Reviews of Physiology, Biochemistry and Pharmacology 143

W0042454

Springer-Verlag Berlin Heidelberg GmbH

Reviews of

143 Physiology Biochemistry and Pharmacology

Editors

E. Bamberg, Frankfurt   M.P. Blaustein, Baltimore
R. Greger, Freiburg   H. Grunicke, Innsbruck
R. Jahn, Göttingen   W.J. Lederer, Baltimore
L.M. Mendell, Stony Brook   A.Miyajima, Tokyo
N. Pfanner, Freiburg   G. Schultz, Berlin
M. Schweiger, Berlin

With 40 Figures and 1 Table

Springer

ISBN 978-3-662-31053-3   ISBN 978-3-540-44510-4 (eBook)
DOI 10.1007/978-3-540-44510-4

Library of Congress-Catalog-Card Number 74-3674

© Springer -Verlag Berlin Heidelberg 2001

Originally published by Springer-Verlag Berlin Heidelberg New York in 2001.
Softcover reprint of the hardcover 1st edition 2001

Production: PRO EDIT GmbH, 69126 Heidelberg, Germany
Printed on acid-free paper – SPIN: 10785458   27/3136wg-5 4 3 2 1 0

Contents

Indexed in Current Contents

Electrophysiological Approaches to the Study of Neuronal Exocytosis and Synaptic Vesicle Dynamics

R. Heidelberger

Department of Neurobiology and Anatomy, W.M. Keck Center for the Neurobiology of Learning and Memory, University of Texas Houston Health Science Center Houston, Texas 77025, USA

Contents

Preview

The cornerstone of synaptic transmission is the calcium-dependent fusion of synaptic vesicles with the plasma membrane and the release of neurotransmitter into the synaptic cleft. Renewed interest in the mechanisms that regulate these important first steps in neuronal communication has been sparked by recent advances in the identification of molecular interactions that underlie these processes and by the development of electrophysiological techniques that permit the direct examination of these processes in real-time. This review will present the emerging view of neuronal exocytosis and endocytosis that has developed as a consequence of these new electrophysiological approaches. Where instructive, results will be discussed in light of other new developments in the field in order to give the reader a comprehensive understanding of our current knowledge. To set the stage, we begin with an overview of the traditional approach to the study of neurotransmitter release, electrical recordings from pairs of synaptically-connected cells and discuss the applicibility of this approach to the study of exocytosis and synaptic vesicle dynamics in the central nervous system. The review will then focus on its main theme, the use of membrane capacitance measurements and whole-terminal patch-clamp recording techniques to directly monitor calcium-dependent synaptic vesicle fusion and retrieval in central neurons. Next, carbon fiber amperometry will be introduced, and information about the mechanics of secretory vesicle fusion and neurotransmitter release obtained with carbon fiber amperometry will be presented. Finally, combinations and refinements of these electrophysiological approaches will be discussed.

1
Paired Recordings

1.0
Introduction

Communication between neurons can occur chemically via synaptic transmission. This communication process begins when an invading depolarization opens voltage-gated calcium channels in the nerve terminal of the presynaptic neuron, allowing calcium from the extracellular fluid to enter into the nerve terminal. The resultant elevation in presynaptic calcium somehow triggers the fusion of synaptic vesicles with the presynaptic plasma membrane and the release of vesicular contents into the synaptic cleft. Secreted neurotransmitter molecules then diffuse passively across the synaptic cleft to the postsynaptic neuron, where they may bind to cell-surface receptors. The postsynaptic receptor with the appropriate number of bound neurotransmitter molecules may either change its conformation to allow ions to cross the postsynaptic membrane and create an electrical signal and/or activate a second-messenger-mediated cascade that might modulate ion channel activity. To complete the signalling process, neurotransmitter is cleared from the synaptic cleft, and the membrane that was added to the presynaptic plasma membrane by the fusion of synaptic vesicles, or its equivalent amount, is retrieved to preserve the surface area of the nerve terminal.

1.1
Introduction to Paired Recordings

Synaptic transmission was first examined at the vesicular level by Katz and his colleagues at the frog neuromuscular junction using intracellular recording techniques to monitor electrical activity. In their classic manuscript, Fatt and Katz described the "chance observations" of randomly occurring spontaneous miniature electric discharges at the endplates of resting muscle fibers (Fatt and Katz 1952). These miniature discharges ,"minis," were found to share a number of important characteristics with end-plate potentials, such as being localized to innervated regions of the muscle fiber and being blocked by curare. More importantly, when the release probability was lowered, the amplitudes of the end-plate potentials were found to vary in a step-like fashion that corresponded to multiples of the mean mini amplitude (Castillo and Katz 1954). Based upon these and other observations, Fatt and Katz proposed that the endplate potential at the neuromoscular junction was comprised of multiple, summed miniature potentials, and they coined the term "quantum" to describe the mean size of these minis (Fatt and Katz 1952). Katz and colleagues went on further to hypothesize that synaptic vesicles were the physical representation of these quanta (Castillo and Katz 1956; Katz 1966; Katz 1969).

1.2
Paired Recordings in the Central Nervous System

In a manner analogous to recording from a presynaptic motor neuron and a postsynaptic muscle fiber, neurotransmitter release can also be studied in synaptically-connected pairs of neurons. In the central nervous system, patch-clamp recording techniques are often preferred over conventional sharp electrode intracellular recordings because of the more favorable signal-to-noise ratio of the patch-clamp technique and because the patch-clamp technique allows one to record from small neurons or parts of neurons, such as dendrites or nerve endings, that would not tolerate impalement with sharp electrodes. Even with these improvements, the application of quantal analysis to synaptic transmission in central neurons has been hampered by the fact that results can be difficult to interpret. One reason for this difficulty is that the release of neurotransmitter from a synaptic vesicle may open only a small number of low conductance ion channels on the postsynaptic neuron (e.g. Bekkers et al. 1990; Edwards et al. 1990; Korn and Faber 1991; Kraszewski and Grantyn 1992; Kullmann and Nicoll 1992; Jonas et al. 1993). Under these circumstances, single quantal events may be below the limit of resolution (Manabe et al. 1992). In addition, distinct peaks may not be apparent in the amplitude histograms of evoked responses (e.g. Bekkers and Stevens 1989; Bekkers et al. 1990). Statistical arguments and deconvolution techniques have been developed to extract quantal information from postsynaptic recordings (e.g. Redman 1990; Korn and Faber 1991), but these methods are indirect, and the outcomes tend to vary depending on the mathematical approach adopted (reviewed in Redman 1990; Korn and Faber 1991; Korn and Faber 1998). Indeed, complex statistical arguments have been used for the past twenty years to argue both for and against a presynaptic origin of long-term potentiation (Korn and Faber 1998).

In addition to small signals, another complication in central neurons arises from that fact that a given neuron may receive multiple synaptic contacts along its dendritic tree, and these inputs may activate more than one kind of postsynaptic receptor. Furthermore, because the synaptic contacts may be located at different places in the dendritic tree, the signals derived from these contacts will be filtered to varying extents by the dendritic cable properties. Therefore, it is highly unlikely that the summated signal recorded at the soma will reveal information about the exocytotic state of any one specific bouton or active zone unless the synaptic inputs can be isolated. The importance of isolating synaptic inputs in order to study quantal release is well-appreciated (e.g. Bekkers and Stevens 1989; Bekkers et al. 1990), and a recent study indicates just how critical this aspect of experimental design can be to the interpretation of the data (Forti et al. 1997). In this beautiful set of experiments, miniature responses were simultaneously recorded from the neuronal soma, which come from multiple boutons located throughout the dendritic tree, and at identified single boutons along the dendritic tree in cultured hippocampal neurons. Comparison of the time course of the minis revealed that the minis from an individual bouton were more consistent in amplitude than those recorded at the soma (Fig. 1.1). Indeed, the minis from a single bouton gave rise to an amplitude histogram that could be fitted with

R. Heidelberger

A

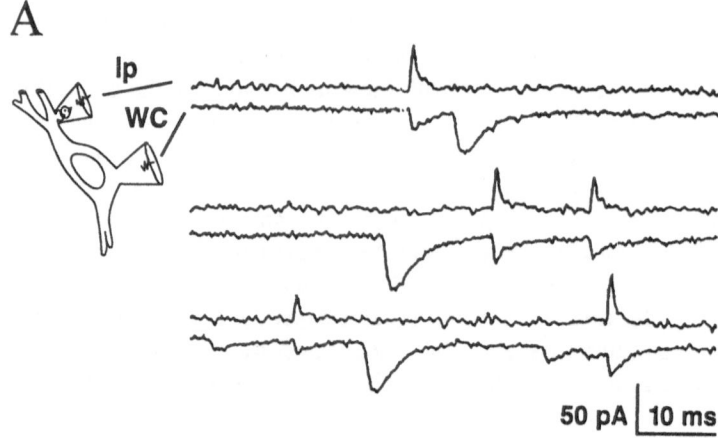

50 pA | 10 ms

B

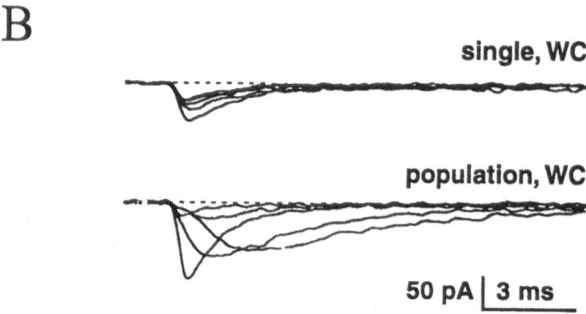

single, WC

population, WC

50 pA | 3 ms

Fig. 1.1. Simultaneous recordings from pairs of identified synapses and somata. A. Simultaneously acquired current traces from a synapse (upper trace in each pair) and soma (lower trace in each pair) of a hippocampal neuron. To make the synaptic recordings, a synapse was visualized by its ability to take up the fluorescent dye FM1-43 in an activity-dependent manner. This fluorescently-labelled region was then sucked into a large patch pipette and a loose-electrical seal (≈ 10 mΩ) was formed. Minis originating from the synapse enclosed in the loose-patch pipette appear synchronously in both traces, whereas minis from other synapses are seen exclusively in the somatic recording. Note the consistency of the waveform of the synaptic minis versus the variability in waveform of the somatic minis. B. Single synapse and population minis recorded at the soma. Minis arising from a single synapse are more similar in waveform than are minis that arise from a population of synapses. (Reprinted by permission from Nature 388:874–878, copyright 1997, Macmillan Magazines Ltd.)

a single gaussian distribution, whereas the minis recorded in the soma, exhibited a high degree of amplitude variability and a broad, skewed amplitude distribution. Broad, skewed amplitude histograms have commonly been reported in the central synapse literature (e.g. Edwards et al. 1990; Manabe et al. 1992; Jonas et al. 1993), and the data of Forti et. al. suggest that at least some of this reported skewing might reflect the spontaneous release of neurotransmitter from multiple boutons rather than the quantal characteristics of any one particular bouton. Thus, these findings highlight the tremendous level of difficulty involved when one attempts to infer information about quantal release from small postsynaptic responses in cells receiving numerous synaptic contacts located at a variety of electronic distances from the site of the recording.

If the goal of an experiment is to extract information about exocytosis by applying the principles of quantal analysis, then the ideal situation would be to record from a postsynaptic neuron that is innervated exclusively on its soma and by a single presynaptic nerve fiber, situation analgous to the neuromuscular junction. This would eliminate concerns about diverse inputs and cable filtering. There are two synapses in the central nervous system where this has been possible, and both are located in the auditory pathway. The first of these is the fast glutamatergic synapse between the auditory nerve and neurons in the anteroventral cochlear nucleus. The second is the fast glutamatergic synapse between neurons in the anteroventral cochlear nucleus and neurons in the medial nucleus of the trapezoid body (the calyx of Held). At both synapses, a single presynaptic axon terminates in a calyceal-style ending that engulfs the soma of a single postsynaptic neuron. Similar to neuromuscular junction, exocytosis at these synapses was observed to be both quantal and stochastic, and the time course of full-sized post-synaptic currents at physiologic external calcium could be recreated by summing the miniature currents at low external calcium (Isaacson and Walmsley 1995; Borst and Sakmann 1996). This indicates that the release probability has the same time course regardless of whether the concentration of external calcium is high or low and is independent of the amount of release. In addition, the synaptic delay, measured as the time between the peak of the calcium influx and the start of a miniature synaptic current, and the temperature sensitivity of release were nearly identical to that reported at the neuromuscular junction (Katz and Miledi 1965; Isaacson and Walmsley 1995; Borst and Sakmann 1996). Thus, the most fundamental properties of exocytosis at these central synapses are comparable to those at the neuromuscular junction. This observation supports the idea that the basic exocytotic machinery is conserved among synapses, as portended by the highly-conserved nature of the individual proteins that comprise the fusion machinery (Rothman and Orci 1992; Sudhof et al. 1993; Rothman 1994; Sudhof 1995; Weber et al. 1998).

Finally, it should be kept in mind that the above approaches to the study of synaptic vesicle fusion and neurotransmitter release rely entirely on the accuracy of postsynaptic receptors to report the timing and magnitude of neurotransmitter release. However, it cannot be assumed that postsynaptic receptors are always accurate reporters of exocytosis. For example, if the postsynaptic receptors desensitize with stimulation or undergo a decrease in number or activity (e.g. Trussell and Fischbach 1989; Trussell et al. 1993; Tang et al. 1994; Liao et al. 1995), this will result in an underestimation of the amount of neurotransmitter that has been released. In addition, not only can the amount of time that it takes to clear the synaptic cleft of neurotransmitter affect the extent to which postsynaptic receptors desensitize, but slow clearance from the synaptic cleft may give the appearance of a delayed or asynchronous component of neurotransmitter release even though exocytosis has ceased. Finally, it has been suggested that at some synapses a single quantum can saturate the postsynaptic response (e.g. Edwards et al. 1990; reviewed in Redman 1990; Stevens 1993), although this interpretation is not without controversy (see Stevens 1993; Forti et al. 1997). The question of whether quantal size is presynaptically or postsynaptically determined is clearly of fundamental importance if one wishes to extrapolate information about neurotransmitter release and synaptic vesicle fusion by analyzing the amplitude of the postsynaptic responses. Under any of these circumstances, an assay of exocytosis that is independent of the postsynaptic response is desirable.

2
Capacitance Measurements

2.0
Introduction

The advent of membrane capacitance measurements as an index of synaptic vesicle fusion and retrieval has allowed for real-time measurements of both exocytosis and endocytosis. In addition to providing a direct method with which to presynaptically monitor membrane fusion and retrieval, membrane capacitance measurements can be readily combined with the simultaneous measurement of intracellular calcium. This combination has led to a new understanding of the secretory process and its regulation by calcium in a variety of secretory cells, including mast cells, eosinophils, adrenal chromaffin cells, pituitary melanotrophs, and pancreatic beta cells. More recently, these techniques have been extended to the study of neuronal exocytosis and endocytosis. This chapter will present data from capacitance experiments that have shaped our understanding of synaptic vesicle fusion and retrieval. Readers interested in the technical details of capacitance measurements are referred to the following excellent sources, (Lindau and Neher 1988) and (Gillis 1995).

2.1
Basic Principles of Capacitance Measurements

The membrane capacitance technique for monitoring synaptic vesicle fusion and retrieval is based upon the premise that the surface area of a cell is proportional to its membrane capacitance. This relationship allows one to use membrane capacitance, an electrically measurable parameter, as an assay to detect changes in cell surface area. In the simplest scenario, an increase in membrane capacitance indicates an increase in cell surface area due to the addition of membrane (i.e., fusion of synaptic vesicles with the plasma membrane), and a decrease in membrane capacitance indicates a decrease in membrane surface area due to membrane retrieval. However, because membrane capacitance measurements report the net membrane surface area, when there is no observed change in membrane capacitance over time, two possible interpretations exist. The first interpretation is that neither exocytosis nor endocytosis has occurred during the observed interval. The second is that both exocytosis and endocytosis have occurred in perfect register, such that the amount of membrane added was precisely balanced by the amount of membrane retrieved. Even when a capacitance increase is discernable, the magnitude and timing of the true exocytotic response can be minimized or exhibit an apparent change in kinetics if endocytosis occurs concurrently. Similarly, studies of endocytosis can be confounded by ongoing exocytosis. This underscores the importance of using independent means to determine the extent to which exocytosis and endocytosis temporally overlap.

Factors other than membrane surface area also contribute to the capacitance of a membrane, and therefore not all changes in membrane capacitance are attributable to changes in membrane surface area. For example, gating charge movements that occur when a voltage-gated ion channel is activated can cause a transient increase in membrane capacitance (Horrigan and Bookman 1994). Fortunately, it is relatively easy to identify and correct for changes due to gating charge movements in the capacitance record (Horrigan and Bookman 1994; Mennerick and Matthews 1996). In addition, spurious changes in membrane capacitance can be caused by changes in the membrane or access conductances (Gillis 1995), and therefore one should monitor these conductances throughout an experiment to determine whether or not they have remained constant. If transient changes in membrane conductance do occur, it may be possible to separate these changes from transient changes in membrane capacitance through the use of a sophisticated voltage command, such as a dual-frequency sine wave, to calculate the membrane capacitance (e.g. Gillis 1995; Barnett and Misler 1997). Finally, not all changes in membrane capacitance that result from membrane addition are indicative of neurotransmitter release. There are several reports of membrane addition that is triggered by calcium but does not result in the exocytosis of neurotransmitter (Coorssen et al. 1996; Ninomiya et al. 1996; Xu et al. 1998). Therefore, one should take care to also verify that an increase in membrane capacitance is due to synaptic vesicle fusion and results in the release of neurotransmitter (e.g. von Gersdorff et al. 1998).

Membrane capacitance measurements are made using a variant of the patch-clamp recording technique, and this dictates that either the nerve terminal be accessible to a patch pipette, or, if the recording is made from the soma, the space-clamp properties of the neuron permit tight voltage control of the nerve terminal. A second consideration is that the secretory response must be large enough to be detectable as a change in capacitance. In mast cells, step-like increases can be observed in the capacitance record that correspond to the fusion of single secretory granules with the plasma membrane (Almers and Neher 1987). In contrast, the small, clear-core synaptic vesicles of neurons are an order of magnitude smaller than the large, dense-core secretory vesicles of the mast cell. Fusion of a single synaptic vesicle will contribute only a small amount, perhaps as little as 26 aF, to the total membrane capacitance (von Gersdorff et al. 1996; Neves and Lagnado 1999). This is below the resolution for capacitance measurements that are made using the whole-cell recording configuration and below the currently achievable resolution using cell-attached recordings (see section 4.0). However, if multiple vesicles fuse in response to a stimulus, which can happen if there are multiple active zones per neuron and/or multiple active release sites per active zone, the summed capacitance change can be sufficiently large to permit detection of neuronal exocytosis. This allows the behavior of a population of vesicles to be examined.

Neurons in which capacitance measurements have been used to successfully study synaptic exocytosis and endocytosis include retinal bipolar neurons (von Gersdorff and Matthews 1994a), retinal photoreceptors (Rieke and Schwartz 1994; Rieke and Schwartz 1996), and hair cells from the sacculus (Parsons et al. 1994) and inner ear (Moser and Beutner 2000). Each of these neurons is accessible to a recording electrode and is electrically compact, and this allows for excellent voltage control of the presynaptic terminal. This is particularly true for the bipolar neuron, in which a whole-cell recording electrode can be placed directly on the synaptic terminal. Each of these neurons also shares the common feature that they contain a specialized structure at their active zones to which small, clear-core synaptic vesicles are tethered. In bipolar neurons and photoreceptors these structures are sheet-like in appearance and are called "synaptic ribbons" (Dowling 1987). In hair cells, these structures are more spherical in shape and are called "synaptic bodies" (Hama and Saito 1977; Lenzi et al. 1999). Multiple synaptic ribbons or synaptic bodies are commonly found in these neurons, depending upon the species, and this presumably indicates that there is more than one region from which secretion of neurotransmitter can occur. The combination of accessibility, electrical compactness, and multiple active zones makes these three neurons of the central nervous system ideal for the study of calcium-dependent membrane addition and retrieval using the capacitance technique.

2.2
The Neuronal Exocytotic Response

The first study of synaptic vesicle fusion in a central neuron was conducted in synaptic terminals of bipolar neurons acutely isolated from the retina of goldfish (von Gersdorff and Matthews 1994a). Capacitance measurements were used to track changes in membrane surface area indicative of synaptic vesicle fusion and retrieval, and intraterminal calcium concentration was monitored using the fluorescent calcium indicator dye Fura-2. To elevate intraterminal calcium and trigger exocytosis, a depolarizing step change in membrane potential of an intensity and duration known to evoke release of the neurotransmitter glutamate from these terminals (Tachibana and Okada 1991) was utilized. In response to membrane depolarization and calcium influx through presynaptic voltage-gated calcium channels (Heidelberger and Matthews 1992; Tachibana et al. 1993), a reversible increase in the membrane capacitance of the synaptic terminal was observed (Fig. 2.1). The addition of membrane was found to absolutely require the elevation of intraterminal calcium, and the amplitude of the capacitance increase was found to closely track the intraterminal calcium. Upon membrane repolarization and closure of calcium channels, the membrane capacitance returned to baseline with a time constant of $\approx$ 1–2 seconds. The simplest interpretation of these data is that the increase in membrane capacitance reflects calcium-dependent membrane addition, presumably due to the calcium-triggered fusion of synaptic vesicles with the plasma membrane during the process of

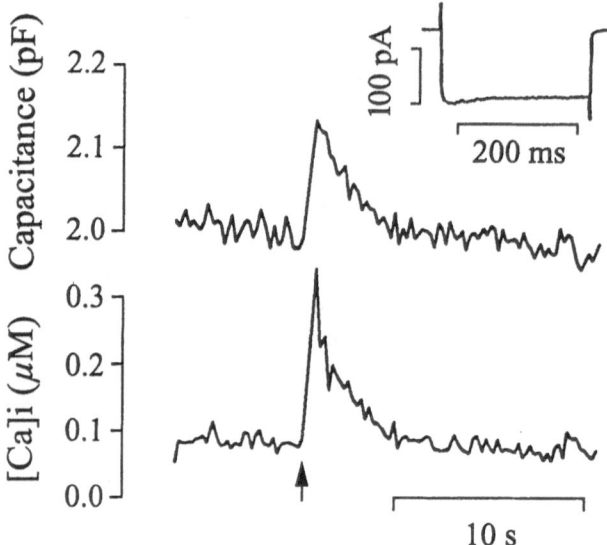

Fig. 2.1. Activation of calcium current increased intraterminal calcium and the membrane capacitance of an isolated synaptic terminal of a retinal bipolar neuron. At the arrow (lower panel), a 250 ms depolarization was given from –60 mV to 0 mV to activate presynaptic calcium channels. The inward current carried by calcium ions is shown on an expanded time scale in the inset. The lower panel shows the intraterminal calcium record calculated from the change in fluorescence of the calcium indicator dye Fura-2. The upper panel shows the corresponding capacitance record. Note that in addition to the capacitance increase seen following calcium entry, the membrane capacitance returns to the prestimulus level following the closure of the presynaptic calcium channels. (Reprinted by permission from Nature 367:735–739, copyright 1994, Macmillan Magazines Ltd.)

exocytosis, and the return of the capacitance to baseline represents membrane retrieval.

This interpretation is supported by two independent lines of evidence. In a very elegant study, the release of excitatory neurotransmitter glutamate from synaptic terminals and the fusion of synaptic vesicles were concurrently monitored (von Gersdorff et al. 1998). These key experiments were performed by acutely positioning a bipolar neuron synaptic terminal on top of a second neuron known to contain functional glutamate receptors. Synaptic terminal membrane capacitance was monitored in parallel with patch-clamp recordings of glutamate-gated currents in the second neuron. The timing and amplitude of a membrane capacitance increase and the timing and amount of glutamate-gated current were found to be in excellent agreement, indicating that in these terminals, increases in membrane capacitance accurately indicate the exocytotic release of the excitatory neuro-transmitter glutamate. In addition, the authors demonstrated that the capacitance increase in bipolar neuron synaptic terminals evoked by brief to moderate intensity depolarizations reflects exocytosis uncontaminated by endocytosis. They also provided evidence to indicate that the capacitance decrease observed upon repolarization solely reflects endocytosis (von Gersdorff et al. 1998). In a separate study, optical imaging techniques were employed to track membrane turnover in bipolar neuron synaptic terminals. Again, the results demonstrated that exocytosis is unlikely to be confounded by endocytosis (Rouze and Schwartz 1998). Thus, the simple interpretation that an increase in membrane capacitance indicates exocytosis and a decrease in membrane capacitance indicates endocytosis holds for the synaptic terminal of the bipolar neuron for standard stimulation protocols. If one uses a more intense, prolonged stimulus or trains of stimuli, this simple interpretation may no longer be valid (Lagnado et al. 1996). Correlations of capacitance changes with neurotransmitter release have not been attempted for the other neurons, but they have been performed for some neuroendocrine cells using carbon-fiber amperometry (see section 3) as the independent detector of release. In addition, capacitance measurements have been combined with optical methods of tracking granule fusion and retrieval to determine the extent to which exocytosis and endocytosis might overlap in chromaffin cells (Smith and Betz 1996).

2.3
The Calcium-Dependence of Exocytosis

The pioneering work of Katz and Miledi clearly demonstrated that the influx of calcium into the presynaptic terminal was necessary for synaptic exocytosis (Katz and Miledi 1967; Miledi 1973). Until recently, however, it has proven very difficult to determine precisely the relationship between the intracellular calcium concentration and the exocytotic response or identify additional calcium-dependent steps along the secretory pathway. The development of calcium-sensitive fluorescent indicator dyes (e.g. Grynkiewicz et al. 1985; Raju et al. 1989; Konishi et al. 1991), capacitance measurements, and neuronal preparations that are accessible to direct biophysical investigation has led to a renewal of the challenge to understand the calcium dependence of synaptic exocytosis and the role of coupling between voltage-gated calcium channels and the fusion apparatus.

2.3.1
The Calcium-Dependence of Synaptic Exocytosis

The first examination of the intraterminal calcium threshold for synaptic exocytosis was also performed in synaptic terminals of retinal bipolar neurons (von Gersdorff and Matthews 1994a). Membrane capacitance measurements were used to track fusion of synaptic vesicle fusion, and intraterminal calcium was monitored with the use of a calcium-sensitive fluorescent indicator dye. To avoid the complications that are associated with attempting to accurately measure the rapidly-changing and spatially-heterogenous calcium signal that results from the activation of voltage-gated calcium channels (e.g. Hernandez-Cruz et al. 1990; Augustine and Neher 1992), intraterminal calcium was homogeneously elevated within the synaptic terminal by either internal dialysis of the cytosol with calcium-buffered solutions or by the external application of a calcium ionophore. Both methods of elevating the intraterminal calcium concentration revealed that the concentration of free calcium required to trigger synaptic vesicle fusion was greater than 20 μM (von Gersdorff and Matthews 1994a). In addition, a second component of membrane addition, triggered by low concentrations of calcium (Lagnado et al. 1996; Rouze and Schwartz 1998) and inhibited by > 20 μM calcium (Rouze and Schwartz 1998), has been reported in optical assays of membrane turnover in synaptic terminals. The significance of this second component for synaptic signaling has not been established, and therefore it will not be discussed further in this review.

The fundamental relationship between intraterminal calcium and the rate of synaptic vesicle fusion, capacitance and calcium measurements has been determined in synaptic terminals by combining flash-photolysis of caged-calcium with capacitance measurements (Heidelberger et al. 1994). Flash-photolysis of caged-calcium can be used to rapidly and homogeneously elevate cytosolic calcium (Naraghi et al. 1998), and this permits the fast acquisition of a spatially-averaged calcium signal to be used as an index of the calcium concentration near release sites. Furthermore, because this photolytic approach does not require a change in membrane potential or the activation of presynaptic voltage-gated conductances, synaptic vesicle fusion can be monitored *during* the stimulus. This permits the rate of synaptic vesicle fusion with respect to intraterminal calcium to be directly determined. This same information is not readily obtainable when a depolarizing voltage step is used to trigger exocytosis due to the confounding effects that the opening of voltage-gated calcium channels changes has on capacitance measurements. One possible caution with this technique arises from reports

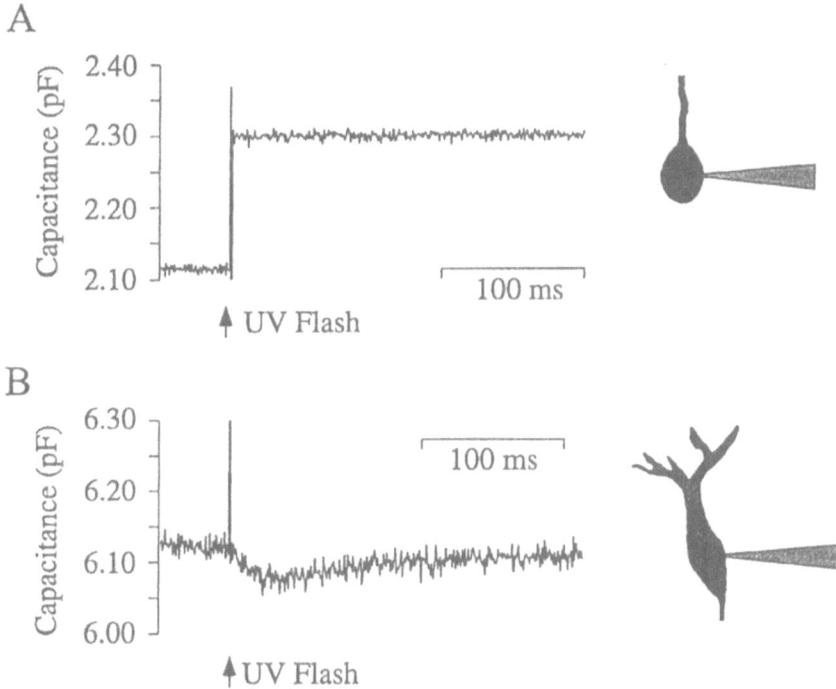

Fig. 2.2. Synaptic terminals, but not somata, respond to the elevation of cytosolic calcium with an increase in membrane capacitance. **A** The capacitance record of an isolated synaptic terminal of a retinal bipolar neuron. Immediately following flash-photolysis of caged-calcium and release of calcium into the cytosol, the membrane capacitance increased by 186 fF. **B** The capacitance record of an isolated bipolar neuron soma that lost its axon and synaptic terminal during the dissociation process. After flash-photolysis of caged-calcium, membrane capacitance slowly decreased by 45 fF, and then slowly returned to baseline. There is no indication of calcium-dependent membrane addition in the soma. In both A and B, timing of the flash is given by the electrical artefact (i.e. spike) in the capacitance record and by the arrow. Flash intensity was near-maximal to elevate internal calcium above 100 µM. (Reproduced from The Journal of General Physiology, 1998, 111, 225–241, by copyright permission of The Rockefeller University Press)

that the global elevation of cytosolic calcium may stimulate fusion of membrane that is not related to exocytosis and release of neurotransmitter (Oberhauser et al. 1995; Oberhauser et al. 1996; Coorssen et al. 1996; Ninomiya et al. 1996; Xu et al. 1998). Control studies in synaptic terminals of bipolar neurons, however, suggest that fusion events that are not related to exocytosis are not a major concern in synaptic terminals (Heidelberger 1998). The exocytotic response evoked by flash-photolysis of caged-calcium was found to be restricted to the synaptic indicating that either the pool of fusion-competent membrane or the fusion machinery, or both, is specifically localized to the synaptic terminal (Fig. 2.2). Furthermore, cross-depletion experiments indicate that the same pool of membrane fuses in response to flash-photolysis of caged-calcium as to the activation of presynaptic calcium channels (Heidelberger 1998).

In synaptic terminals, elevation of the intraterminal calcium concentration to greater than 10–20 µM via the flash-photolysis of caged-calcium was

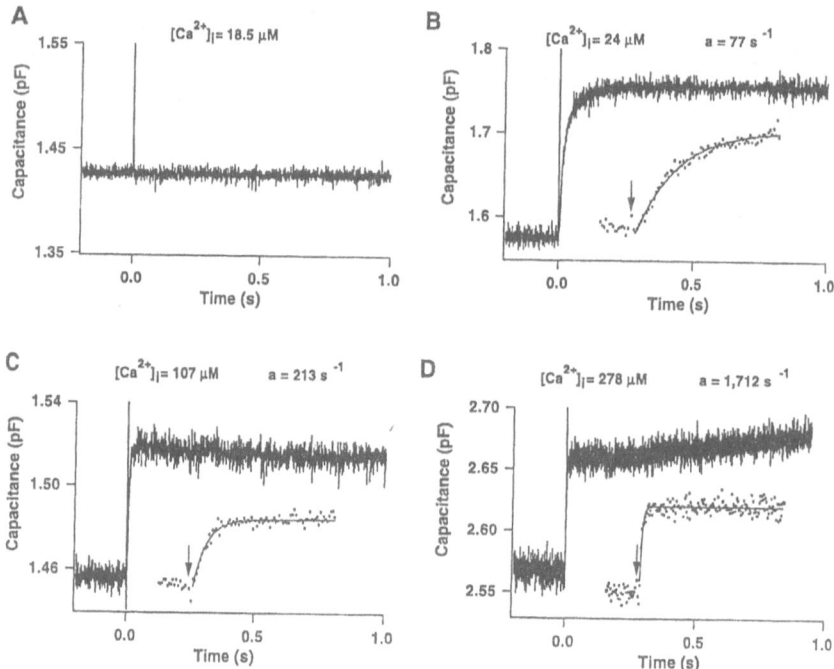

Fig. 2.3. An increase in membrane capacitance is observed when intraterminal calcium is elevated above 20 µM by flash-photolysis of caged-calcium. The timing of the flash is given by the flash artefact at time = 0 or by arrows (insets). The numbers above each trace show the average intraterminal calcium concentration 100 ms after the flash calculated using the ratiometric calcium-indicator dye Furaptra. The rate constant, a, was obtained for each response by fitting a single exponential to the rising phase of the capacitance trace. Data in A–C were obtained with a 1,600 Hz sine wave and a 190 V flash. Flash intensity was controlled by the insertion of neutral density filters into the light path. Trace D was obtained with a 3,200 Hz sine wave and an unattenuated 225 V flash. Insets show 60 ms of the trace on an expanded time scale. Curve (insets) shows the exponential fit to the data. Note that the rate of membrane addition increases with higher intraterminal calcium. (Data used with permission from Nature 371:513–515, copyright 1994, Macmillan Magazines Ltd.)

observed to trigger a rapid increase in membrane capacitance (Heidelberger et al. 1994). This calcium threshold for synaptic vesicle exocytosis is in good agreement with the estimate of > 20 μM obtained in the dialysis experiments of von Gersdorff and Matthews (von Gersdorff and Matthews 1994a). The somewhat lower estimate obtained in the flash studies is likely to be the more accurate estimate because in a flash-photolysis experiment the calcium rise is concerted rather than protracted, making it easier to detect a small capacitance response. The rising phase of a capacitance response in synaptic terminals was typically well-described by a single exponential function (Fig. 2.3). This suggests that the population of synaptic vesicles that fuses in response to a global elevation in intraterminal calcium is homogeneous with respect to the calcium-dependent kinetics of fusion (Heidelberger et al. 1994; Heidelberger 1998;).

The rate of the capacitance rise was steeply dependent upon the post-flash intraterminal free calcium concentration (Fig. 2.3), particularly for intraterminal calcium concentrations between 10 and 200 μM. Within this calcium range, the rate of membrane addition increased by nearly four log units for a one log unit increase in intraterminal calcium (Fig. 2.4B). This observation is in agreement with the fourth-power relationship between presynaptic calcium and neurotransmitter release that has been proposed for synaptic release, based upon indirect methods, at neuromuscular junction (Dodge and Rahamimoff 1967), squid giant synapse (Augustine et al. 1985; Augustine and Charlton 1986), and cerebellar synapses (Mintz et al. 1995). Between 100–200 μM calcium, the rate of membrane addition was half-maximal, and above ≈ 200 μM, the rate of membrane addition abruptly saturated with a rate constant of ≈ 3000/s (Fig. 2.4B) (see also (Heidelberger 1998). In this latter range of calcium concentrations, the calcium-dependent delay between the elevation of intraterminal calcium and the start of the capacitance response was less than one millisecond (Heidelberger et al. 1994). These results indicate that in nerve terminals, synaptic vesicles can exocytose rapidly, synchronously, and with minimal delay when the local calcium concentration rises above ≈ 100 μM.

To more quantitatively evaluate the calcium-sensitivity of the synaptic release process and approximate the affinities of the individual calcium binding sites, the relationship between the rate of membrane addition and the intraterminal free calcium concentration was mathematically modeled (Heidelberger et al. 1994). The form of the model (see Fig. 2.4A) was borrowed by analogy from earlier work in neuroendocrine cells in which calcium-dependent secretion was proposed to occur by a series of sequential steps (Heinemann et al. 1993; Neher and Zucker 1993; Thomas et al. 1993b).

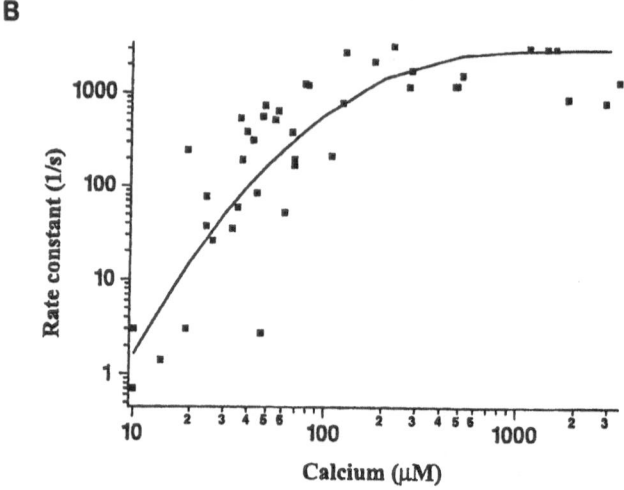

Fig. 2.4. The calcium-dependence of the rate of synaptic vesicle fusion. A. A model of calcium-dependent exocytosis in synaptic terminals. The model details the sequential binding of 4 calcium ions followed by a final, calcium-independent fusion step. The vesicle pool B_0 represents the population of synaptic vesicles that are fusion-competent and require only the elevation of intraterminal in order to proceed towards fusion. The starting size of B_0, 77 fF, was obtained from the average size of the capacitance increases. Pools B_1, B_2, B_3 and B_4 represent populations of vesicles with one, two, three, or four calcium ions bound. The rate constant for the fusion step, γ, was set to 3000 s^{-1}, a value close to the maximum measured rate of membrane addition. The forward and backwards rate constants, α and β, were set to 14 x 10^6 $M^{-1}s^{-1}$ and 2,000 s^{-1} respectively. The cooperativity factor for the fowards direction, a, was set to 1, and the cooperativity factor for the backwards reaction, b, was set to 0.4. Thus, the dissociation constant for the first calcium binding site was 143 µM, 57 µM for the second binding site, 23 µM for the third, and for the last one, 9.14 µM. B. Exponential rate constants for the rise of each capacitance response are plotted against the post-flash intraterminal calcium concentration. The curve represents the predicted relationship between rate of membrane addition and intraterminal calcium that is based upon the model and using the parameters indicated in (A). (Reprinted by permission from Nature 371:513–515, copyright 1994, Macmillan Magazines Ltd.)

In the chromaffin cell model, the transition from the docked and primed state, 'B_0,' to the fused state, 'C,' was proposed to require three independent calcium-binding reactions that preceded a final, calcium-independent, fusion step (Heinemann et al. 1994). While this type of model describes the chromaffin cell date quite well, it does not adequately describe the data from synaptic terminals. In synaptic terminals, if no cooperativity between binding sites is assumed, then the binding of at least six calciums ions has to be proposed in order to match the steep rise and abrupt saturation of the rate versus calcium relationship. Because the initial slope suggested a Hill coefficient of four, a fourth calcium binding step was added to the model, and a cooperative interaction between the binding sites was added to accommodate the steep rise and abrupt saturation. The resultant synaptic model fits the terminal data quite well (Fig. 2.4B, curve) and predicts that the binding of each of four calcium ions to the calcium sensor (be it a single protein or several proteins) occurs independently of each other. However, once the four ions are bound, the model predicts that it would be difficult to remove a calcium ion from the four-calcium bound state. Conceptually, one might imagine that a change occurs in the calcium sensor when four calcium ions are bound and that this conformational switch favors the reaction proceeding towards vesicle fusion. However, once the first calcium ion is lost from the sensor, the model predicts that it will become progressively easier to remove each of the remaining calcium ions. Thus, the model might be considered "doubly" cooperative in the sense that it requires the binding of more than one calcium ion in order to trigger exocytosis, and it calls for cooperativity between binding sites in the unbinding direction. The model estimates that the overall Kd of the fusion reaction for calcium is 194 µM, and that the Kd's for each of the four calcium-binding sites range from 143 µM (for the first calcium ion to bind) to 9.14 µM (for the last of the four ions to bind) (Heidelberger et al. 1994). The relatively low affinities indicate that in synaptic terminals, the calcium-sensor for exocytosis is likely to release calcium from its binding sites quickly when calcium channels close and calcium gradients begin to dissipate.

2.3.2
The Calcium-Dependence of Exocytosis in Other Secretory Cells

2.3.2.1
Photoreceptors

Quantitative data about the calcium-dependence of the rate of exocytosis is not available for other synaptic terminals. However, the calcium-threshold of release has been directly determined for the rod photoreceptor. In this tonically-active neuron, flash-photolysis experiments with caged-calcium suggest that the calcium threshold for release is $\approx 2\ \mu M$ (Rieke and Schwartz 1996). This contrasts sharply with the 10–20 μM calcium threshold of the bipolar neuron. Indeed, this low threshold is reminiscent of the calcium threshold reported for secretory cells that contain dense-core vesicles (see below). A rationale for a lower calcium threshold in the photoreceptor may lie in its operation. Photoreceptors differ from most other neurons in that they tonically release neurotransmitter (Rieke and Schwartz 1996), and information about light onset is signaled by a reduction in the release neurotransmitter rather than by a switch from quiescence to secretion. Furthermore, the light response is not particularly fast, taking as long as 200 ms for the change in membrane potential to reach its peak following absorption of a photon (Rodieck 1998). Thus, one might argue that speed of the offset of neurotransmitter release is not a critical feature of synaptic signaling in the rod photoreceptor, and therefore a low-affinity calcium sensor is not mandated. One might also argue that a 10-fold lower calcium-threshold could be advantageous to a tonically-secreting neuron because it would help to contain the average cytosolic calcium concentration during on-going secretion. From these perspectives, a calcium binding protein with a low threshold for release and a high affinity for calcium might be desirable for a neuron like the rod photoreceptor. This interpretation remains purely speculative at present because full characterization of the calcium dependence of glutamate exocytosis in photoreceptors has not yet been achieved.

2.3.2.2
Large Dense-Core Granule Fusion

Exocytosis of dense-core granules in neurosecretory cells has a markedly different calcium-dependence than the exocytosis of small, clear-core vesicles in bipolar neuron synaptic terminals. In contrast to the 10–20 μM threshold observed in synaptic terminals (Heidelberger et al. 1994; von Gersdorff and Matthews 1994a), the threshold for non-synaptic release is typically estimated to be less than 1 μM. Cells examined include neuroendocrine cells, neurohypophysial endings and neuronal somata (Thomas et al. 1990; Lindau et al. 1992; Okano et al. 1993; Huang and Neher 1996). In addition, PC12 cells, transformed cells isolated from rat pheochromcytomas that contain both small and large vesicles, exhibit two distinct components of release (Kasai et al. 1996). The faster component, thought to be due to the fusion of small, synaptic-like vesicles, has a calcium threshold > 10 μM, and the slower component has a lower threshold and is thought to be associated with large dense-core granule fusion (Kasai et al. 1996).

The calcium-dependence of the rate of release also appears to be substantially different between synaptic vesicle and dense-core granule fusion. For example, secretion from neuroendocrine cells is less steeply dependent upon internal calcium, exhibiting a third-order dependence upon intracellular calcium. Furthermore, unlike in synaptic terminals, there is no evidence to support a cooperative interaction between the binding sites (Thomas et al. 1993b; Heinemann et al. 1994; Huang and Neher 1996). In addition, the rate of release can be an order of magnitude or more slower in neurosecretory (Thomas et al. 1993b; Heinemann et al. 1994; Ninomiya et al. 1997) than in synaptic terminals (ie, Heidelberger et al. 1994; Mennerick and Matthews 1996; Heidelberger 1998). A faster component of release that approaches the rates measured in synaptic terminals has been observed in a population of bovine adrenal chromaffin cells, but it is unclear what other factors differentiate these cells from their slower counterparts (Heinemann et al. 1994).

Based upon the above comparisons, the following generalizations may be made. The first is that there can be striking differences in the calcium-binding properties of the calcium sensor for exocytosis used by secretory cells. These differences may reflect the functional requirements of the cells. Neurons specialized for the rapid release of neurotransmitter are likely to possess a calcium sensor with four or more low-affinity calcium-binding sites. Secretory cells with lower requirements for speed are apt to use a calcium sensor with a lower number of binding sites, with each site having a relatively high affinity for calcium. Secondly, there is more likely to be cooperativity between the calcium binding sites of the calcium sensor at synapses, where the timing of release may be critical, than in cells that secrete in a humoral fashion. Thirdly, the maximum rate of exocytosis seems to be faster at synapses than for neuroendocrine cells. These features all have important implications for the timing of exocytosis (section 2.4.2). In addition, they suggest that there are important differences in the calcium-sensitivity of the secretory machinery at the protein level, between the calcium sensor for synaptic exocytosis and the calcium sensor used to trigger the release of hormones from neuroendocrine cells.

2.3.3
Is There Heterogeneity in the Calcium-Sensitivity of Fusion?

Recent studies in squid giant synapse and in sea urchin egg fertilization have provided evidence to support the hypothesis that not all secretory vesicles have the same calcium sensitivity for fusion (Hsu et al. 1996; Blank et al.

1998a; Blank et al. 1998b). In one instance this has been attributed to a difference in the intrinsic calcium-sensitivity of granules (Blank et al. 1998a; Blank et al. 1998b). In another instance, the difference is thought to be due to a calcium-dependent adaption of the secretory apparatus (Hsu et al. 1996). These experiments are particularly intriguing because the methods employed to elevate intracellular calcium (flash-photolysis of caged-calcium and injection of calcium-buffered solutions) eliminate concerns about the possible contributions of activity-dependent changes in calcium influx and arguments about the location of the secretory granules relative to calcium channels and point strongly towards the possibility that something downstream from calcium entry is different. In vertebrate secretory cells, while some populations of vesicles fuse with different calcium-dependent kinetics than others, data obtained with the capacitance technique do not support either a differential or a changing calcium sensitivity of the calcium-binding trigger protein for exocytosis. Rather, the common interpretation has been that vesicles with slower release kinetics are located more distantly from the calcium channels than those with faster release kinetics (see section 2.5). The strongest evidence for this interpretation comes from the results of experiments that use flash-photolysis of caged-calcium to uniformly elevate intraterminal calcium and trigger exocytosis. If individual vesicles were to have different calcium affinities *a priori*, one would imagine that the amplitude of the capacitance response should be graded with respect to intraterminal calcium such that a larger flash should trigger a larger capacitance jump. Such a relationship has not been observed in either neuroendocrine cells or synaptic terminals (Gillis 1995; Heidelberger 1998). Furthermore, in synaptic terminals, which exhibit two kinetically distinct components of vesicle fusion in response to membrane depolarization (von Gersdorff and Matthews 1994a; Mennerick and Matthews 1996; Sakaba et al. 1997; von Gersdorff et al. 1998), the capacitance increase elicited by flash-photolysis of caged-calcium is typically well-described by a single exponential, consistent with the fusion of a single kinetic population of vesicles (Heidelberger et al. 1994; Heidelberger 1998). Thus, these data argue against the idea of fusion-competent synaptic vesicles that have different intrinsic calcium sensitivities in bipolar neurons synaptic terminals. In addition, the data support the hypothesis that location of synaptic vesicles relative to the sites of calcium influx underlies the observed kinetic components of fusion reported in depolarization experiments. A recent study has indicated that vesicles that newly join the release-ready pool are more reluctant to fuse than other vesicles at the calyx of Held (Wu and Borst 1999), but additional studies such as those described above will be necessary to determine whether this is due to an intrinsic difference in the ability of these newly-arrived vesicles to fuse or whether newly-arriving vesicles may be located more distantly from calcium channels.

2.4
The Location of Calcium Channels With Respect to Release Sites

2.4.1
Theoretical and Experimental Evidence

The data from the flash-photolysis experiments suggest that when the calcium concentration near release sites is 200 µM or higher, the rate of synaptic vesicle fusion is at its maximum. This raises the question of whether a calcium concentration of 200 µM or higher is actually achieved at the plasma membrane under normal physiological conditions. From a theoretical perspective, the answer to this question may be "yes." Mathematical models have estimated that the calcium ion concentration achieved near the mouth (< 50 nm) of an open calcium channel, a calcium microdomain, can reach several hundred micromoles (Chad and Eckert 1984; Simon and Llinas 1985; Yamada and Zucker 1992). The existence of a high calcium concentration under the plasma membrane when calcium channels are open is also supported by experimental evidence. Current through calcium-activated potassium channels has been very cleverly used to estimate the submembrane calcium concentration. Using this approach, the submembrane calcium concentration is believed to reach 1 mM in response to membrane depolarization in hair cells (Roberts et al. 1990) and 60 µM in chromaffin cells (Prakriya et al. 1996). Estimates at the squid giant synapse using optical methods have reported that presynaptic calcium may rise to 200–300 µM in discrete regions in response to membrane depolarization (Llinas et al. 1992). Thus, it seems reasonable to postulate that if the calcium-sensitive neuronal trigger for synaptic release is located within molecular proximity of an open calcium channel, it may indeed sense a free calcium concentration of a few hundred micromoles and be able to fuse with high probability within a fraction of a millisecond as proposed (e.g. Heidelberger et al. 1994).

The second question raised by the results of the flash-photolysis studies is whether synaptic release sites are located close enough to the calcium channels to detect microdomains of elevated calcium. The answer is likely to be "yes." Several classes of voltage-gated calcium channels have been found to interact at the molecular level with syntaxin (Bennett et al. 1992; Sheng et al. 1994; Catterall 1998; Wiser et al. 1999; Yang et al. 1999; Mochida 2000), a component of the minimal fusion machinery (Weber et al. 1998). This molecular interaction appears to be necessary for rapid, calcium-dependent synchronous release of neurotransmitter (Mochida et al. 1996). In addition, when the interaction between calcium channels and syntaxin was altered, both calcium channel function (Bezprozvanny et al. 1995; Stanley and Mirotznik 1997; Catterall 1998; Zhong et al. 1999) and the calcium-sensitivity of exocytosis was affected (Rettig et al. 1997). Bipolar neurons, similar to photoreceptors and hair cells, trigger glutamate release via calcium influx through a presynaptic voltage-gated calcium channel of the L-type (Heidelberger and Matthews 1992; Tachibana et al. 1993). This particular calcium channel has not been molecularly characterized, and so it is not yet known whether it too interacts with syntaxin. However, the isoform of syntaxin used in bipolar neurons has been identified as syntaxin 3 (Morgans et al. 1996), and it is tempting to speculate that this isoform specifically interacts with the presynaptic L-type calcium channel.

Experimental evidence in synaptic terminals also favors the interpretation that presynaptic calcium channels and the secretory machinery are co-localized. As illustrated in Fig. 2.1, when calcium-influx through voltage-gated calcium channels is used to trigger exocytosis, the corresponding spatially-averaged intraterminal calcium concentration is in the 1–2 µM range (von Gersdorff and Matthews 1994a). However, elevating the bulk cytosolic calcium to 1–2 µM is ineffective at triggering synaptic vesicle fusion. In order to trigger exocytosis when the intraterminal calcium concentration is elevated in a spatially homogeneous manner, the intraterminal calcium concentration must rise to at least 10–20 µM (Heidelberger et al. 1994; von Gersdorff and Matthews 1994a; Heidelberger 1998). These apparently disparate findings can be reconciled if one embraces the hypothesis that local microdomains of elevated calcium concentration exist in restricted regions under the plasma membrane near open calcium channels and considers that such rapidly rising and falling microdomains are not detectable with the standard fluorescence techniques that have been employed for the measurement of intracellular calcium (e.g. Augustine and Neher 1992). From this perspective, one must then assume that in order to sense these microdomains and be governed by changes in local calcium rather than by the average cytosolic calcium, the release machinery must also be located close to the presynaptic calcium channels.

The question of whether presynaptic calcium channels are colocalized with release sites in synapses has also been experimentally addressed by comparing the efficacies of exogeneous calcium buffers to suppress exocytosis. The underlying assumption is that the calcium sensor for exocytosis and the exogeneous buffer will compete for calcium ions. If the sites of calcium entry are located very near to the calcium sensor for exocytosis, an exogeneous buffer that has fast calcium binding kinetics will have a better chance at competing with the endogeneous sensor for calcium ions compared to an exogeneous buffer with an identical dissociation constant but slower on-rate (Stern 1992). If the sites of calcium entry are remote from the calcium sensor, then the on-rate of the exogeneous buffers becomes less important, and a calcium buffer with the appropriate dissociation constant will be able to intercept calcium ions before they reach the sensor. At squid giant synapse, calcium buffers with fast calcium binding kinetics are more efficacious in blocking neurotransmission than buffers with similar Kd's but slower calcium binding kinetics (Adler et al. 1991). Similarly in synaptic terminals and hair cells of the inner ear, the fast calcium buffer BAPTA is a far better blocker of exocytosis than EGTA, a calcium chelator with a dissociation constant similar to BAPTA but with a slower on-rate (von Gersdorff and Matthews 1994a; Mennerick and Matthews 1996; Sakaba et al. 1997; Moser and Beutner 2000). These results are consistent with a close co-localization of calcium channels and release sites at active zones.

Similar results are observed in peptidergic nerve terminals of the neuro-hypophysis. In these honorary neurons, BAPTA, with its faster on-rate, was found to be a better blocker of fusion than EGTA (Seward et al. 1995). Furthermore, closer scrutiny revealed that the previously reported inhibition of secretion by EGTA (Lim et al. 1990) may actually be attributable to the prevention of an obligatory calcium-dependent preparatory step rather than to inhibition of calcium-triggered fusion (Seward et al. 1995). At the calyx of Held, EGTA appreciably dampens the secretory response (Borst and Sakmann 1996), and this has been interpreted to indicate that the fusion of a synaptic vesicle at this synapse requires the opening of multiple calcium channels rather than a single calcium channel. In contrast to nerve terminals, secretion is substantially reduced or even blocked by millimolar EGTA in endocrine cells and dorsal root ganglion cell somata (Neher and Marty 1982; Augustine and Neher 1992; Ammala et al. 1993; Huang and Neher 1996). This suggests that in these secretory cells, calcium channels and release sites are not as closely co-localized as they are in nerve terminals. Taken together, the results of the experiments described in this and the preceding sections support the hypothesis that in nerve terminals, relative to neuroendocrine cells, exocytosis is typically driven by the high levels of calcium found in close proximity to calcium channels.

2.4.2
Implications for the Synchronicity of Synaptic Release

The ideal relay synapse is one that faithfully and in a time-locked manner transmits information about an incoming stimulus to a postsynaptic neuron. At the neuromuscular junction and squid giant synapse, the exocytotic response begins with a minimal delay following calcium entry (Llinas et al. 1981) and terminates within a few milliseconds after calcium entry ceases (Katz and Miledi 1965; Augustine et al. 1985). One way to achieve this high level of synchrony would be to have calcium channels and release machinery be located in the molecular proximity of each other and to have the calcium sensor for exocytosis be a calcium-binding protein with a low affinity for calcium. This would allow the calcium sensors to quickly bind calcium ions and trigger release when calcium channels open. When the channels close and the calcium microdomains begin to collapse, the calcium sensors would be able to quickly release the bound calcium ions due to their low-affinity binding sites, and exocytosis would cease in a timely manner. In contrast, if the calcium sensors were to have a high affinity for calcium, these calcium-binding proteins would hold on to the calcium ions for a longer period of time following the closure of calcium channels and this could contribute to secretion that is asynchronous with stimulus offset.

The previously presented mathematical model of the calcium-dependence of synaptic exocytosis can be used to estimate the idealized off-rate of the exocytotic response. In synaptic terminals, if one makes the assumption that there is an instantaneous drop in the calcium concentration below threshold, the synaptic terminal model predicts that secretion should decay to zero within 1 ms. Making the same idealized assumption about the drop in cytosolic calcium in chromaffin cells, secretion would be expected to take three times as long to decay (e.g. Heinemann et al. 1994). Thus, the models suggest that the high-affinity calcium sensor of these neuroendocrine cells contributes to the asynchrony of humoral secretion, whereas the low-affinity synaptic trigger aids in time-locking the synaptic exocytotic response to the stimulus. These estimates of the cessation time most likely represent the lower limit of the actual time course of the termination of release because calcium under the membrane does not instantaneously drop to zero, but will return to baseline with a time course that is determined by the mechanisms of calcium clearance. These include calcium buffering and the diffusion of buffered calcium away from release sites, sequestration of calcium into intracellular organelles, and extrusion of calcium from the cytosol.

In the retina, bipolar neurons provide the sole pathway by which information from the photoreceptors of the outer retina is relayed to neurons of the inner retina, which provide the output signal to the brain (Dowling 1987). Physiologically, bipolar neurons have been shown to possess several attributes of a good relay neuron. For example, synaptic vesicle fusion can occur extremely rapidly upon elevation of intraterminal calcium (Heidelberger et al. 1994; Mennerick and Matthews 1996). Additionally, the rate of neurotransmitter release is steeply dependent upon intraterminal calcium (Heidelberger et al. 1994), which may help to synchronize the exocytotic response. Furthermore, capacitance measurements indicate that exocytosis triggered by standard stimulation protocols ceases quickly following the closure of presynaptic calcium channels (see von Gersdorff and Matthews 1994a; von Gersdorff et al. 1996; Mennerick and Matthews 1996). However, the capacitance technique is not sufficiently sensitive to detect the fusion of an individual or a limited number of synaptic vesicle(s), and therefore the sensitive bioassay of glutamate release described in section 2.2. has also been employed to search for asynchronous release. These studies did not reveal any asynchronous release when the depolarizing voltage step was less than 200 ms in duration and was to a voltage that did not maximally activate calcium influx (von Gersdorff et al. 1998). A small amount of asynchronous release was detected, however, in response to stronger or longer depolarizations, but the amount of asynchronous release was not copious and ceased within a few hundred milliseconds (Sakaba et al. 1997; von Gersdorff et al. 1998). Thus, when calcium entry is not excessive, both the onset and offset of exocytosis are tightly coupled to calcium entry in synaptic terminals. This result is consistent with the hypothesis that synaptic exocytosis is driven by a calcium-sensitive protein (or proteins) with low-affinity calcium-binding sites.

The small component of asynchronous release that follows long, strong depolarizations in synaptic terminals contrasts with the pronounced component of asynchronous release that is observed in secretory cells that lack synaptic specializations and active zones. In adrenal chromaffin cells, for example, the onset of secretion can begin as late as 50 ms after the start of the stimulus (Chow et al. 1992). Furthermore, asynchronous secretion in chromaffin cells can last as long as several hundred milliseconds after a relatively brief depolarization (Chow et al. 1992; Chow et al. 1996). A contributing factor to this impressive asynchronous release is the chromaffin cell's high-affinity calcium sensor for exocytosis (Chow et al. 1996). The major contributor to asynchronous release in chromaffin cells, however, appears to be the lack of co-localization between calcium channels and release sites (Chow et al. 1994; Chow et al. 1996) (see section 3.3).

The calcium-binding affinities of the calcium sensors for exocytosis have not been directly characterized in other secretory cells, but it is clear from paired-recordings that in addition to a large synchronous component of release, some central neurons exhibit a second, small component of release that occurs after the cessation of a stimulus (Gleason et al. 1993; Goda and Stevens 1994; Geppert et al. 1994; Borges et al. 1995; Atluri and Regehr 1998). In most instances, the underlying mechanism of this asynchronous release has not been definitively established, but studies in cerebellar granule cell to stellate cell synapses indicate that calcium is a very important regulator of this delayed release (Atluri and Regehr 1998). Whether calcium ions bind to a different trigger molecule for asynchronous release than for synchronous release is not directly known, but the experimentally-derived model for delayed release suggests that a calcium-dependent process with slow kinetics can account for delayed release (Atluri and Regehr 1998). Interestingly, when the putative calcium binding protein that triggers neuronal exocytosis, synaptotagmin I, was genetically "knocked-out," the synchronous component of release was lost in hippocampal neurons, but the asynchronous component remained and if anything, was slightly enhanced (Geppert et al. 1994). The favored interpretation of these data is that the asynchronous component of release is triggered by a molecule that is genetically distinct from the molecule that is used to trigger synchronous release (Geppert et al. 1994). An intriguing alternative interpretation is that coupling between the presynaptic calcium channels and the fusion machinery is reduced when synaptotagmin I is not present (Neher and Penner 1994). The latter interpretation seems plausible because there are documented interactions between calcium channels, synaptotagmin and the minimal fusion machinery (Kim and Catterall 1997; Catterall 1998; Seagar et al. 1999) and because the calcium-dependent kinetics of release change when one of the components of the minimal fusion machinery is cleaved (Xu et al. 1998). Taken as a whole, the experimental evidence supports the hypothesis that the binding-kinetics of the calcium sensor(s) for release and the degree of co-localization of sites of calcium elevation and the release machinery will help to determine whether release is synchronous with the stimulus or whether it has an asynchronous component. Additionally, other factors, such as the time course of clearance of intraterminal calcium and the time course of clearance of neurotransmitter from the synaptic cleft, will also influence the off-set of the exocytotic response.

2.5
Synaptic Vesicle Pools

The concept that there might be a pool of neurotransmitter molecules within a presynaptic nerve terminal that is limited in size and is released preferentially over another pool of neurotransmitter molecules has been around almost as long as the quantal release hypothesis (e.g. Takeuchi 1958; Birks and MacIntosh 1961; Elmqvist and Quastel 1965; Betz 1970). However, our knowledge of these putative pools, the exchange between them, and their significance for short-term plasticity has been incomplete, particularly with respect to central neurons. The advent of the use of capacitance measurements to directly track synaptic vesicle fusion, combined with measurements on intracellular calcium, has permitted kinetically-distinct populations of vesicles to be defined. This section will present data that represents the current view of vesicle pools and movement between these pools, and the implications of synaptic vesicle pool dynamics for facilitation and depression of a synapse.

While reading this section, the reader should keep in mind that the descriptive names given to these functionally-defined populations of vesicles in the literature can be somewhat arbitrary. Typically the first pool studied, by virtue of its ease of fusion, has been granted the title "release-ready pool." The "ultrafast pool," also called the "immediately-releasable pool" or the "rapidly-releasable pool," is generally defined relative to the release-ready pool based upon its apparently faster kinetics of fusion. The term ultrafast will be used here to avoid confusion between the similar-sounding release-ready pool and rapidly-releasable pool, both of which have been abbreviated "RRP" in the literature. To bring uniformity to the nomenclature, in this review, the term "release-ready pool" will refer to those granules or vesicles that are independent of ATP and are fully fusion-competent upon the elevation of internal calcium. The term "ultrafast" pool will be used to denote a population of granules or vesicles that has been primed by ATP but has faster apparent fusion kinetics than the release-ready pool and is thought to be located nearer the calcium channels that drive release. This small, fast pool is released upon brief activation of voltage-gated channels and is not blocked by EGTA and only minimally affected, if at all, by BAPTA. The present definitions works quite well for chromaffin cells and synaptic terminals, where pools meeting these criteria have been well-described. Vesicle pools in other neurons that are less well-defined will be discussed in the section that seems most appropriate, given the available information.

2.5.1
Pools of Large Dense-Core Granules in Neuroendocrine Cells

In neuroendocrine cells, exocytosis triggered by calcium influx typically proceeds with a brief, rapid initial burst that is followed by a slower and gradually diminishing rate of secretion (Lim et al. 1990; Augustine and Neher 1992; Bittner and Holz 1992; Lindau et al. 1992). Neuroendocrine cell release in response to flash-photolysis of caged-calcium also shows a multi-phasic response characterized by a rapid initial burst followed by slower rates of secretion (Neher and Zucker 1993; Thomas et al. 1993a; Heinemann et al. 1994). These observations have led investigators to propose that there is a pool of vesicles available for rapid release, named the release-ready pool and that there is one or more pools of reserve vesicles that feed sequentially into this release-ready pool in a calcium-dependent manner in order replen-ish it (e.g. Heinemann et al. 1993; Neher and Zucker 1993; Thomas et al. 1993a; Heinemann et al. 1994). The release-ready pool is postulated to con-tain vesicles that are fully fusion competent and available for rapid release upon the elevation of intracellular calcium (Neher and Zucker 1993). The observed fast exocytotic response would represent fusion of granules in the release-ready pool in this model, and the slower rates of release would be due to the ongoing refilling of the release-ready pool and the subsequent fusion of these newly added granules.

In neuroendocrine cells, the physiologically described release-ready pool has a correlate pool in biochemical and morphological studies. Studies in permeabilized chromaffin cells and PC12 have shown that exocytosis can be separated into two components based upon the requirement for MgATP (Holz et al. 1989; Bittner and Holz 1992; Hay and Martin 1992). Granules in the ATP-independent pool are thought to be fully fusion competent, having been primed previously by ATP. Similar to the release-ready pool, this population of granules requires only the elevation of internal calcium in order for fusion to occur. In addition, the number of granules in this ATP-independent pool corresponds quite well to the number of granules physi-cally located at the plasma membrane (Parsons et al. 1994). This implies that the release-ready pool of granules are those granules that have been primed by ATP and are docked at the plasma membrane. However, only as small fraction of the morphologically docked granules are able to participate in the most rapid component of the exocytotic response (Parsons et al. 1994; Steyer et al. 1997). It is presently unclear whether after docking, further matura-tional steps of the granules are required in order to truly become fully-fusion competent or whether the subset of granules with the fastest release kinetics are simply more closely associated with the calcium channels and fusion machinery. Given that neuroendocrine cells lack synaptic zones, it seems reasonable to assume that only a subset of all docked and primed granules find themselves situated near calcium channels (e.g. Chow et al. 1994; Chow et al. 1996). This interpretation is supported by the discovery of a small pool of chromaffin granules that have faster fusion kinetics than the larger release-ready pool (Horrigan and Bookman 1994). This small pool was revealed only when exocytosis was triggered with very brief depolarizing voltage-steps, which is consistent with the idea that this smaller population of primed granules are located very near to calcium channels (i.e., an ultra-fast pool).

It should be noted that not all investigators favor a simple model of se-cretion that involves distinct populations of vesicles and the sequential movement of vesicles between these populations. For example, Seward and colleagues have suggested a different interpretation of the phases of exocy-tosis observed in neurohypophysial nerve endings (Seward et al. 1995). In their model, there is no release-ready pool *per se*, but there is a calcium-dependent preparatory step that occurs at low, sub-fusion-threshold calcium levels that must be completed in order for calcium-dependent fusion to take place. This preparatory step could conceivably act to create a pool of func-tional or "release-ready" granules or to prepare release sites for the next round of release (e.g. Weis et al. 1999). Reminiscent of this action of calcium is a recent report from the calyx of Held, in which intraterminal calcium is suggested to convert newly-acquired vesicles in the release-ready pool from those that are reluctant to fuse to those that will fuse with a high probability in response to calcium (Wu and Borst 1999). Another model, developed at the squid giant synapse to explain short-term depression, is that the release-machinery's sensitivity to calcium changes in an activity-dependent manner (Hsu et al. 1996). This model fits their data better than a pool-depletion model. Finally, it has been suggested that there may be intrinsic differences in the calcium-sensitivity of release of secretory granules (e.g. Blank et al. 1998a; Blank et al. 1998b). In this review, data will be interpreted primarily from the simple perspective of the vesicle pool model set forth in the preced-ing paragraphs. By selecting this perspective, it does not imply nor is it in-tended to imply that activity-dependent changes in the movement of vesicles between pools (section 2.5.5), in the size (von Ruden and Neher 1993; Gillis et al. 1996) or capacity (Stevens and Wesseling 1999) of vesicle pools, or in the release machinery are precluded.

2.5.2
Vesicle Pools in Synaptic Terminals

2.5.2.1
The Release-Ready Pool

In synaptic terminals, three different approaches have been used to identify and characterize the release-ready pool of synaptic vesicles. Each method has revealed that the release-ready pool is finite in size and can be depleted.

In the first method (Fig. 2.5A), capacitance increases were evoked by fixed amplitude voltage steps that varied in length from 20 ms to 2 sec. The amplitude of the capacitance increase was found to increase with pulse duration until it reached a plateau value of 150 fF (von Gersdorff and Matthews 1994a; Neves and Lagnado 1999). The size of the readily releasable pool has also been estimated by holding the pulse duration constant and varying the amplitude of the voltage step (Fig. 2.5B). Again, the amplitude of the capacitance jump was observed to saturate at $\approx$ 150 fF (von Gersdorff and Matthews 1997), consistent with a finite pool size. The size of the release-ready pool has also been estimated from the amplitude of the capacitance jump in flash-photolysis experiments. In these experiments, the elevation of internal calcium is extremely rapid (<<1 ms), and therefore there is virtually no opportunity for refilling of the release-ready pool to occur during either calcium elevation or the exocytotic response. Estimates of the size of the release ready pool in synaptic terminals in flash-photolysis experiments when MgATP is present are consistent with estimates obtained with voltage steps, exhibiting an average capacitance increase of $\approx$ 120 fF (Heidelberger 1998). Furthermore, a plateau is clearly observed in the individual exocytotic

---

Fig. 2.5. Estimates of the release-ready and ultrafast pool sizes. A. The relationship between the amplitude of the average capacitance increase is plotted against the duration of a voltage step from -60 mv to 0 mV. Each point represents the mean of results from 5–17 terminals $\pm$ sem. For pulse durations of 100 ms or less, the response to 5–25 pulses were averaged for each terminal in order to reduce the noise in the capacitance record. The solid line was fitted to the results for durations < 200 ms by a least squares criterion and had a slope of 730 fF/s. The dashed line is drawn at the plateau level of 150 fF, which represents the maximum average response attaned at durations > 200 ms. (Reprinted by permission from Nature 367:735–739, copyright 1994, Macmillan Magazines Ltd.) B. The average capacitance jump is plotted against the calcium current amplitude. Depolarizations were 250 msec from a holding potential of -60 mV to -20 mV, -10 mV, or 0 mV. Each point is the average from 4–11 responses, and the vertical lines show $\pm$ s.e.m. The internal pipette solution contained standard solution with either 0.5 mM BAPTA (filled circles) or EGTA (open circles). (von Gersdorff and Matthews 1997). Reprinted with permission from the Society for Neuroscience) C. Summary data from five terminals showing the average capacitance increase to six brief depolarizations from -60 mV to 0 mV. The solid line is the least-squares monoexponential fit to the first four data points. The fit is characterized by a time constant of 1.5 ms and an amplitude of 33 fF. Note the deviation of the final point from the fit suggesting a slower component of release elicited by longer pulses. (Mennerick and Matthews 1996, copyright Cell Press)

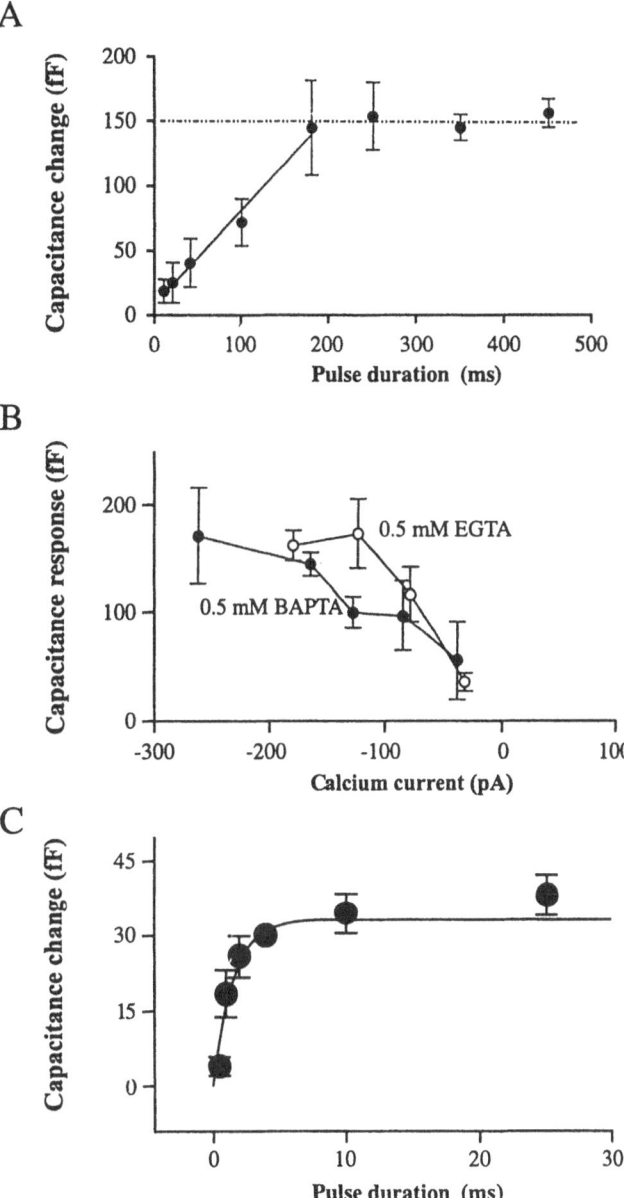

responses in flash-photolysis experiments (e.g. Fig. 2.3), despite the fact that the presynaptic calcium concentration may remain above threshold for many seconds (Heidelberger et al. 1994). Again, this observation is consistent with the interpretation that a pool of synaptic vesicles has been depleted and under the conditions and time scale of these experiments, refilling of this pool has not occurred.

A depletable pool of release-ready synaptic vesicles has also been observed in neurohypophysial nerve terminals (Lindau et al. 1994; Hsu and Jackson 1996; Giovannucci and Stuenkel 1997), and a depletable pool of vesicles has been postulated for the calyx of Held (von Gersdorff et al. 1997; Wang and Kaczmarek 1998; Wu and Borst 1999; Schneggenburger et al. 1999) and for hippocampal neurons (Stevens and Tsujimoto 1995; Rosenmund and Stevens 1996). In hair cells, the situation is somewhat more complex. Rapid depletion of a release-ready pool was not reported in hair cells of the frog sacculus (Parsons et al. 1994), and in these neurons, high rates of release were maintained for up to two seconds (Parsons et al. 1994). Within this time frame more vesicles fused than are thought to be morphologically docked at the plasma membrane at dense bodies (Parsons et al. 1994; Lenzi et al. 1999). In hair cells of the inner ear, two components of release have been observed (Moser and Beutner 2000). The fast component is depletable and may represent an ultrafast pool. The slow component exhibited maintained release for up to 1 second (Moser and Beutner 2000). At present, no explanation exists to account for the prolongued secretion observed in hair cells, but it could be that during stimulation, the rate of refilling of the release-ready pool matches the depletion rate to allow for sustained transmission. Consistent with this hypothesis is the observation that electron tomography has failed to reveal depletion of vesicles at the synaptic body of saccular hair cells following depolarization (Lenzi et al. 1999). In addition to the possibility of rapid refilling, the component of release that can be sustained in hair cells of the inner ear might result from the fusion of additional vesicles in a release-ready pool that are located at sites distant to the calcium channels (e.g. Moser and Beutner 2000). Morphologically-docked vesicles that are not located at synaptic bodies have been reported in vestibular hair cells (Lenzi et al. 1999). Clearly, more data are needed to resolve this very interesting situation in these sensory neurons.

As in chromaffin cells, the physiologically-defined release-ready pool in synaptic terminals also seems to have biochemical and morphological correlates. Experiments in which intraterminal cytosolic ATP has been very greatly reduced or replaced with a slowly-hydrolyzable analog indicate that the release-ready pool of vesicles in synaptic terminals is that pool of vesicles that has already undergone the last ATP-dependent priming step

(Heidelberger 1998). Thus, it is fair to argue that because the fusion of vesicles in the release-ready pool is ATP-independent, requiring only the elevation of intraterminal calcium to trigger exocytosis, use of the term "release-ready" to describe this physiologically-defined 150 fF pool is quite appropriate. The average amplitude of the release-ready pool of $\approx$ 150 fF corresponds to the fusion of $\approx$ 6000 vesicles, assuming an average vesicle diameter of 29 nm and a specific membrane capacitance of $1\mu F/cm^2$ (von Gersdorff et al. 1996). This number of vesicles agrees surprisingly well with the number of vesicles that are located at and tethered to the synaptic ribbons (von Gersdorff et al. 1996). Furthermore, a nice correlation exists between the size of a synaptic terminal, the number of synaptic ribbons it contains, and the amplitude of the capacitance response (von Gersdorff et al. 1996). While not definitive, these data are highly suggestive of the possibility that the release-ready pool of synaptic vesicles may be that population of vesicles that are fully-fusion competent and tethered to the synaptic ribbons, placing these vesicles within several vesicle diameters of the fusion machinery. Furthermore, there is no evidence in the literature of active zones in these particular neurons other than those at the synaptic ribbons.

2.5.2.2
The Ultrafast Pool

In addition to a release-ready pool, there is strong evidence for an ultrafast pool of vesicles in synaptic terminals (Fig. 2.5C). This small pool, whose $\approx$ 30–45 fF size corresponds to the fusion of $\approx$ 1100–1700 vesicles, was detected when very brief (1–30 ms duration) depolarizations were used to probe the exocytotic response (Mennerick and Matthews 1996; Gomis et al. 1999). The rate of fusion extrapolated from a plot of amplitude versus pulse duration was estimated to be $\approx$ 670/sec or $\approx$ 20,000 fF/s (Mennerick and Matthews 1996). This is significantly faster than the extrapolated fusion rate for the release-ready pool by more than an order of magnitude (von Gersdorff and Matthews 1994a). Indeed, the rate of fusion of this small pool is so rapid that it is limited by the activation kinetics of the presynaptic calcium channels (Mennerick and Matthews 1996). When calcium influx is evoked through pre-activated calcium channels, the rate of fusion of the ultrafast pool is even faster than reported above and is comparable to the maximum rate observed in flash-photolysis experiments (Heidelberger et al. 1994; Heidelberger 1998).

The rapid rate of fusion reported for the ultrafast pool suggests that this vesicle pool might be situated molecularly near to the presynaptic calcium channels. This interpretation is supported by studies that compare the relative efficacies of EGTA and BAPTA to block this component of release. EGTA, the slower calcium buffer, was unable to block release from the ultra-fast pool, but it could damp release to some extent from the release-ready pool (Mennerick and Matthews 1996; Sakaba et al. 1997). The fast calcium buffer BAPTA depressed release from both pools (Mennerick and Matthews 1996). These results suggest that relative to the release-ready pool, the ≈ 30 fF ultrafast pool is located nearer to the presynaptic calcium channels. Supporting this conclusion are the results of a study in which serial recon-structions of bipolar neuron ribbon synapses were performed at the EM level (von Gersdorff et al. 1996). The morphological estimate of 1000–1400 docked vesicles (von Gersdorff et al. 1996) corresponds quite nicely with the reported size (≈ 30 fF or ≈ 1100 vesicles) and the hypothesized position of the ultrafast pool (at the active zone membrane) suggested by the physio-logical experiments. The existence of a small population of synaptic vesicles that are located nearer the presynaptic calcium channels than other fusion-competent vesicles and have faster fusion kinetics may not be unique to synaptic ribbons. Hair cells of the inner ear (Moser and Beutner 2000) and neurohypophysial nerve endings (Hsu and Jackson 1996, Giovannucci and Stuenkel 1997) are also reported to have a small pool of fusion-competent vesicles that is not effectively blocked by EGTA and may represent an ultra-fast pool. Preliminary evidence in cone photoreceptors also suggests the presence of an ultrafast pool of synaptic vesicles (DeVries 2000). An ultrafast component of release has not yet been sought in saccular hair cells.

The identification of an ultrafast pool in synaptic terminals is significant for several reasons. First, the rapid rate of fusion provides evidence that a physiological stimulus can trigger the fusion of synaptic vesicles with rates that approximate those achieved by flash-photolysis of caged-calcium. If one makes the assumption that the calcium-dependence of the rate of release is similar regardless of whether release is driven by membrane depolarization or by global elevation of intraterminal calcium via the flash-photolysis of caged-calcium, then one can use the rate of release derived from a depolari-zation experiment and the relationship between the rate of membrane addi-tion and intraterminal calcium determined in flash-photolysis experiments to estimate the calcium concentration seen by the release machinery during membrane depolarization. From this type of calculation, it appears that in synaptic terminals, the average synaptic vesicle in the rapidly-releasable pool senses a calcium concentration of at least 100 µM and possibly as high as 325 µM during a brief membrane depolarization (Fig. 2.6). These data therefore provide additional support for the hypothesis that the calcium concentration near release sites can be very high when presynaptic calcium

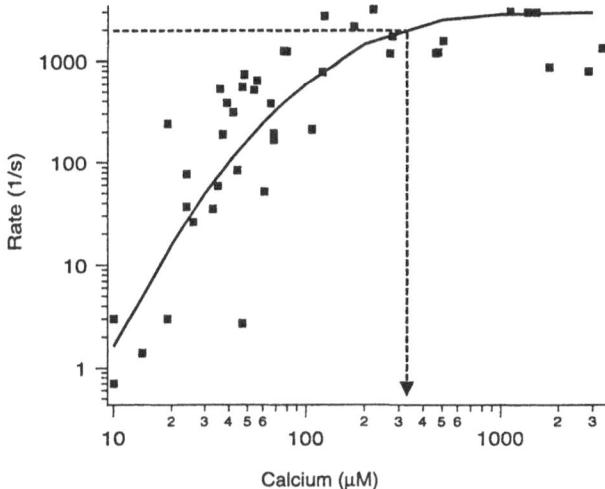

Fig. 2.6. The relationship between rate of membrane addition and intraterminal calcium can be used to back-calculate the calcium concentration at the release site. Synaptic vesicles in the ultrafast pool can fuse with a rate of ≈ 2000 s^{-1}, provided that the calcium-channel kinetics are not rate-limiting. A rate of 2000 s^{-1} corresponds to a calcium concentration of ≈ 325 μM near release sites (dotted line), according to to the synaptic terminal secretion model (curve). The data points (squares), independently of the model, indicate that the intraterminal calcium must rise to more than 100 μM or more in order to achieve such a high rate of fusion. (Graph is reprinted by permission from Nature 371:513–515, copyright 1994, Macmillan Magazines Ltd)

channels open. Secondly, the identification of the rapidly-releasable pool resolves the mystery of why the average rate of fusion of the release-ready pool is slower in a depolarization experiment than in a flash-photolysis experiment (e.g. compare Heidelberger et al. 1994; von Gersdorff and Matthews 1994a). One possibility is that in response to membrane depolarization, the average calcium concentration near release sites rises no higher than ≈ 20–50 μM due to imperfect coupling between calcium channels and release sites. The discovery of a rapidly-releasable pool suggests that this is not so, but rather, it is the calcium concentration near the population of release-ready vesicles that averages 20–50 μM. Given the structure of bipolar neuron ribbon synapses and proposed arrangement of the release-ready pool (von Gersdorff et al. 1996), it is reasonable to postulate that some vesicles in the fusion-competent pool are closer to the sites of calcium entry then others and therefore experience a higher calcium concentration (e.g. the rapidly-releasable pool). Similarly, vesicles located at a distance may see lower concentrations of calcium. The end result is that the population fusion rate of the release-ready pool is retarded relative to the fusion rate of the very fastest vesicles. Finally, the identification of the rapidly-releasable fast pool is significant because in characterizing this pool, it has become clear that the activation time course of the presynaptic calcium channel is able to influence the rate of neurotransmitter release (Mennerick and Matthews 1996). This raises the possibility that modulation of the activation kinetics of presynaptic calcium channels will also modulate the rate of exocytosis and possibly contribute to the plasticity of a synapse.

2.5.3
Depletion and Refilling of Synaptic Vesicle Pools

2.5.3.1
The Rate of Depletion

The rate of depletion of a release-ready pool of vesicles has been estimated for several neurons. For bipolar neuron synaptic terminals, the average rate at which vesicles in the release-ready pool are exocytosed in response to activation of voltage-gated channels is $\approx$ 28,000 vesicle/sec (Matthews 1996b), assuming a vesicle diameter of 29 nm (von Gersdorff et al. 1996). This rate is of similar magnitude to the depletion rate of the release-ready pool in saccular hair cells (Parsons et al. 1994) and the slow component of release in inner ear hair cells (Moser and Beutner 2000) and virtually identical to the rate of depletion of the fast component of release (Moser and Beutner 2000). There are multiple synaptic areas within each of these neurons, and therefore another way to express the rate of release is as a function of the number of synaptic ribbons. For the bipolar neuron synaptic terminal, which has approximately 60 synaptic ribbons (von Gersdorff et al. 1996), the depletion rate can be expressed as 500 vesicles/sec/synaptic ribbon (Matthews 1996b). For a saccular hair cell with 20 synaptic bodies (Roberts et al. 1990), a similar rate (500 vesicles/sec/synaptic body) of depletion is revealed. If one assumes that all of the vesicles in the rapidly-releasable pool represent vesicles that are sitting at release sites, the rate of release in a synaptic terminal then falls to $\approx$ 21 vesicles/release site/sec. This rate is comparable to the rate of 22 vesicles/sec/release site reported for the sustained component of exocytosis in amacrine cell synapses (Gleason et al. 1993; Borges et al. 1995) and for the rate of release reported in hippocampal neurons (Stevens and Tsujimoto 1995). The number of active zones and docked vesicles has not yet been established for inner ear hair cells.

2.5.3.2
Refilling of the Release-Ready Pool

Another question that has been addressed using capacitance techniques is the length of time it takes to refill the release-ready pool of vesicles following depletion. The answer to this question provides valuable information about short-term depression of the synapse. In response to the standard depolarization, synaptic terminals of bipolar neurons exhibit paired-pulse depression (Mennerick and Matthews 1996; von Gersdorff and Matthews 1997). Recovery of the release-ready pool from paired-pulse depression proceeds with a time constant of ≈ 8 seconds (Fig. 2.7), taking slightly more than 20 seconds to be completely refilled under standard conditions (von Gersdorff and Matthews 1997). Similar estimates of the time course of refilling have been obtained in cultured hippocampal neurons using optical techniques. In these studies, the reported time constants of refilling ranged between 10–40 seconds. (Ryan et al. 1993; Ryan and Smith 1995; Liu and Tsien 1995; Stevens and Tsujimoto 1995). These measurements reflect the refilling time of a population of vesicles, rather than the time required to refill an individual release site. The latter may be an order of magnitude faster (Stevens and Wang 1995).

The population of synaptic vesicles that serves to refill the release-ready pool has not been identified, but several possibilities exist. The first is that the reserve pool is comprised of all the vesicles in the synaptic terminal that are not docked and primed. The synaptic terminals of bipolar neurons typically contain 700,000–900,000 synaptic vesicles (von Gersdorff et al. 1996), of which only 6,000 are thought to comprise the release-ready pool (Matthews 1996a; Matthews 1996b). This leaves a putative reserve pool that could be as large as 894,000 vesicles. It is not known definitively whether the vesicles in this pool are equivalent with respect to maturational stage and mobility, but physiological results obtained in synaptic terminals suggest that they are not. For example, when synaptic terminals are stimulated with a train of depolarizations to evoke release, the cumulative increase in membrane capacitance saturates at 300 fF (von Gersdorff and Matthews 1997). Given that the amplitude of the release-ready pool is 150 fF, this result suggests that the release-ready pool can be reloaded one time before the synapse becomes fully depressed. These data also raise the possibility that there may be a small, depletable reserve pool that is used preferentially to fill the release-ready pool. If such a small reserve pool does exist and refilling of the release-ready pool occurs preferentially via this small reserve pool, then recovery from depletion of the release-ready pool would be expected to take

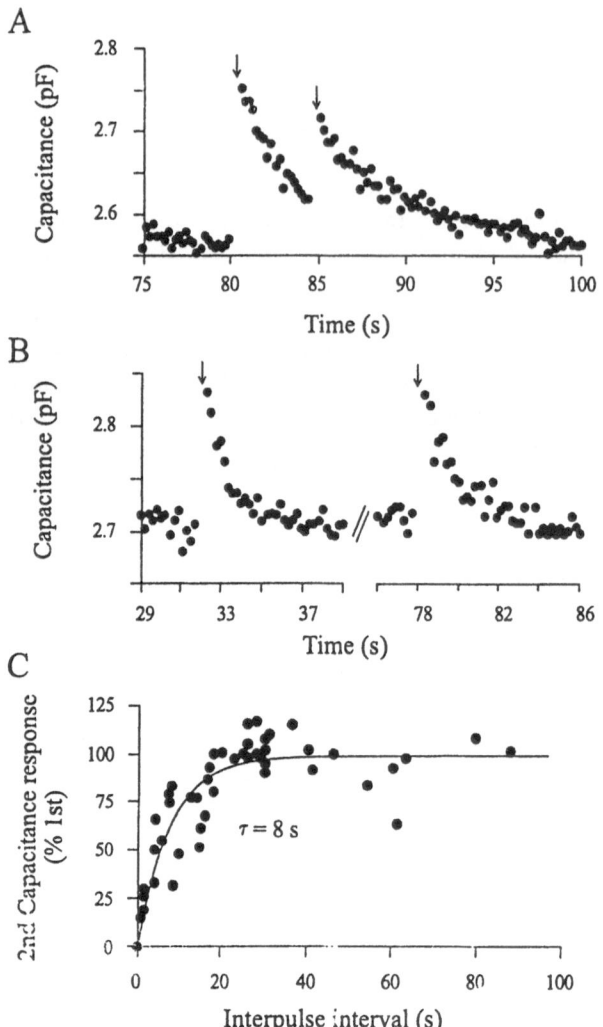

Fig. 2.7. Exocytosis in synaptic terminals exhibits paired-pulse depression. **A** Two capacitance responses were elicited about 4 s apart. The arrows show the timing of 250 ms pulses from –60 mV to 0 mV. The indicated time is relative to break-in. **B** Two capacitance responses were elicited about 46 s apart. The arrows indicate the timing of 250 ms pulses from –60 mV to 0 mV. **C** Summary of experiments like those shown in A and B. To compare results across cells, the second capacitance response is expressed as a percentage of the first response. The interval between pulses is shown on the abscissa. The curve is the best-fit single exponential, which has a time constant $\tau = 8$ s. (von Gersdorff and Matthews 1997). Reprinted with permission from the *Society for Neuroscience*

longer if this reserve pool is also depleted. Consistent with this prediction, the time course of recovery of the release-ready pool after a pulse train was found to take more than the usual 20 seconds required to refill the release-ready pool after depletion by a single stimulus (von Gersdorff and Matthews 1997). In addition to the pool-depletion model, alternative interpretations of these data exist. These include the possibility that some intracellular factor necessary for pool-refilling or vesicle fusion is depleted during a pulse train or that some factor that inhibits pool-refilling or vesicle fusion accumulates during the pulse trains. Another possibility, is that there an activity-dependent decrease in the capacity of the release-ready pool (Stevens and Wesseling 1999).

2.5.3.3
Refilling of the Rapidly-Releasable Pool

Similar to the release-ready pool, the rapidly-releasable pool also appears to be finite in size (Fig. 2.5C) and able to be depleted (Mennerick and Matthews 1996; Gomis et al. 1999). The time constant of refilling of the ultrafast pool after paired-pulse depression was estimated to be of the same order of magnitude as the release-ready pool (≈ 4 s vs ≈ 8 s) when refilling occurred between depolarizations (Mennerick and Matthews 1996). Other investigators have reported an additional, fast component of refilling ($\tau_1 \approx 640$ ms) (Gomis et al. 1999). Fast and slow components of refilling have also been reported for the fast component of release observed in hair cells of the inner ear ($\tau_1 \approx 140$ ms; $\tau_2 \approx 3$s) (Moser and Beutner 2000) and for refilling at the calyx of Held (Wu and Borst 1999). The significance of these two components remains unclear, but they could represent the kinetics of refilling at different release sites (Moser and Beutner 2000). It has also been postulated that the fast component of refilling might possibly reflect a poststimulus acceleration of refilling due to residual calcium (Gomis et al. 1999; Moser and Beutner 2000). If so, this implies that in bipolar cells, the location of residual calcium and the sites it acts on to speed the functional refilling of the rapidly-releasible pool must be located near to each other as this fast component of release is not sensitive to EGTA (Gomis et al. 1999). In contrast, calcium-dependent acceleration of the fast component of refilling does not appear to be present at calyx of Held (Wu and Borst 1999). The functional significance of these two components of pool refilling remains obscure.

The population of vesicles that is used to replenish the ultrafast pool has not yet been firmly established. Interestingly, although one might anticipate that the vesicles in the ultrafast pool are supplied by vesicles in the release-ready pool, the time course of refilling of the ultrafast pool appears to be independent of the state of the release-ready pool. For example, the time course of refilling of the ultrafast pool was not altered by the prior depletion of the release-ready pool (Mennerick and Matthews 1996). In addition, if 5 mM EGTA was added to the internal solution to preferentially block release from the release-ready pool, there was no effect on the refilling rate of the ultrafast pool (Mennerick and Matthews 1996). These data seem to suggest that at basal calcium, refilling of the ultrafast pool does not occur via the release-ready pool. This raises the question of which pool acts as the reserve pool for the ultrafast component. It would be interesting to determine whether the small reserve pool described in the preceding paragraph subserves this function or whether vesicles come from the reserve pool at large. Finally, it has been noted that during periods of elevated calcium, the refilling of the ultrafast pool seems to be more an order of magnitude more rapid than at resting calcium (Mennerick and Matthews 1996). The role of calcium in pool refilling will be discussed in section 2.5.4.2.

2.5.4
Conversion of Paired-Pulse Depression Into Facilitation

As stated earlier, the synaptic terminal of the bipolar neuron exhibits paired-pulse depression under standard whole-cell recording conditions (Mennerick and Matthews 1996; von Gersdorff and Matthews 1997). The depressed second response in a pair could not be attributed to a reduction in calcium influx, and this suggests that a locus downstream of calcium entry is responsible for synaptic depression. When calcium entry was reduced by decreasing the stimulus intensity or when the calcium-buffering capacity of a terminal was increased by the addition of exogeneous calcium buffer, paired-pulse depression was converted into paired-pulse facilitation (von Gersdorff and Matthews 1997). Certainly, these results show that paired-pulse depression and facilitation can have a purely presynaptic component, since the capacitance technique is a purely presynaptic technique and the isolated synaptic terminal preparation rules out changes in postsynaptic receptor activity and retrograde signaling as mediating mechanisms. These results also indicate that activity, via changes in intraterminal calcium, can modify some aspect of the release machinery that is downstream of the presynaptic calcium channels and lead to a change in synaptic efficacy.

Perhaps the simplest way to conceptualize paired-pulse facilitation and depression in synaptic terminals is with a simple pool-depletion model. The basic premise is that there is a release-ready pool of fixed size, and if the first stimulus evokes the fusion of the majority of synaptic vesicles in the release-ready pool, fewer vesicles will be available to be released in response to an identical, second stimulus. (This presupposes that the second stimulus is given shortly after the first so that pool-refilling does not occur between stimuli.) Similarly, if the first stimulus is weak, only a few vesicles are released, leaving a large release-ready pool available for release by a second stimulus. If this second stimulus comes on the heels of the first such that there is a summated calcium signal, then release in response to the second stimulus will be greater than that in response to the first. An analogous interpretation has been proposed for hippocampal neurons, in which synapses with a high probability of release have more vesicles available for fusion and tend to exhibit paired-pulse depression, whereas synapses with a low release probability often have less vesicles available for release and show paired-pulse facilitation (Dobrunz and Stevens 1997; Murthy et al. 1997). Similar relationships between release probability and facilitation and depression have also been reported in the cerebellum (Dittman and Regehr 1998; Dittman et al. 2000) and in neurons of the anteroventral cochlear nucleus (Bellingham and Walmsley 1999; Oleskevich et al. 2000). At these synapses, modification of the release sites by calcium, rather than pool depletion, was the favored interpretation (Dittman and Regehr 1998; Bellingham and Walmsley 1999). Physiologically, however, it may be difficult to distinguish between pool depletion and refractory release sites, given that the size and rate of pool refilling is likely to also be regulated by intraterminal calcium (see section 2.5.5).

These reports indicate that a better understanding of the role that presynaptic calcium plays in regulating synaptic transmission at other steps in the secretory pathway, in addition to its role in triggering exocytosis, is needed. Future directions of investigation should undoubtedly include an examination of the role of presynaptic calcium-binding proteins in modulating release and a study of the calcium-dependent modulation of synaptic proteins, particularly those that comprise or interact with the fusion machinery. Pairing these studies with ultrastructural analysis of facilitated and depressed synapses would aid in determining whether these proteins influence the number of synaptic vesicles docked at the active zones, in addition to possibly altering the secretory machinery.

2.5.5
Movement Between Pools

2.5.5.1
Neuroendocrine Cells

In one model of secretion proposed for chromaffin cells, the transitions from the reserve pool to the release-ready pool and from the release-ready pool to the fused state are hypothesized to be calcium-dependent (Heinemann et al. 1993). While this model may be considered simplistic in that it assumes that calcium is uniform throughout the cell and that at any instant the calcium concentration acting at each of the calcium-dependent steps is the same, the model is a reasonable first approximation. This assumption seems valid from the perspective that it probably takes longer to refill the release-ready pool than it does to achieve spatial homogeneity of intracellular calcium. The appeal of this model is that it has stimulated the formulation of testable predictions about the role of calcium in regulating the size of the release-ready pool and in determining the time course of pool refilling in chromaffin cells (e.g. Heinemann et al. 1993; von Ruden and Neher 1993; Smith et al. 1998).

To test the hypothesis that intracellular calcium can facilitate the refilling of the release-ready pool in chromaffin cells, augmentation and depression of the secretory response were examined with respect to intracellular calcium (von Ruden and Neher 1993). In these experiments, a brief train of depolarizing pulses was given. The capacitance response to successive pulses in the train typically showed a pronounced depression that could not be attributed to a reduction in calcium influx or to changes in calcium handling within the cell. If the cell was allowed to rest for about 1 minute and then rechallenged, the capacitance response amplitudes recovered. The interpretation is that the depression of the secretory response represents the depletion of a pool of granules and the recovery of the secretory response respresents replenishment of this vesicle pool. If the intracellular calcium was raised modestly between pulse trains by either membrane depolarization or by the release of calcium from internal stores, the recovery of the capacitance response to a depolarizing train was enhanced both with respect to the rate of recovery and the amplitude of the post-recovery capacitance response (von Ruden and Neher 1993). More recently, it has been shown that the stimulatory effect of calcium on refilling of the release-ready pool of granules is reduced but not blocked by the inhibition of protein kinase C (PKC). This suggests that calcium acts via two distinct pathways to regulate the supply of fusion competent granules (Smith et al. 1998). Since both cytosolic calcium and PKC activity are affected by synaptic stimulation, a logical conclusion is that the recent stimulus history of a synapse will influence the number of secretory granules that are available for release, and therefore affect the probability of release (e.g. Rosenmund and Stevens 1996).

2.5.5.2
Synaptic Terminals

The calcium-dependence of pool refilling has not been as rigorously studied in synaptic terminals, but several pieces of evidence suggest that calcium-dependent pool-refilling exists at these synapses. For instance, compare the capacitance jump evoked by a 2 second depolarization with the capacitance jump evoked by a 250 ms depolarization (Fig. 2.8) (von Gersdorff and Matthews 1997). If one considers that the time constant of refilling of the release-ready pool is ≈ 8 seconds, one would predict that the capacitance increase evoked by the 2 second depolarization should be 20% larger than that evoked by the 250 ms depolarization. However, the measured amplitude of the capacitance response evoked by the second depolarization is more than

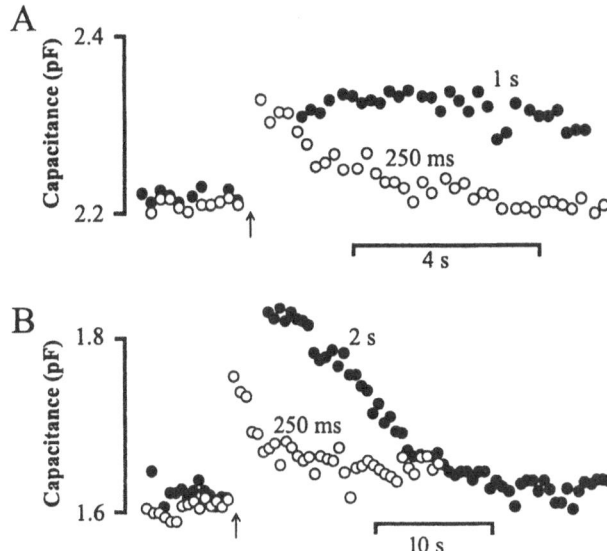

Fig. 2.8. Recovery from paired-pulse depression can occur during, and may be accelerated, during membrane depolarization. Two responses from the same terminal are superimposed. Both responses were elicited at the arrow by depolarizations from a holding potential of –60 mV to 0 mV. The first lasted 250 ms (open circles) and the second, given 179 (s) afterwards, lasted 2 s. (von Gersdorff and Matthews 1997). Reprinted with permission from the Society for Neuroscience

20% greater than the response to the 250 ms depolarization. This enhancement of the amplitude of the capacitance response can be interpreted as an indication that during the calcium entry, not only do calcium ions trigger release, but they can also accelerate the refilling rate of the release-ready pool. Consistent with these data is the finding that refilling of the release-ready pool is sensitive to exogenous calcium buffer (Gomis et al. 1999). An argument for calcium-dependent refilling of the ultrafast pool has also been made (Mennerick and Matthews 1996). Recovery from paired-pulse depression of the ultrafast pool has a time constant of ≈ 4 seconds. In contrast, an estimate of the refilling rate extrapolated from the relationship between the amplitude of the capacitance increase and the pulse duration yields a time constant of only 350 ms. These data indicate that refilling of the ultrafast pool is an order of magnitude faster when calcium channels are open than when calcium channels are closed.

The above experiments are highly indicative of the possibility that calcium can enhance replenishment of synaptic vesicle pools. These results do not preclude other interpretations, such as that membrane depolarization itself can influence mobilization of synaptic vesicles or that the calcium-dependent step(s) is/are acting to prepare vesicles (Seward et al. 1995) or release sites (Weis et al. 1999) for release. However, there is additional experimental evidence to support a direct role for calcium in mobilizing vesicles to the active zone. In the bioassay for glutamate release from synaptic terminals described in previous sections, the authors observed that when the presynaptic calcium current is less than 50 pA/pF in magnitude, the pseudo post-synaptic response contains two distinct components (von Gersdorff et al. 1998). When calcium entry is greater than 50 pA/pF, the two components of the post-synaptic response meld together to give a single component (von Gersdorff et al. 1998). The interpretation is that the first component represents fusion of vesicles in the ultrafast pool, and the second component represents the fusion of vesicles that belong to the pool that feeds into the ultrafast pool in a calcium-dependent manner. To specifically test whether the second component of release was sensitive to intraterminal calcium, EGTA was introduced into the synaptic terminal via the patch-pipette (von Gersdorff et al. 1998). Under these conditions, the separation between the two components of glutamate release was increased, as would be predicted if calcium accelerates the mobilization of synaptic vesicles from a second pool to an ultrafast pool, and the chelation of calcium by EGTA impedes this mobilization. These results suggest that when calcium is high, mobilization is rapid and this causes a blending of the two components. Conversely when calcium is low, mobilization proceeds more slowly, and this allows the two components to be temporally distinguished. Alternatively, the first component may represent fusion of vesicles nearest the calcium channels and the second component may represent fusion of vesicles located a little further away. For a ribbon synapse, the two possibilities go hand in hand because the vesicles tethered near the base of the ribbon, believed to comprise the ultrafast fast pool, are those which are closest to the plasma membrane and calcium channels. The vesicles on the upper rows may serve as the second pool while at the same time being located more distantly from calcium channels. Additional evidence to support a direct role for calcium in synaptic vesicle mobilization comes from the drosophila neuromuscular junction. In this preparation, raising the external calcium concentration results in an increase in the number of vesicles that are docked at release sites (Koenig et al. 1993). Calcium-dependent refilling of vesicle pools also been postulated for other central neurons (Dittman and Regehr 1998; Stevens and Wesseling 1998; Wang and Kaczmarek 1998), in addition to invertebrate neurons (Gingrich and Byrne 1985).

2.5.6
ATP and Refilling of the Release-Ready Pool

Studies of secretion in permeabilized neuroendocrine cells have established a functional role for ATP in the priming of large dense-core granules for calcium-dependent fusion and the maintenance of exocytosis (e.g. Baker and Knight 1978; Holz et al. 1989; Bittner and Holz 1992; Martin et al. 1995; Banerjee et al. 1996). Recently, electrophysiological evidence has established a role for ATP in functionally replenishing the supply of release-ready of small clear-core synaptic vesicles in synaptic terminals (Heidelberger 1998). In these experiments, the release-ready pool was first depleted by flash-photolysis of caged-calcium and then probed after allowing sufficient time for pool replenishment to occur. When MgATP was omitted from the internal recording solution or replaced with the non-hydrolyzable analog ATP-γ-S, the exocytotic responses following a pool-depleting stimulus were greatly attenuated (Fig. 2.9). This was true even when the interval between stimulations was increased to six times the standard time constant of refilling to allow for slow refilling of the release-ready pool (Heidelberger 1998). Importantly, such a large reduction in the amplitude in the exocytotic response was not observed when MgATP was included in the internal solution (Heidelberger 1998). Because the intraterminal calcium was elevated via flash-photolysis of caged-calcium and measured with a calcium-indicator dye, the reduction in the amplitude of subsequent capacitance responses cannot be attributed to a change in calcium. A requirement for MgATP in replenishing the release-ready pool in synaptic terminals has also been re-

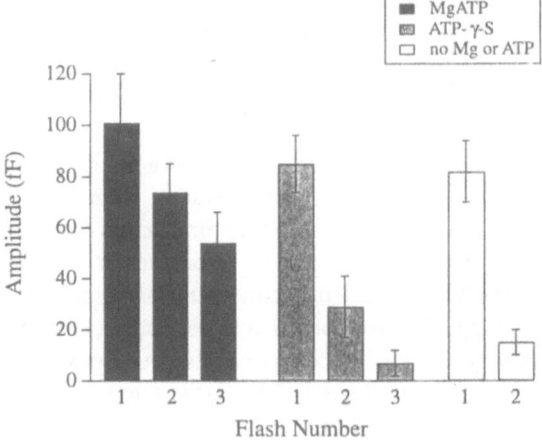

Fig. 2.9. Terminals with MgATP maintained the ability to respond to stimulation better than those without. The average amplitude of the capacitance response to the first, second and third UV-flash, given at a super-threshold intensity, is compared across three groups of synaptic terminals. Flashes were separated by a minimum of ≈ 50 s, and typically > 60 s, to allow refilling of the release-ready pool of synaptic vesicles to occur between stimulations. The solid bars indicated data from terminals dialyzed with an internal solution containing MgATP. In the second group, indicated by the gray bars, terminals were dialyzed with an ATP–γ–S-containing solution. In the final group, indicated by the hatched bars, terminals were dialyzed with solution that which contained neither Mg^{2+} nor nucleotide. Error bars indicate ± 1 s.e.m. (Reproduced from The Journal of General Physiology, 1998, 111, 225–241, by copyright permission of The Rockefeller University Press)

ported under conditions in which intraterminal calcium is elevated by the activation of presynaptic voltage-gated calcium channels (Heidelberger and Matthews 1997). Thus, functional replenishment of the release-ready pool in synaptic terminals appears to be dependent upon the presence of cytosolic MgATP. Therefore, these results suggest that without ATP, the synapse will exhibit a profound and lasting depression following the depletion of the release-ready pool. The ATP-dependence of synaptic vesicle pool-refilling and maintained release has not been probed in other central neurons, but there is strong evidence from the drosophilia neuromuscular junction to indicate that maintenance of the release-ready pool of synaptic vesicles requires the hydrolysis of ATP by the protein NSF (N-ethylmaleimide-sensitive fusion protein) (Kawasaki et al. 1998; Tolar and Pallanck 1998).

Although the above results convincingly demonstrate that ATP is required to maintain release in synaptic terminals, they do not answer the question of where in the secretory pathway this ATP requirement lies. For example, it could be that ATP is required for localizing vesicles to the release sites and that in the ATP-depleted state, vesicles are simply not docked at the active zones. A second possibility is that without ATP the vesicles are correctly docked at active zones, but the vesicles are incapable of calcium-dependent fusion. A third possibility is that the vesicles are properly docked at active zones, but it is the release sites that are not able to accommodate another round of fusion. The first hypothesis has only been examined at the ultrastructural level with respect to the function of the ATPase NSF, and it appears that hydrolysis of ATP by NSF does not play a significant role in the docking of synaptic vesicles (Kawasaki et al. 1998; Schweizer et al. 1998). Additional putative roles for ATP in vesicle docking have not yet been addressed. For the second hypothesis, several reactions have been identified which implicate ATP in vesicle priming. The first is that ATP is required as a phosphate donor to support kinase activity. For example, phosphorylation by myosin light-chain kinase has been proposed to be important in ATP-dependent priming in chromaffin cells (Kumakura et al. 1994). In addition, a requirement for ATP in the priming of large dense-core granules has been shown to be mediated via the phosphorylation of phosphatidyl inositol (Hay and Martin 1993; Hay et al. 1995; Martin 1998). The evidence suggests that these phosphorylation reactions are quite important for the priming of large dense-core granules, but it is not yet clear whether they play also a role in the fusion of small clear-core synaptic vesicles (e.g. Khvotchev and Sudhof 1998; but see Wiedemann et al. 1998). The second and third hypotheses share some overlap in the sense that it can be difficult to distinguish between a vesicle that is reluctant to fuse and a release site that is not able to accommodate fusion. Locus aside, the evidence suggests that the hydrolysis of ATP by NSF is necessary for priming of the fusion machinery. Indeed, hydrolysis of ATP by NSF has most recently been proposed to either rearrange SNARE complexes following fusion so that they can engage in subsequent rounds of release (Kawasaki et al. 1998; Jahn and Sudhof 1999) or to prime vesicles (Banerjee et al. 1996; Kawasaki et al. 1998) or the release machinery (Tolar and Pallanck 1998) for fusion. In synaptic terminals, the available data does not allow a requirement for ATP in a hydrolysis reaction to be distinguished from a phosphorylation step for which ATP-γ-S is a poor donor, and therefore each of these mechanisms must be considered.

2.5.7
Synthesis of Synaptic Vesicle Pool Information

The results presented in the following sections suggest that the sequential models of exocytosis developed for neuroendocrine cells need to be modi-

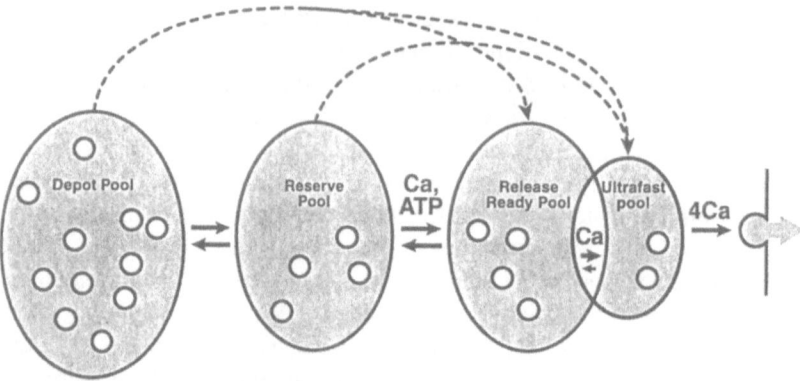

Fig. 2.10. A schematic model of exocytosis in synaptic terminals. Solid arrows indicate those transitions which are very likely to exist. Dashed lines indicate those transitions which may be possible, but for which there is no direct evidence

fied in order to account for the data on exocytosis and synaptic vesicle pools in synaptic terminals. One important deviation from earlier models is the possibility that vesicle pools can be refilled from more than one source (Fig. 2.10). In addition, this revision of the neuroendocrine cell models highlights the possibility that in synaptic terminals, as in neuroendocrine cells, there may be more than one calcium-dependent step in the refilling pathway. Calcium- and activity-dependent steps in replenishing synaptic vesicles pools have not always been considered when modeling synaptic depression and facilitation in the central nervous sytem, and the results presented earlier suggest that they should be. Finally, the revised model provides a simple framework with which to conceptualize the new synaptic data.

First, consider the ultrafast pool of vesicles. This pool presumably represents those docked and primed vesicles nearest the presynaptic calcium channels, given the evidence presented earlier. This small pool can proceed directly to calcium-dependent fusion with minimal delay. Next, consider the release-ready pool. This larger pool can fuse with the similar rates as the ultrafast pool, provided that calcium is uniformly elevated within the terminal. This, along with the comparative actions of EGTA and BAPTA on release from this pool and its ATP-independent release, suggests that the release-ready pool may differ from the ultrafast pool only in that the release-ready pool is located more distantly from the calcium channels. Therefore, in this model, the ultrafast pool is depicted as a subset of the release-ready pool. The present schematic further assumes that under physiological conditions, vesicles must pass through the ultrafast pool in order to fuse.

The evidence suggests that the rate of refilling of the release-ready pool is enhanced by the elevation of intraterminal calcium. It also seems that there may be a particular reserve pool that can refill the release-ready pool one time before it, too, is depleted. In this model, transitions from this reserve pool to the release-ready pool are depicted as being calcium-dependent, but it could also be that transitions from the depot pool are calcium-dependent. In addition to calcium, the functional replenishment of the release-ready pool was shown to require ATP. The actual site or sites of ATP action have not been established, but presumably if the requirement lay at the depot pool to reserve pool transition, terminals without ATP would have exhibited a standard amplitude second, and this was not typically the case. Therefore, in this model ATP is shown to be required for refilling of the release-ready pool from the reserve pool. It should be kept in mind that there be more than one step in the secretory pathway that requires ATP, including those ATP-dependent reactions that are discussed above and that the actual locus of ATP action has not yet been determined.

Refilling of the ultrafast pool was also shown to have a calcium-dependent component. A calcium-dependent translocation step along the ribbon could conceivably account for this reaction, and therefore, a calcium-dependent exchange of vesicles between the release-ready pool and the ultrafast is postulated. However, a calcium-dependent refilling step may occur from one of the other pools. In contrast to what happens during a stimulus, at basal calcium refilling of the ultrafast pool may occur independently of the release-ready pool. The identity of the reserve pool that subserves this function is unknown, so in this model, this slower form of refilling of the ultrafast pool is suggested to come from both a small reserve pool and a larger, ill-defined depot pool.

2.6
Endocytosis

Endocytosis serves the dual purpose of maintaining the surface area of a cell and retrieving and recycling vesicular components. Endocytosis can be critical to ongoing activity at a synapse, particularly if replenishment of the release-ready pool depends to a large extent on the retrieval and local recycling of endocytosed vesicles. Under such conditions, suppression of endocytosis leads to synaptic depression (Koenig et al. 1993). Therefore, modulation of endocytosis must be considered as a mechanism by which the strength of a synapse might potentially be regulated (Klingauf et al. 1998). Endocytosis can be examined with excellent temporal resolution using the capacitance technique. A potential problem may arise if exocytosis and endocytosis overlap temporally, but for synaptic terminals of retinal bipolar neurons, overlap does not appear to occur with the standard stimulation protocols (von Gersdorff et al. 1998). Chromaffin cells, on the other hand, will have some contamination of endocytosis with on-going exocytosis, but probably only for the first 300 ms after the depolarization (Engisch and Nowycky 1998). The extent to which endocytosis and exocytosis occur concurrently in other secretory cells is unknown.

2.6.1
Endocytosis in Synaptic Terminals

In bipolar neuron synaptic terminals, endocytosis follows exocytosis and restores the membrane surface area precisely to the prestimulus baseline. For a single depolarization from –60 to 0 mV for 500 ms, the time constant of membrane retrieval is typically a few seconds ("fast endocytosis", see Fig. 2.1). Optical methods in hippocampal neurons have also reported a component of endocytosis that has a time constant in the 1–10 sec range, depending on the experimental conditions (Klingauf et al. 1998; Kavalali et al. 1999). In synaptic terminals, the rate of fast endocytosis can be predicted from the time course of the recovery of intracellular calcium in synaptic terminals, where it has been shown to follow a fourth-order dependence on cytosolic calcium (von Gersdorff and Matthews 1994a). With more intense stimulation protocols, such as a train of depolarizations, the rate of membrane recovery exceeds that predicted by the recovery time course of internal calcium. In these instances, endocytosis can have a time constant in the range of 10's of seconds. This is similar to the ≈ 14 second time constant of endocytosis reported in hair cells in response to a 1 second depolarization (Parsons et al. 1994) and the estimate in hippocampal neurons that endocytosis is half-completed within 20 seconds following a train of action potentials (Ryan and Smith 1995) or within 60 seconds in response to a maintained depolarization (Ryan et al. 1993). The dissociation between calcium recovery and endocytosis in response to strong stimulation may indicate that a factor, in addition to intraterminal calcium, regulates the rate of endocytosis. For example, a component necessary for fast endocytosis may be depleted or have its activity modulated following strong stimulation. Equally plausible is that idea that some molecule that serves as a braking mechanism for endocytosis is activated by a strong stimulation paradigm. Alternatively, the two rates of endocytosis may reflect two distinct types of membrane retrieval. For example, the faster form of endocytosis might be associated with a slow type of "kiss and run" phenomenon in which the fused vesicle does not fully collapse into the plasma membrane, but instead is retrieved intact (e.g.

Ceccarelli and Hurlbut 1980b; Meldolesi and Ceccarelli 1981). A longer endocytosis time course, which exceeds the time course predicted by the recovery rate of intraterminal calcium, might arise when a vesicle collapses fully into the plasma membrane following fusion. This latter form of retrieval might require clathrin, whereas the former might not. Ultrastructural studies of endocytosis at the neuromuscular junction have also suggested two types of endocytosis (see Meldolesi and Ceccarelli 1981), with one type most prominent in the first few seconds following stimulation and not mediated by clathrin, and the second type peaking about 30 s after the stimulus and involving the formation of clathrin-coated pits (Miller and Heuser 1984). Similar results have been reported in nerve terminals of drosophila photoreceptors (Koenig and Ikeda 1996). Further investigation that include ultrastructural analyses is needed to determine whether the faster and slower forms of endocytosis observed in synaptic terminals represent distinct forms of endocytosis or activity-dependent modulation of a single process.

2.6.2
Neuroendocrine Cells

Two modes of endocytosis that differ with respect to their time courses have been reported in neuroendocrine cells (Henkel and Almers 1996; Burgoyne 1998); see also (Heinemann et al. 1994; Thomas et al. 1994; Hsu and Jackson 1996) and in pancreatic beta-cells (Proks and Ashcroft 1995). One recently described mode of endocytosis in chromaffin cells, termed compensatory endocytosis, restores the capacitance record following stimulation precisely back to baseline and has a time constant on the order of seconds (Engisch and Nowycky 1998). Thus, with respect to time course and the precise restoration of membrane surface area, compensatory endocytosis resembles the fast endocytosis observed in synaptic terminals. However, some important differences exist. Compensatory exocytosis does not occur in all chromaffin cells, but the probability that it will occur is increased with an increase in the size of the capacitance jump and with an increase in the amount of calcium influx (Engisch and Nowycky 1998). In synaptic terminals, there is no correlation between fast endocytosis and the amplitude of the capacitance response, and fast endocytosis does not have an obvious calcium influx minimum. Conceivably, a calcium minimum might exist in synaptic terminals, but due to the high calcium concentration required to trigger exocytosis in nerve terminals, this threshold is always exceeded and therefore never observed. A further difference between compensatory endocytosis in neuroendocrine cells and fast endocytosis in synaptic terminals is that compensatory endocytosis is typically preserved in the perforated-patch configuration, but lost in the whole-cell recording configuration (Smith and Neher 1997), suggesting that compensatory endocytosis requires a freely diffusible cytosolic molecule. In contrast, fast endocytosis in synaptic terminals is maintained equally well in either whole-cell or perforated patch configurations (e.g. von Gersdorff and Matthews 1994a; von Gersdorff and Matthews 1994b; Sakaba et al. 1997; but see Neves and Lagnado 1999).

2.6.3
Excess Retrieval

A second type of endocytosis that has been described in neuroendocrine cells is termed excess retrieval. It is characterized by a rapid, calcium-triggered retrieval of membrane that typically undershoots the baseline capacitance, followed by a slow recovery of membrane capacitance to the prestimulus value. Like compensatory and rapid endocytosis, excess retrieval is also a rapid process. The time constant of retrieval can typically be fitted by the sum of two exponentials with a fast component (seconds) and an even faster component (1 ms–100 ms) (Artalejo et al. 1995; Engisch and Nowycky 1998). Excess retrieval is triggered by an intense stimulation that results in a large increase in intracellular calcium (Thomas et al. 1994; Artalejo et al. 1995; Proks and Ashcroft 1995; Smith and Neher 1997; Engisch and Nowycky 1998). Indeed, the likelihood of observing this type of rapid endocytosis appears to increase when internal calcium is elevated to 50 μM or more (Heinemann et al. 1994). Once excess retrieval is triggered, however, the rate at which excess retrieval proceeds is independent of cytosolic calcium (Smith and Neher 1997). The significance of excess retrieval is unknown, but it has been postulated that it serves to endocytose membrane that has been previously added but not retrieved (Thomas et al. 1994; Smith and Neher 1997; Engisch and Nowycky 1998).

Excess retrieval has not been observed in synaptic terminals. Even when cytosolic calcium is raised to >> 100 μM and a rapid form of endocytosis is triggered, the membrane capacitance still does not undershoot the baseline (Heidelberger 1998; Figs. 2, 3). Thus, well-balanced and well-timed endocytosis may represent a distinct feature of the synaptic release and retrieval machinery. Interestingly, in contrast to the synaptic terminals of bipolar neurons, a slow form of calcium-triggered endocytosis that undershoots the resting membrane capacitance has been observed in the cell bodies of isolated bipolar neurons (see Fig. 2.2; also Heidelberger 1998). This excess retrieval was evoked by a very intense stimulation that elevated the cytosolic calcium well above 100 μM (Heidelberger 1998) and slowly recovers to baseline. The fact that this excess endocytosis is observed exclusively in somata and not in terminals raises the possibility that excess retrieval may involve the retrieval of membrane that has been added constitutively, rather than membrane added via synaptic vesicle fusion and the release of neurotransmitter. A similar interpretation has been recently postulated for excess retrieval in chromaffin cells (Smith and Neher 1997; Engisch and Nowycky 1998).

2.6.4
Calcium Regulation of Endocytosis

Calcium ions have long been proposed to be an important factor for endocytosis (Ceccarelli and Hurlbut 1980a). However, consensus about the effect of calcium ions on endocytosis has not been reached. This may be due in part to other factors that contribute to the regulation of endocytosis and in part, to the existence of more than one mode or type of endocytosis. In synaptic terminals of bipolar neurons, rapid endocytosis is slowed by the elevation of internal calcium, with half-maximal inhibition achieved at $\approx$ 500 nM calcium (von Gersdorff and Matthews 1994b). The effect of intraterminal calcium concentrations above $\approx$ 2 μM, however, were not examined. Consistent with a calcium-dependent inhibition of endocytosis, optical assays of endocytosis in synaptic terminals have shown that endocytosis does not occur during periods of calcium influx (Rouze and Schwartz 1998). In nerve terminals of the posterior pituitary and cortical synaptosomes, endocytosis also appears to be inhibited by elevated internal calcium (Hsu and Jackson 1996; Branchaw et al. 1998; Cousin and Robinson 2000). In chromaffin cells, tetany, with its strong elevation of internal calcium, has been reported to result in either a lack of endocytosis or delayed or slowed endocytosis (Artalejo et al. 1995), and dialysis of a chromaffin cell interior with 50 μM calcium has been shown to delay the onset of endocytosis for a few minutes (Smith and Betz 1996). Results obtained in PC12 cells, clonal cells of the adrenal medulla, also support the observation that high internal calcium can inhibit endocytosis. Here, endocytosis associated with large dense-core granules occurs only when the average cytosolic calcium falls below 10 μM (Kasai et al. 1996), suggesting that high calcium inhibits endocytosis. A perplexing additional finding is that this slow endocytosis is also observed when the average cytosolic calcium exceeds 40–50 μM. The authors have intrepreted this to indicate that the initiation of slow endocytosis is blocked by high calcium but that the appearance of slow endocytosis is facilitated by high calcium.

In contrast to the reports that elevation of intracellular calcium inhibits endocytosis are the reports that calcium is either required for triggering endocytosis or has no effect on endocytosis. In chromaffin cells, the time constant of compensatory endocytosis was found to decrease with increasing calcium influx until it reached a plateau at ≈ 6 seconds (Engisch and Nowycky 1998). In addition, rapid endocytosis associated with excess retrieval has been demonstrated to require an elevation in intracellular calcium (Artalejo et al. 1995). Consistent with this finding is the observation that transient fusion events are detected with greater frequency following an increase in external calcium (Ales et al. 1999). The rate of fast endocytosis was also reported to increase with an increase in internal calcium in PC12 cells (Kasai et al. 1996). Calcium-triggered endocytosis was revealed in somata of dorsal root ganglion neurons when exocytosis was blocked with EGTA (Huang and Neher 1996). In sea urchin eggs, optical methods have demonstrated that the endocytosis that follows cortical granule exocytosis requires calcium influx through voltage-gated channels (Vogel et al. 1999). Optical studies of vesicle cycling in hippocampal neurons have indicated that fast endocytosis is spurred on by an elevation in internal calcium (Klingauf et al. 1998). In contrast, calcium seems to have little effect on the slower mode of endocytosis in hippocampal neurons (time constant ≈ 20 seconds) (Ryan and Smith 1995) or on endocytosis at neuromuscular junction (Ramaswami et al. 1994; Wu and Betz 1996).

From the above data, one must conclude that calcium is an important and complex regulator of endocytosis. However, at present there does not appear to be a simple way to reconcile these strikingly different reports. The calcium-dependence of endocytosis cannot be classified according to vesicle type (clear-core vs dense core). Neither can it be readily grouped according to cell-type (neuron vs. neuroendocrine), nor by time course of endocytosis (fast vs. slow). It may simply be that there is more than one type of calcium-regulated endocytosis. Furthermore, if the relationship between internal calcium and endocytosis is non-linear or bell-shaped, as has been reported for continuous vesicle cycling (Rouze and Schwartz 1998), this could also contribute to the complexity. A non-linear relationship might allow a role for calcium in both initiating endocytosis and in inhibiting it when calcium rises above a certain level in a particular intracellular location. The matter becomes even more complex if the dose-response curve can be shifted along the calcium axis in a cell-specific or activity-dependent manner. A hypothesis that contains elements of some of these speculations has recently been proposed (Cousin and Robinson 2000). In this model, there are two modes of calcium-regulated endocytosis that are ultimately mediated by the protein dynamin. It is the difference in the binding-affinities of the receptors to which calcium ions bind and the subcellular location of these receptors that provide the switch between calcium-dependent inhibition of endocytosis and calcium-driven endocytosis. Each of these hypotheses is intriguing and has merit. What is needed now is for these hypotheses to be carefully considered and experimentally evaluated. It would be most helpful if the regulation of endocytosis by calcium would be reexamined across secretory cell types using standardized stimulation protocols and precise measurements of the relevant intracellular calcium concentration (e.g. global vs. local; somatic vs synaptic).

3
Amperometry

3.0
Introduction

The addition of carbon fiber amperometry to our box of tools for studying presynaptic release mechanisms has advanced our knowledge of exocytosis, particularly with respect to the fusion process itself. This technique has excellent temporal resolution and can be used to monitor the release of oxidizable substances from a single secretory cell in real-time. The signal resolution is such that single quantal release events can be detected. Amperometry has the advantage that it is not plagued by the non-linearities that can confound the use of postsynaptic receptors as reporters of neurotransmitter release. Additionally, carbon fiber amperometry does not necessitate that the cell of interest be held under whole-cell voltage clamp (although this combination of techniques is often desirable), and therefore the requirements for cell morphology are less stringent than those of capacitance measurements. Furthermore, the externally-positioned carbon fiber provides a non-invasive way to monitor release. For all of these reasons, carbon fiber amperometry provides an excellent adjunct to paired pre- and post-synaptic recordings and capacitance measurements for the study of exocytosis.

The carbon fiber amperometric technique is based upon the principle of chemical reduction-oxidation reactions. The carbon fiber electrode (2–5 μm radius) is placed very near to a secretory cell and is held at a potential that is at or above the redox potential of the secreted, oxidizable molecules of interest. (The rate of oxidation is speeded by exceeding the redox potential). When molecules are released from the cell during exocytosis, they diffuse away from the cell, and some fraction will strike the carbon fiber and become oxidized. When oxidized, these molecules give up electrons, and these electrons contribute to an electrical signal that is detected by the carbon fiber microelectrode. The amperometric current is directly related to the amount of released oxidizable substance according to Faraday's Law: $Q = zFM$, where Q is the measured amperometric spike charge, F is Faraday's constant, z is the number of electrons/molecule that are given up in the electrochemical oxidation, and M is the number of moles of the secreted, oxidizable substance. Neurotransmitter molecules that can be electrochemically detected with this technique include catecholamines such as norepinephrine, epinephrine, and dopamine, and indolamines like serotonin. Because the oxidation reaction occurs very quickly, amperometry has proved very useful in studying the time course of secretion from individual granules. Further refinements in the manufacture of these electrodes have improved the signal-to-noise such that the quantal release attributed to the fusion of small synaptic vesicles can also be detected (Zhou and Misler 1995a; Zhou et al. 1996). Unfortunately, there is no time-resolved method for the electrochemical detection of amino acid neurotransmitters such as glutamate or GABA at present. However, a coating for the carbon fiber has been devised to catalyze the oxidation of insulin and allow insulin release from single granules of pancreatic beta-cells to be detected (Huang et al. 1995), and a calcium-indicator dye has been added to the surface of carbon fibers to permit the release of calcium that occurs during catecholamine secretion to be monitored (Xin and Wightman 1998). It may be that additional modifications of the carbon fiber surface will allow other released substances to be electrochemically detected in a time-resolved manner. Readers interested in more details about the electrochemical detection of secreted substances are referred to two excellent reviews (Chow and von Ruden 1995) and (Travis and Wightman 1998). Cyclic voltometry, an electrochemical technique that can be used to electrochemically identify a secreted substance based upon its oxidation profile, is also discussed in these sources.

3.1
The Basic Response

As with capacitance measurements, amperometry was first used to study the secretion of large dense-core granules in neuroendocrine cells and has only more recently been applied to the study of neurotransmitter release. From the large dense-core granule work, it has become clear that a single calcium-dependent release event can be resolved as a single spike in the amperometric trace, particularly when the probability of release is reduced (e.g. Wightman et al. 1991; Chow et al. 1992), a situation analogous to the conditions for detecting miniature end-plate potentials at the neuromuscular

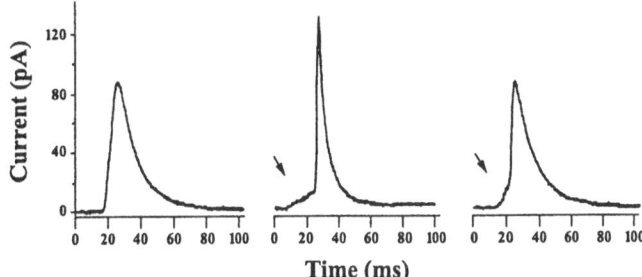

Fig. 3.1. Single amperometric current transients in bovine adrenal chromaffin cells. Each transient is thought to correspond to the exocytotic release of an oxidizable substance from an individual chromaffin granule. The fast rising phase in two of the three examples shown is preceded by a small pedestal or "foot," indicated by the arrows. The foot is thought to denote the development of the fusion pore and the leakage of granule contents into the bath. The spike is thought to correspond to the rapid dilation of the pore and the expulsion of granule contents. (Reprinted by permission from Nature 356:60–63, copyright 1992, Macmillan Magazines Ltd)

junction. Quantal release events that occur near to the carbon fiber are represented as brief, spike-like signals in the amperometric record (Fig. 3.1). Quantal release events that occur at a distance are subject to diffusional delays and are thus represented by longer duration, shorter amplitude (i.e., temporally smeared) spikes. In response to strong stimulation, such as flash-photolysis of caged-calcium, these spikes can merge to form a plateau upon which other spikes may ride (Haller et al. 1998).

The amperometric spikes have structure to them and in order to place this structure in context, it is important to introduce the concept of the fusion pore. The fusion pore was first described and studied in mast cells of the beige mouse because the secretory granules of these cells are extraordinarily large, and this permitted single fusion events to detected in the whole-cell capacitance record (Fernandez et al. 1984; Breckenridge and Almers 1987; Zimmerberg et al. 1987; Spruce et al. 1990). From these investigations it was hypothesized that a small pore is formed upon the fusion of a secretory granule with the plasma membrane. This fusion pore was further hypothesized to permit electrical and biochemical access between the granule interior and the extracellular space (Lindau and Almers 1995). A small amount of release of oxidizable substance through this fusion pore could be detected amperometrically by a closely-positioned carbon fiber (Alvarez de Toledo et al. 1993). The fusion pore was also observed to abruptly and rapidly dilate (Breckenridge and Almers 1987; Spruce et al. 1990) to give rise to a large amperometric spike (Alvarez de Toledo et al. 1993). Consistent with these observations, the majority of single quantal release events in chromaffin cells were found to be preceded by a small "foot" signal that most likely corresponds to the trickle of oxidizable substance through a fusion pore and into the extracellular bath (see arrows, Fig. 3.1; Chow et al. 1992; Zhou et al. 1996). In addition, rapid amperometric spikes were observed that could represent the rapid dilation of the fusion pore and the rapid, concerted release of oxidizable material (Chow et al. 1992; Zhou et al. 1996). Occasionally, however, a long foot signal was observed in the amperometric record that was not followed by a spike. In these instances, it is believed that the fusion pore did not fully dilate and therefore that the granule did not collapse fully into the plasma membrane. Alternatively, the granule may have released all of its catecholamines during the long foot signal such that when the rapid dilation of the fusion pore finally did occur, there were no catecholamine molecules left to be released. Indeed, it has been estimated that a stand-alone foot can release almost 80% of the releasable content of the granule (Zhou et al. 1996)

3.2
Amperometric Responses in Neurons

At present, amperometry has been used to study synaptic exocytosis in only a limited number of neurons. One of these is the large Retzius cell of the leech. This neuron releases serotonin as a fast neurotransmitter from both small clear-core synaptic-like vesicles and large dense-core granules. Consistent with two types of secretory vesicles, two types of amperometric spikes have been observed (Bruns and Jahn 1995). The amperometric spikes attributed to the fusion of small synaptic vesicles occurred within the first few milliseconds following a single action potential and were not preceded by a foot-signal. The second type of response was a larger spike that followed the action potential with a delay of several milliseconds or more and was often preceded by a foot-signal. These latter spikes are similar to the signals observed in amperometric records from chromaffin cells. For both types of amperometric spikes, release proceeded very rapidly once initiated, such that the rise times of the amperometric spikes were less than 60 microseconds. The rapidity of this response suggests that there is an abrupt opening of a pre-assembled fusion pore. The lack of a foot-signal and the fast time course of release from the small vesicles might be attributable, at least in part, to the discharge of synaptic vesicle contents through an undilated fusion pore. If true, this would have important implications for vesicle recycling. A small synaptic vesicle that retains its vesicle matrix because it has only opened a small fusion pore (rather than dilating fully and collapsing into the plasma membrane), may need only to be filled with neurotransmitter before it can rejoin the secretory pathway (see discussion in section 4.2).

In the vertebrate kingdom, quantal catecholamine exocytosis from presynaptic neurons has been successfully detected using carbon fiber amperometry in cultured superior cervical ganglion neurons (Zhou and Misler 1995b; Koh and Hille 1997) and cultured retinal amacrine cells (Hochstetler et al. 2000a). As might be predicted for neuronal exocytosis, amperometric spikes had rapid rise times and were of short duration (Zhou and Misler 1995b). For both neurons, the typical vesicle released ≈ 30,000 molecules of neurotransmitter (Zhou and Misler 1995b; Hochstetler et al. 2000b). This is similar to the estimate of ≈ 10,000 acetylcholine molecules that comprise a quantum at the vertebrate neuromuscular junction (Kuffler and Yoshikami 1975). In contrast to this number, the number of molecules released from a large dense-core chromaffin granule can be two orders of magnitude larger (e.g. Wightman et al. 1991; Jankowski et al. 1992). This is not entirely surprising, given the larger volume of the typical chromaffin cell granule relative to a small synaptic vesicle. For example, biochemical estimates place the concentration of norepinephrine within a dense-core granule to be between 0.2 and 0.5 M (e.g. Hillarp 1959; Lagercrantz 1976; Klein 1982; Klein and Lagercrantz 1982; Johnson 1988). If one assumes an average catecholamine concentration of 0.35 M, then a single quantal release event that liberates 30,000 molecules would correspond to the fusion of a single vesicle with a diameter of ≈ 65 nm, assuming that the vesicle is fluid-filled and neglecting the thickness of the vesicle membrane. This estimate falls within the range of vesicle diameters observed in neurons of the superior cervical ganglia (Zhou and Misler 1995b). With the same assumptions, the number of molecules contained within a single 300 nm chromaffin granule would be approximately two orders of magnitude greater, which is the difference in magnitude that is observed.

In addition to the above central neurons, carbon fiber amperometry has also been used to monitor neurotransmitter release from synaptic vesicles of dopaminergic neurons in the midbrain and from the substantia nigra. In cultures of midbrain neurons, the ability to detect single fusion events allowed quantal size to be analyzed before and after experimental treatments. The surprising result was that quantal size, as determined by the number of molecules of dopamine released, was not invariant (Pothos et al. 1998a). Rather, quantal size was subject to modulation by growth factors and alterations in neurotransmitter synthesis (Pothos et al. 1998a). This finding may turn out to be extremely important because it suggests that in addition to the sites that have been previously postulated, quantal size may represent an additional locus of synaptic plasticity (Pothos and Sulzer 1998; Pothos et al.

1998a; Pothos et al. 1998b). Furthermore, these results raise the possibility that the variations in quantal size and broad amplitude histograms that have been reported for central neurons might also be due to the selective modulation of particular boutons.

In substantia nigra neurons (Jaffe et al. 1998), a key question was to determine whether the somatic release of dopamine occurs via exocytosis or a non-vesicular mechanism. To address this, amperometric recordings were made from single neurons in substantia nigra brain slices (Jaffe et al. 1998). The amperometric record showed spontaneous discrete release events, and the frequency of spikes increased with depolarization and could be blocked by cadmium, consistent with a calcium-regulated secretory process. Some of the amperometric spikes had amplitudes of nearly 20 pA, indicating the passage of at least 6×10^7 molecules per second. This is slightly faster than the 2.5×10^7 molecules per second that was estimated for the rate of catecholamine passage through a 500 nm diameter fusion pore in chromaffin cells (Albillos et al. 1997). Based upon this information, the authors concluded that the release of dopamine from these neuronal cell bodies most likely occurs via exocytosis, rather than by a slower transport mechanism. The data also indicated that there may be more than one population of vesicles that contain oxidizable substances in these neurons. As in retzius neurons of the leech (Bruns and Jahn 1995) and snail neurons (Clark and Ewing 1997), these two populations may represent large dense-core and small synaptic-like vesicles, but this remains to be determined. Another exciting aspect of this work is that it demonstrates that carbon fiber amperometry can be adapted to the study of neurotransmitter release in the brain slice preparation.

3.3
Combined with Capacitance Measurements

As indicated in the previous section, the application of carbon fiber amperometry to the study of exocytosis has revealed several new aspects about the quantized release of oxidizable substances, and this technique becomes even more powerful when combined with other approaches, such as capacitance measurements or measurement of intracellular calcium. For example, amperometry best detects those release events that occur close to the tip of the carbon fiber, which may represent only 10% of the total events, depending on the geometry of the cell and the carbon fiber (Chow et al. 1992; Chow et al. 1994; Haller et al. 1998). This feature can be an advantageous under some circumstances. Indeed, a small carbon fiber can be moved along the surface of a cell to map the location of release sites (e.g. Robinson et al.

1995). In contrast to amperometry, whole-cell capacitance technique detects exocytosis from the entire cell and is not limited to oxidizable substances. In addition, of the two, only capacitance measurements can be used to track endocytosis. However, amperometry can be used to detect the release of oxidizable substances during a voltage step, when capacitance measurements may not be valid due to the changes in membrane conductance evoked by the voltage step. Thus, these two techniques can provide mutually beneficial information about synaptic vesicle dynamics. An excellent example of the benefits of utilizing both techniques simultaneously comes studies of catecholamine release from chromaffin cells.

Chow and colleagues have exploited the mutually beneficial attributes of these approaches to determine why the time course of exocytosis from neuroendocrine cells is slower than the time course of release at synapses (Chow et al. 1994; Chow et al. 1996). These authors reasoned that the integral of the latency histogram of amperometric spikes should be proportional to the capacitance record, provided that the observed capacitance increase is attributable only to secretion. By scaling the integral of the amperometric latency histogram appropriately with respect to the capacitance record, they derived an estimate for the rate of exocytosis that was achieved in response to a depolarizing voltage step. This rate estimate was compared against the relationship between rate and exocytosis obtained in flash-photolysis experiments with caged-calcium to yield an estimate of the calcium concentration seen by the release machinery during the depolarization. This exercise revealed that in response to the opening of voltage-gated calcium channels, the release machinery of chromaffin cells sees an average calcium concentration that is $< 10 \mu M$ (Chow et al. 1994). This is approximately an order of magnitude lower than the average calcium concentration seen by the release machinery in synaptic terminals (Fig. 2.6). Whereas the calcium concentration is thought to rise and fall within microseconds at the mouth of a calcium channel (reviewed in (Neher 1998), additional calculations suggested that the calcium concentration seen by the release machinery in chromaffin cells is not only much slower, but that it decays over tens of milliseconds (Chow et al. 1994). Additionally, it was shown that the addition of exogenous mobile calcium buffers, including low millimolar concentrations of EGTA, could reduce the amount of delayed release (Chow et al. 1996). This in turn implies that the sites of exocytosis and the location of calcium channels are not molecularly near to each other in chromaffin cells, consistent with arguments presented in section 2.4.2.

Carbon fiber amperometry can also be an important adjunct to whole-cell capacitance measurements when information either about the relative timing of membrane fusion and release of neurotransmitter is desired (e.g.

Chow et al. 1996) or about the size of the vesicles that underlie the whole-cell
exocytotic response is desired (e.g. Moser and Neher 1997). To obtain either
piece of information, capacitance and amperometric signals from stimulated
secretory cells are simultaneously recorded. Amperometric spikes are then
used to align the whole-cell capacitance records, and the capacitance records
are averaged to permit single fusion events to be detected above the noise.
Such careful cross-correlations studies have estimated the mean size of fusion-competent mouse chromaffin granules to be between 0.6 and 2 fF
(Moser and Neher 1997). In addition, cross-correlation studies have revealed
the important finding that membrane fusion precedes release of vesicle
contents by a few milliseconds (Chow et al. 1996). The reason for this brief
delay is not known, but it could be that the earliest fusion pore that is elec-
trically-detectable does not readily pass catecholamines or that it conducts
too little to be detected by the carbon fiber (Chow et al. 1996). Alternatively,
this brief delay could indicate that there is another step between the formation of the fusion pore and before catecholamine release that must occur in
order to for release to occur. One postulated postfusion step entails the
swelling of the secretory granule matrix and the exchange of charged neurotransmitter in the vesicle with counter ions in the extracellular fluid or the
co-release of neurotransmitter and counter ions from the vesicle (reviewed
in Rahamimoff and Fernandez 1997). A postfusion, calcium-dependent
regulation of the fusion pore has also been suggested (Zhou et al. 1996). At
present, it is not clear whether either of these mechanisms contributes to the
the brief delay between fusion and the actual release of neurotransmitter
described above. Furthermore, because this combination of techniques has
not been applied to the fusion of small, clear-core synaptic vesicles, it is not
yet known whether synaptic release also exhibits a brief postfusion delay in
neurotransmitter release.

4
Refinements in the Presynaptic Detection
of Exocytosis and Endocytosis

4.0
Capacitance Measurements

Detailed analysis of single fusion events would be expected to provide new
insights into the mechanisms of synaptic exocytosis and endocytosis much
the way that fluctuation analysis and the study of single channel properties
has provided crucial information about the elementary signals that underlie
the whole-cell macroscopic current. For cells in which carbon fiber am-

perometry is not applicable, accurate determination of the underlying size of the single fusion event may be obtained from the whole-cell capacitance record by analyzing trial-to-trial fluctuations of the capacitance increase evoked by a stimulus (Moser and Neher 1997). Alternatively, limiting the size of the patch of membrane from which capacitance measurements are made will improve the signal-to-noise ratio and increase the potential to detect single events (Neher and Marty 1982). Lindau and his colleagues have applied this logic to study large dense-core granule fusion in neutrophils. Using a high-frequency sine wave stimulation in the cell-attached configuration, they were able to resolve both spontaneous and calcium-dependent step increases in membrane capacitance that were as small as 0.1 fF, corresponding to the fusion of a single 60 nm secretory granule (Lollike et al. 1995; Lollike et al. 1998). In addition, additional refinements of the cell-attached capacitance technique have permitted the fusion pore of secretory vesicles as small as $\approx$ 110 nm diameter to be measured (Debus and Lindau 2000). Unfortunately, the study of small synaptic vesicle fusion remains below these current resolution limits.

Single fusion events have been studied in neuroendocrine cells using the cell-attached refinement of the capacitance technique (Lollike et al. 1995; Kreft and Zorec 1997; Lollike et al. 1998). From these studies, it is clear that large-dense core granules also form a fusion pore when they fuse with the plasma membrane during exocytosis, as was reported for the giant granules of mast cells from the beige mouse. The fusion pore was quite stable and sometimes expanded only very slowly over the 100 ms following fusion (Lollike et al. 1995). In other instances, the fusion pore was observed to rapidly expand. At still other times, the fusion pore was observed to flicker and close without achieving full expansion (Albillos et al. 1997; Ales et al. 1999). While the average conductance of secretory granule fusion pores in giant granules of mast cells has been estimated to be between 100–300 pS (Breckenridge and Almers 1987; Spruce et al. 1990), the conductance of the fusion pore in neuroendocrine cells can be as low as 35 pS (Lollike et al. 1995). This range of these conductances is similar to the ranges of conductances reported for gap junctions and ion channels, raising the possibility that the fusion pore may share some features with ion channels. Additional characteristics of large dense-core granule fusion pores are discussed in section 4.2.

4.1
Amperometry

The detection of single fusion events via changes in membrane capacitance are not the only techniques that have benefitted from the use of techniques previously applied to the study of single channels. For example, fluctuation analysis of amperometric records has also been used to determine the relative size of a secretory quantum in chromaffin cells (Haller et al. 1998). This analysis was performed under experimental conditions in which the release probability was high, and therefore the detection of single quantal events in the amperometric record was difficult. The experimentally collected data were compared with simulations that were designed to test assumptions about the size of the underlying elementary signal. The results indicated that under these experimental conditions, the secretory quantum was smaller than would be predicted (Haller et al. 1998). Possible explanations for this finding include an abundance of foot-only secretory events, secretory events that occur at heterogeneous locations, or a heterogeneous size distribution of granules (Haller et al. 1998).

4.2
Combined Patch-Amperometry and Patch-Capacitance

To more directly analyze single amperometric events, Albillos and colleagues placed a carbon fiber microelectrode within a patch-clamp recording electrode in order to electrochemically detect the exocytotic release of oxidizable substances from a region of membrane limited to that region contained within the patch pipette. Fusion events in this same patch of membrane were simultaneously monitored using cell-attached capacitance measurements. The combined use of patch-amperometry and patch-capacitance measurements permitted these investigators to simultaneously observe the opening of the fusion pore and the release of catecholamine from a single dense-core granule in a chromaffin cell in real-time (Albillos et al. 1997). With this approach, they estimated the size of the fusion pore to be ≈ 2.5 nm in diameter and showed that as previously surmised, catecholamine could indeed be released through the undilated fusion pore. Transient dilations of the fusion pore were also observed. In both rat and bovine chromaffin cells, catecholamine release through the undilated fusion pore was nearly complete and occurred approximately 10% of the time (Albillos et al. 1997; Ales et al. 1999). This result suggests that a dense-core granule need not collapse fully into the plasma membrane in order to release its contents, but rather might actually be able to fuse transiently with the plasma membrane and then be retrieved as an intact vesicle. Furthermore, given that a fusion pore may be less than 3 nm in diameter, it seems unlikely that the large molecules that comprise the dense-core would be lost through a small fusion pore (Albillos et al. 1997; Ales et al. 1999). These results, combined with the work of others (reviewed in Palfrey and Artalejo 1998), have led to the novel suggestion that in addition to recycling through an endosomal compartment, recycling of large dense core vesicles can occur locally, depending upon the stimulation protocol.

Even with the improved signal-to-noise ratio, the patch-amperometry technique is not yet sensitive enough to study the fusion pore and its characterists in small synaptic vesicles. However, other lines of evidence suggest that there may be local recycling of synaptic vesicles. One piece of evidence comes from a recent optical study of synaptic vesicle dynamics in cultured hippocampal neurons (Murthy and Stevens 1998). Earlier tracer studies have also raised the possibility that synaptic vesicles might be locally recycled (reviewed in (Ceccarelli and Hurlbut 1980b). This evidence is particularly convincing for ribbon synapses and dense bodies, where tracer-labelled vesicles have been found at these synaptic active zone structures (Schacher et al. 1974; Schaeffer and Raviola 1978; Evans et al. 1981; Cooper and McLaughlin 1983; Siegel and Brownell 1986). Local recycling of synaptic vesicles would be consistent with the hypothesis that important vesicle contents are not lost during the process of synaptic exocytosis and that there is little sorting of vesicle surface molecules that needs to be done following the fusion event. These features would not be incompatible with the possibility that a synaptic vesicle may be able to release all of its contents through an undilated fusion pore (e.g. Neher 1993; Bruns and Jahn 1995). Hopefully, future refinements of these techniques will be developed that will allow this hypothesis to be more directly tested.

Concluding Comments

Great strides have been made in recent years in our ability to study exocytosis and endocytosis in living nerve terminals in real-time. A basic understanding of synaptic vesicle fusion, neurotransmitter release, membrane retrieval and synaptic vesicle recycling is now well on its way to being fully achieved. The future is bright, and further refinements of the electrophysiological approaches presented in this review are expected to reveal ever more detailed information about these fundamental aspects of neuronal synaptic communication. In addition, these approaches can be combined with other tools, such as optical assays of vesicle recycling, ultrastructural analyses of active zones, and molecular biological probes, to examine further the role of the presynaptic terminal and synaptic vesicle dynamics in synaptic signaling and its modulation.

Acknowledgements. Supported by a grant from the National Institutes of Health and by awards from the Alfred P. Sloan Foundation and the Esther A. and Joseph Klingenstein Fund. The author wishes to thank Dr. Steve DeVries for critical reading of this manuscript and Katherine P. West for her assistance in its preparation.

References

Adler EM, Augustine GJ, Duffy SN, Charlton MP (1991) Alien intracellular calcium chelators attenuate neurotransmitter release at the giant squid synapse. J Neuroscience 11:1496-1507

Albillos A, Dernick G, Horstmann H, Almers W, Alvarez de Toledo G, Lindau M (1997) The exocytotic event in chromaffin cells revealed by patch amperometry. Nature 389:509-12

Ales E, Tabares L, Poyato JM, Valero V, Lindau M, Alvarez de Toledo G (1999) High calcium concentrations shift the mode of exocytosis to the kiss-and-run mechanism [see comments]. Nature Cell Biology 1:40-4

Almers W, Neher E (1987) Gradual and stepwise changes in the membrane capacitance of rat peritoneal mast cells. J Physiol (Lond) 386:205-17

Alvarez de Toledo G, Fernandez-Chacon R, Fernandez JM (1993) Release of secretory products during transient vesicle fusion. Nature 363:554-8

Ammala C, Eliasson L, Bokvist K, Larsson O, Ashcroft FM, Rorsman P (1993) Exocytosis elicited by action potentials and voltage-clamp calcium currents in individual mouse pancreatic B-cells. Journal of Physiology 472:665-88

Artalejo CR, Henley JR, McNiven MA, Palfrey HC (1995) Rapid endocytosis coupled to exocytosis in adrenal chromaffin cells involves Ca2+, GTP, and dynamin but not clathrin. Proc Natl Acad Sci USA 92:8328-32

Atluri PP, Regehr WG (1998) Delayed release of neurotransmitter from cerebellar granule cells. J Neurosci 18:8214-27

Augustine GJ, Charlton MP (1986) Calcium dependence of presynaptic calcium current and post-synaptic response at the squid giant synapse. Journal of Physiology 381:619-40

Augustine GJ, Charlton MP, Smith SJ (1985) Calcium entry and transmitter release at voltage-clamped nerve terminals of squid. J Physiol (Lond) 367:163-81

Augustine GJ, Neher E (1992) Calcium requirements for secretion in bovine chromaffin cells. J Physiol (Lond) 450:247-71

Baker PF, Knight DE (1978) Calcium-dependent exocytosis in bovine adrenal medullary cells with leaky plasma membranes. Nature 276:620-2

Banerjee A, Barry VA, DasGupta BR, Martin TFJ (1996) N-Ethylmaleimide-sensitive factor acts at a prefusion ATP-dependent step in Ca2+-activated exocytosis. J Biol Chem 271:20223-6

Barnett DW, Misler S (1997) An optimized approach to membrane capacitance estimation using dual-frequency excitation. Biophysical Journal 72:1641-58

Bekkers JM, Richerson GB, Stevens CF (1990) Origin of variability in quantal size in cultured hippocampal neurons and hippocampal slices. Proceedings of the National Academy of Sciences of the United States of America 87:5359-62

Bekkers JM, Stevens CF (1989) NMDA and non-NMDA receptors are co-localized at individual excitatory synapses in cultured rat hippocampus. Nature 341:230-3

Bellingham MC, Walmsley B (1999) A novel presynaptic inhibitory mechanism underlies paired pulse depression at a fast central synapse. Neuron 23:159-70

Bennett MK, Calakos N, Scheller RH (1992) Syntaxin: a synaptic protein implicated in docking of synaptic vesicles at presynaptic active zones. Science 257:255-9

Betz WJ (1970) Depression of transmitter release at the neuromuscular junction of the frog. Journal of Physiology 206:629-44

Bezprozvanny I, Scheller RH, Tsien RW (1995) Functional impact of syntaxin on gating of N-type and Q-type calcium channels. Nature 378:623-6

Birks J, MacIntosh FC (1961) Acetylcholine Metabolism of a Synaptic Ganglion. Can. J. Biochem. Physiol. 39:787-827

Bittner MA, Holz RW (1992) Kinetic analysis of secretion from permeabilized adrenal chromaffin cells reveals distinct components. J Biol Chem 267:16219-25

Blank PS, Cho MS, Vogel SS, Kaplan D, Kang A, Malley J, Zimmerberg J (1998a) Submaximal responses in calcium-triggered exocytosis are explained by differences in the calcium sensitivity of individual secretory vesicles. Journal of General Physiology 112:559-67

Blank PS, Vogel SS, Cho MS, Kaplan D, Bhuva D, Malley J, Zimmerberg J (1998b) The calcium sensitivity of individual secretory vesicles is invariant with the rate of calcium delivery. Journal of General Physiology 112:569-76

Borges S, Gleason E, Turelli M, Wilson M (1995) The kinetics of quantal transmitter release from retinal amacrine cells. Proc Natl Acad Sci USA 92:6896-900

Borst JG, Sakmann B (1996) Calcium influx and transmitter release in a fast CNS synapse [see comments]. Nature 383:431-4

Branchaw JL, Hsu SF, Jackson MB (1998) Membrane excitability and secretion from peptidergic nerve terminals. Cell Mol Neurobiol 18:45-63

Breckenridge LJ, Almers W (1987) Currents through the fusion pore that forms during exocytosis of a secretory vesicle. Nature 328:814-7

Bruns D, Jahn R (1995) Real-time measurement of transmitter release from single synaptic vesicles. Nature 377:62-5

Burgoyne RD (1998) Two forms of triggered endocytosis in regulated secretory cells. Journal of Physiology 506:589

Castillo J, Katz B (1954) Quantal Components of The End-Plate Potential. J. Physiol. 124:560-573

Castillo J, Katz B (1956) Biophysical aspects of neuro-muscular transmission. In: Butler JAV (ed) Progess in Biophysics. Pergamon Press, London, pp 122-170 vol 6)

Catterall WA (1998) Structure and function of neuronal Ca2+ channels and their role in neurotransmitter release. Cell Calcium 24:307-23

Ceccarelli B, Hurlbut WP (1980a) Ca2+-dependent recycling of synaptic vesicles at the frog neuromuscular junction. Journal of Cell Biology 87:297-303

Ceccarelli B, Hurlbut WP (1980b) Vesicle hypothesis of the release of quanta of acetylcholine. Physiological Reviews 60:396-441

Chad JE, Eckert R (1984) Calcium domains associated with individual channels can account for anomalous voltage relations of CA-dependent responses. Biophys J 45:993-9

Chow RH, Klingauf J, Heinemann C, Zucker RS, Neher E (1996) Mechanisms determining the time course of secretion in neuroendocrine cells. Neuron 16:369-76

Chow RH, Klingauf J, Neher E (1994) Time course of Ca2+ concentration triggering exocytosis in neuroendocrine cells. Proc Natl Acad Sci USA 91:12765-9

Chow RH, von Ruden L (1995) Electrochemical detection of secretion from single cells. In: Sakmann B, Neher E (eds) Single-channel recording, second edn. Plenum Press, New York, pp 245-276

Chow RH, von Ruden L, Neher E (1992) Delay in vesicle fusion revealed by electrochemical monitoring of single secretory events in adrenal chromaffin cells. Nature 356:60-3

Clark RA, Ewing AG (1997) Quantitative measurements of released amines from individual exocytosis events. Molecular Neurobiology 15:1-16

Cooper NG, McLaughlin BJ (1983) Tracer uptake by photoreceptor synaptic terminals. I. Dark-mediated effects. J Ultrastruct Res 84:252-67

Coorssen JR, Schmitt H, Almers W (1996) Ca2+ triggers massive exocytosis in Chinese hamster ovary cells. Embo J 15:3787-91

Cousin MA, Robinson PJ (2000) Ca(2+) influx inhibits dynamin and arrests synaptic vesicle endocytosis at the active zone. Journal of Neuroscience 20:949-57

Debus K, Lindau M (2000) Resolution of patch capacitance recordings and of fusion pore conductances in small vesicles. Biophysical Journal 78:2983-97

DeVries SH (2000) Selective distribution of AMPA and kainate receptors at the mammalian cone to Off bipolar cell synapse. Invest Ophthalmol Vis Sci 41:S621. Abstract nr 3294

Dittman JS, Kreitzer AC, Regehr WG (2000) Interplay between facilitation, depression, and residual calcium at three presynaptic terminals. Journal of Neuroscience 20:1374-85

Dittman JS, Regehr WG (1998) Calcium dependence and recovery kinetics of presynaptic depression at the climbing fiber to Purkinje cell synapse. Journal of Neuroscience 18:6147-62

Dobrunz LE, Stevens CF (1997) Heterogeneity of release probability, facilitation, and depletion at central synapses. Neuron 18:995-1008

Dodge FA, Jr., Rahamimoff R (1967) Co-operative action a calcium ions in transmitter release at the neuromuscular junction. J Physiol (Lond) 193:419-32

Dowling JE (1987) Wiring of the retina The retina: an approachable part of the brain. The Belknap Press of Harvard University Press, Cambridge, Mass., pp 42-80

Edwards FA, Konnerth A, Sakmann B (1990) Quantal analysis of inhibitory synaptic transmission in the dentate gyrus of rat hippocampal slices: a patch-clamp study. J Physiol (Lond) 430:213-49

Elmqvist D, Quastel DM (1965) A quantitative study of end-plate potentials in isolated human muscle. Journal of Physiology 178:505-29

Engisch KL, Nowycky MC (1998) Compensatory and excess retrieval: two types of endocytosis following single step depolarizations in bovine adrenal chromaffin cells. J Physiol (Lond) 506:591-608

Evans JA, Liscum L, Hood DC, Holtzman E (1981) Uptake of horseradish peroxidase by presynaptic terminals of bipolar cells and photoreceptors of the from retina. Journal of Histochemistry & Cytochemistry 29:511-8

Fatt P, Katz B (1952) Spontaneous subthreshold activity at motor nerve endings. J Physiol 117:109-28

Fernandez JM, Neher E, Gomperts BD (1984) Capacitance measurements reveal stepwise fusion events in degranulating mast cells. Nature 312:453-5

Forti L, Bossi M, Bergamaschi A, Villa A, Malgaroli A (1997) Loose-patch recordings of single quanta at individual hippocampal synapses. Nature 388:874-8

Geppert M, Goda Y, Hammer RE, Li C, Rosahl TW, Stevens CF, Sudhof TC (1994) Synaptotagmin I: a major Ca2+ sensor for transmitter release at a central synapse. Cell 79:717-27

Gillis KD (1995) Techniques for membrane capacitance measurements. In: Sakmann B, Neher E (eds) Single-Channel Recordings, second edn. Plenum Press, New York, pp 155-198

Gillis KD, Mossner R, Neher E (1996) Protein kinase C enhances exocytosis from chromaffin cells by increasing the size of the readily releasable pool of secretory granules. Neuron 16:1209-20

Gingrich KJ, Byrne JH (1985) Simulation of synaptic depression, posttetanic potentiation, and presynaptic facilitation of synaptic potentials from sensory neurons mediating gill-withdrawal reflex in Aplysia. Journal of Neurophysiology 53:652-69

Giovannucci DR, Stuenkel EL (1997) Regulation of secretory granule recruitment and exocytosis at rat neurohypophysial nerve endings. J Physiol (Lond) 498:735-51

Gleason E, Borges S, Wilson M (1993) Synaptic transmission between pairs of retinal amacrine cells in culture. J Neurosci 13:2359-70

Goda Y, Stevens CF (1994) Two components of transmitter release at a central synapse. Proc Natl Acad Sci USA 91:12942-6

Gomis A, Burrone J, Lagnado L (1999) Two actions of calcium regulate the supply of releasable vesicles at the ribbon synapse of retinal bipolar cells. Journal of Neuroscience 19:6309-17

Grynkiewicz G, Poenie M, Tsien RY (1985) A new generation of Ca2+ indicators with greatly improved fluorescence properties. J Biol Chem 260:3440-50

Haller M, Heinemann C, Chow RH, Heidelberger R, Neher E (1998) Comparison of secretory responses as measured by membrane capacitance and by amperometry. Biophysical Journal 74:2100-13

Hama K, Saito K (1977) Fine structure of the afferent synapse of the hair cells in the saccular macula of the goldfish, with special reference to the anastomosing tubules. Journal of Neurocytology 6:361-73

Hay JC, Fisette PL, Jenkins GH, Fukami K, Takenawa T, Anderson RA, Martin TF (1995) ATP-dependent inositide phosphorylation required for Ca(2+)-activated secretion. Nature 374:173-7

Hay JC, Martin TF (1992) Resolution of regulated secretion into sequential MgATP-dependent and calcium-dependent stages mediated by distinct cytosolic proteins. Journal of Cell Biology 119:139-51

Hay JC, Martin TF (1993) Phosphatidylinositol transfer protein required for ATP-dependent priming of Ca(2+)-activated secretion. Nature 366:572-5

Heidelberger R (1998) Adenosine triphosphate and the late steps in calcium-dependent exocytosis at a ribbon synapse. J Gen Physiol 111:225-41

Heidelberger R, Heinemann C, Neher E, Matthews G (1994) Calcium dependence of the rate of exocytosis in a synaptic terminal. Nature 371:513-5

Heidelberger R, Matthews G (1992) Calcium influx and calcium current in single synaptic terminals of goldfish retinal bipolar neurons. J Physiol (Lond) 447:235-56

Heidelberger R, Matthews G (1997) A requirement for MgATP in endocytosis and pool refilling, but not in late steps of exocytosis. Biophys J 72:A228

Heinemann C, Chow RH, Neher E, Zucker RS (1994) Kinetics of the secretory response in bovine chromaffin cells following flash photolysis of caged Ca2+. Biophys J 67:2546-57

Heinemann C, von Ruden L, Chow RH, Neher E (1993) A two-step model of secretion control in neuroendocrine cells. Pflugers Arch 424:105-12

Henkel AW, Almers W (1996) Fast steps in exocytosis and endocytosis studied by capacitance measurements in endocrine cells. Current opinion in Neurobiology 6:350-7

Hernandez-Cruz A, Sala F, Adams PR (1990) Subcellular calcium transients visualized by confocal microscopy in a voltage-clamped vertebrate neuron. Science 247:858-62

Hillarp NA (1959) Further observations on the state of thecatecholamones stored in the adrenal medullary granules. Acta Physiol Scand 47:271-279

Hochstetler SE, Puopolo M, Gustincich S, Raviola E, Wightman RM (2000a) Real-time amperometric measurements of zeptomole quantities of dopamine released from neurons. Analytical Chemistry 72:489-96

Hochstetler SE, Puopolo M, Gustincich S, Raviola E, Wightman RM (2000b) Real-time amperometric measurements of zeptomole quantities of dopamine released from neurons. Anal Chem 72:489-96

Holz RW, Bittner MA, Peppers SC, Senter RA, Eberhard DA (1989) MgATP-independent and MgATP-dependent exocytosis. Evidence that MgATP primes adrenal chromaffin cells to undergo exocytosis. J Biol Chem 264:5412-9

Horrigan FT, Bookman RJ (1994) Releasable pools and the kinetics of exocytosis in adrenal chromaffin cells. Neuron 13:1119-29

Hsu SF, Augustine GJ, Jackson MB (1996) Adaptation of Ca(2+)-triggered exocytosis in presynaptic terminals. Neuron 17:501-12

Hsu SF, Jackson MB (1996) Rapid exocytosis and endocytosis in nerve terminals of the rat posterior pituitary. J Physiol (Lond) 494:539-53

Huang L, Shen H, Atkinson MA, Kennedy RT (1995) Detection of exocytosis at individual pancreatic beta cells by amperometry at a chemically modified microelectrode. Proceedings of the National Academy of Sciences of the United States of America 92:9608-12

Huang LY, Neher E (1996) Ca(2+)-dependent exocytosis in the somata of dorsal root ganglion neurons. Neuron 17:135-45

Isaacson JS, Walmsley B (1995) Counting quanta: direct measurements of transmitter release at a central synapse. Neuron 15:875-84

Jaffe EH, Marty A, Schulte A, Chow RH (1998) Extrasynaptic vesicular transmitter release from the somata of substantia nigra neurons in rat midbrain slices. J Neurosci 18:3548-53

Jahn R, Sudhof TC (1999) Membrane fusion and exocytosis. Annual Review of Biochemistry 68:863-911

Jankowski JA, Schroeder TJ, Holz RW, Wightman RM (1992) Quantal secretion of catecholamines measured from individual bovine adrenal medullary cells permeabilized with digitonin. Journal of Biological Chemistry 267:18329-35

Johnson RG, Jr. (1988) Accumulation of biological amines into chromaffin granules: a model for hormone and neurotransmitter transport. Physiological Reviews 68:232-307

Jonas P, Major G, Sakmann B (1993) Quantal components of unitary EPSCs at the mossy fibre synapse on CA3 pyramidal cells of rat hippocampus. J Physiol (Lond) 472:615-63

Kasai H, Takagi H, Ninomiya Y, Kishimoto T, Ito K, Yoshida A, Yoshioka T, Miyashita Y (1996) Two components of exocytosis and endocytosis in phaeochromocytoma cells studied using caged Ca2+ compounds. J Physiol (Lond) 494:53-65

Katz B (1966) Nerve, muscle and synapse. McGraw-Hill, New York. 193 pp

Katz B (1969) The release of neural transmitter substance. Charles C. Thomas, Springfield, Illinois. 60 pp

Katz B, Miledi R (1965) The measurement of synaptic delay, and the time course of acetylcholine release at the neuromuscular junction. Proc. Roy. S. B. 161:483-95

Katz B, Miledi R (1967) The timing of calcium action during neuromuscular transmission. J Physiol 189:535-544

Kavalali ET, Klingauf J, Tsien RW (1999) Properties of fast endocytosis at hippocampal synapses. Philosophical Transactions of the Royal Society of London - Series B: Biological Sciences 354:337-46

Kawasaki F, Mattiuz AM, Ordway RW (1998) Synaptic physiology and ultrastructure in comatose mutants define an In vivo role for NSF in neurotransmitter release. J Neurosci 18:10241-9

Khvotchev M, Sudhof TC (1998) Newly synthesized phosphatidylinositol phosphates are required for synaptic norepinephrine but not glutamate or gamma-aminobutyric acid (GABA) release. Journal of Biological Chemistry 273:21451-4

Kim DK, Catterall WA (1997) Ca2+-dependent and -independent interactions of the isoforms of the alpha1A subunit of brain Ca2+ channels with presynaptic SNARE proteins. Proceedings of the National Academy of Sciences of the United States of America 94:14782-6

Klein RL (1982) Chemical composition of the large noradrenergic vesicles. In: Klein RL, Lagercrantz H, Zimmermann H (eds) Neurotransmitter vesicles. Academic Press, London, pp 133-174

Klein RL, Lagercrantz H (1982) Insights into the functional role of the norandrenergic vesicles. In: Klein RL, Lagercrantz H, Zimmermann H (eds) Neurotransmitter vesicles. Academic Press, London, pp 219-240

Klingauf J, Kavalali ET, Tsien RW (1998) Kinetics and regulation of fast endocytosis at hippocampal synapses. Nature 394:581-5

Koenig JH, Ikeda K (1996) Synaptic vesicles have two distinct recycling pathways. Journal of Cell Biology 135:797-808

Koenig JH, Yamaoka K, Ikeda K (1993) Calcium-induced translocation of synaptic vesicles to the active site. J Neurosci 13:2313-22

Koh DS, Hille B (1997) Modulation by neurotransmitters of catecholamine secretion from sympathetic ganglion neurons detected by amperometry. Proceedings of the National Academy of Sciences of the United States of America 94:1506-11

Konishi M, Hollingworth S, Harkins AB, Baylor SM (1991) Myoplasmic calcium transients in intact frog skeletal muscle fibers monitored with the fluorescent indicator furaptra. Journal of General Physiology 97:271-301

Korn H, Faber DS (1991) Quantal analysis and synaptic efficacy in the CNS. Trends Neurosci 14:439-45

Korn H, Faber DS (1998) Quantal analysis and long-term potentiation. Comptes Rendus de l Academie des Sciences - Serie Iii, Sciences de la Vie 321:125-30

Kraszewski K, Grantyn R (1992) Unitary, quantal and miniature GABA-activated synaptic chloride currents in cultured neurons from the rat superior colliculus. Neuroscience 47:555-70

Kreft M, Zorec R (1997) Cell-attached measurements of attofarad capacitance steps in rat melanotrophs. Pflugers Archiv - European Journal of Physiology 434:212-4

Kuffler SW, Yoshikami D (1975) The number of transmitter molecules in a quan-
    tum: an estimate from iontophoretic application of acetylcholine at the neuro-
    muscular synapse. Journal of Physiology 251:465-82
Kullmann DM, Nicoll RA (1992) Long-term potentiation is associated with increases
    in quantal content and quantal amplitude. Nature 357:240-4
Kumakura K, Sasaki K, Sakurai T, Ohara-Imaizumi M, Misonou H, Nakamura S,
    Matsuda Y, Nonomura Y (1994) Essential role of myosin light chain kinase in the
    mechanism for MgATP- dependent priming of exocytosis in adrenal chromaffin
    cells. J Neurosci 14:7695-703
Lagercrantz H (1976) On the composition and function of large dense cored vesicles
    in sympathetic nerves. Neuroscience 1:81-92
Lagnado L, Gomis A, Job C (1996) Continuous vesicle cycling in the synaptic termi-
    nal of retinal bipolar cells. Neuron 17:957-67
Lenzi D, Runyeon JW, Crum J, Ellisman MH, Roberts WM (1999) Synaptic vesicle
    populations in saccular hair cells reconstructed by electron tomography. Journal
    of Neuroscience 19:119-32
Liao D, Hessler NA, Malinow R (1995) Activation of postsynaptically silent synapses
    during pairing-induced LTP in CA1 region of hippocampal slice. Nature 375:400-
    4
Lim NF, Nowycky MC, Bookman RJ (1990) Direct measurement of exocytosis and
    calcium currents in single vertebrate nerve terminals. Nature 344:449-51
Lindau M, Almers W (1995) Structure and function of fusion pores in exocytosis and
    ectoplasmic membrane fusion. Current Opinion in Cell Biology 7:509-17
Lindau M, Neher E (1988) Patch-clamp techniques for time-resolved capacitance
    measurements in single cells. Pflugers Arch 411:137-46
Lindau M, Rosenboom H, Nordmann J (1994) Exocytosis and endocytosis in single
    peptidergic nerve terminals. Adv Second Messenger Phosphoprotein Res 29:173-
    87
Lindau M, Stuenkel EL, Nordmann JJ (1992) Depolarization, intracellular calcium
    and exocytosis in single vertebrate nerve endings. Biophys J 61:19-30
Liu G, Tsien RW (1995) Properties of synaptic transmission at single hippocampal
    synaptic boutons. Nature 375:404-8
Llinas R, Steinberg IZ, Walton K (1981) Relationship between presynaptic calcium
    current and postsynaptic potential in squid giant synapse. Biophys J 33:323-51
Llinas R, Sugimori M, Silver RB (1992) Microdomains of high calcium concentration
    in a presynaptic terminal. Science 256:677-9
Lollike K, Borregaard N, Lindau M (1995) The exocytotic fusion pore of small gran-
    ules has a conductance similar to an ion channel. J Cell Biol 129:99-104
Lollike K, Borregaard N, Lindau M (1998) Capacitance flickers and pseudoflickers of
    small granules, measured in the cell-attached configuration. Biophysical Journal
    75:53-9
Manabe T, Renner P, Nicoll RA (1992) Postsynaptic contribution to long-term po-
    tentiation revealed by the analysis of miniature synaptic currents. Nature 355:50-
    5
Martin TF (1998) Phosphoinositide lipids as signaling molecules: common themes
    for signal transduction, cytoskeletal regulation, and membrane trafficking. An-
    nual Review of Cell & Developmental Biology 14:231-64
Martin TF, Hay JC, Banerjee A, Barry VA, Ann K, Yom HC, Porter BW, Kowalchyk
    JA (1995) Late ATP-dependent and Ca++-activated steps of dense core granule
    exocytosis. Cold Spring Harb Symp Quant Biol 60:197-204

Matthews G (1996a) Neurotransmitter release. Annu Rev Neurosci 19:219-33

Matthews G (1996b) Synaptic exocytosis and endocytosis: capacitance measurements. Curr Opin Neurobiol 6:358-64

Meldolesi J, Ceccarelli B (1981) Exocytosis and membrane recycling. Philosophical Transactions of the Royal Society of London – Series B: Biological Sciences 296:55-65

Mennerick S, Matthews G (1996) Ultrafast exocytosis elicited by calcium current in synaptic terminals of retinal bipolar neurons. Neuron 17:1241-9

Miledi R (1973) Transmitter release induced by injection of calcium ions into nerve terminals. Proc R Soc Lond B Biol Sci 183:421-5

Miller TM, Heuser JE (1984) Endocytosis of synaptic vesicle membrane at the frog neuromuscular junction. Journal of Cell Biology 98:685-98

Mintz IM, Sabatini BL, Regehr WG (1995) Calcium control of transmitter release at a cerebellar synapse. Neuron 15:675-88

Mochida S (2000) Protein-protein interactions in neurotransmitter release. Neuroscience Research 36:175-82

Mochida S, Sheng ZH, Baker C, Kobayashi H, Catterall WA (1996) Inhibition of neurotransmission by peptides containing the synaptic protein interaction site of N-type Ca2+ channels. Neuron 17:781-8

Morgans CW, Brandstatter JH, Kellerman J, Betz H, Wassle H (1996) A SNARE complex containing syntaxin 3 is present in ribbon synapses of the retina. J Neurosci 16:6713-21

Moser T, Beutner D (2000) Kinetics of exocytosis and endocytosis at the cochlear inner hair cell afferent synapse of the mouse. Proceedings of the National Academy of Sciences of the United States of America 97:883-8

Moser T, Neher E (1997) Estimation of mean exocytic vesicle capacitance in mouse adrenal chromaffin cells. Proc Natl Acad Sci USA 94:6735-40

Murthy VN, Sejnowski TJ, Stevens CF (1997) Heterogeneous release properties of visualized individual hippocampal synapses. Neuron 18:599-612

Murthy VN, Stevens CF (1998) Synaptic vesicles retain their identity through the endocytic cycle. Nature 392:497-501

Naraghi M, Muller TH, Neher E (1998) Two-dimensional determination of the cellular Ca2+ binding in bovine chromaffin cells. Biophysical Journal 75:1635-47

Neher E (1993) Cell physiology. Secretion without full fusion. Nature 363:497-8

Neher E (1998) Vesicle pools and Ca2+ microdomains: new tools for understanding their roles in neurotransmitter release. Neuron 20:389-99

Neher E, Marty A (1982) Discrete changes of cell membrane capacitance observed under conditions of enhanced secretion in bovine adrenal chromaffin cells. Proceedings of the National Academy of Sciences of the United States of America 79:6712-6

Neher E, Penner R (1994) Mice sans synaptotagmin. Nature 372:316-7

Neher E, Zucker RS (1993) Multiple calcium-dependent processes related to secretion in bovine chromaffin cells. Neuron 10:21-30

Neves G, Lagnado L (1999) The kinetics of exocytosis and endocytosis in the synaptic terminal of goldfish retinal bipolar cells. Journal of Physiology 515:181-202

Ninomiya Y, Kishimoto T, Miyashita Y, Kasai H (1996) Ca2+-dependent exocytotic pathways in Chinese hamster ovary fibroblasts revealed by caged-Ca2+ compound. J Biol Chem 271:17751-4

Ninomiya Y, Kishimoto T, Yamazawa T, Ikeda H, Miyashita Y, Kasai H (1997) Kinetic diversity in the fusion of exocytotic vesicles. Embo J 16:929-34

Oberhauser AF, Robinson IM, Fernandez JM (1995) Do caged-Ca2+ compounds mimic the physiological stimulus for secretion? Journal of Physiology, Paris 89:71-5

Oberhauser AF, Robinson IM, Fernandez JM (1996) Simultaneous capacitance and amperometric measurements of exocytosis: a comparison. Biophys J 71:1131-9

Okano K, Monck JR, Fernandez JM (1993) GTP gamma S stimulates exocytosis in patch-clamped rat melanotrophs. Neuron 11:165-72

Oleskevich S, Clements J, Walmsley B (2000) Release probability modulates short-term plasticity at a rat giant terminal. Journal of Physiology 524:513-523

Palfrey HC, Artalejo CR (1998) Vesicle recycling revisited: rapid endocytosis may be the first step. Neuroscience 83:969-89

Parsons TD, Lenzi D, Almers W, Roberts WM (1994) Calcium-triggered exocytosis and endocytosis in an isolated presynaptic cell: capacitance measurements in saccular hair cells. Neuron 13:875-83

Pothos EN, Davila V, Sulzer D (1998a) Presynaptic recording of quanta from midbrain dopamine neurons and modulation of the quantal size. Journal of Neuroscience 18:4106-18

Pothos EN, Przedborski S, Davila V, Schmitz Y, Sulzer D (1998b) D2-Like dopamine autoreceptor activation reduces quantal size in PC12 cells. J Neurosci 18:5575-85

Pothos EN, Sulzer D (1998) Modulation of quantal dopamine release by psychostimulants. Adv Pharmacol 42:198-202

Prakriya M, Solaro CR, Lingle CJ (1996) [Ca2+]i elevations detected by BK channels during Ca2+ influx and muscarine-mediated release of Ca2+ from intracellular stores in rat chromaffin cells. J Neurosci 16:4344-59

Proks P, Ashcroft FM (1995) Effects of divalent cations on exocytosis and endocytosis from single mouse pancreatic beta-cells. Journal of Physiology 487:465-77

Rahamimoff R, Fernandez JM (1997) Pre- and postfusion regulation of transmitter release. Neuron 18:17-27

Raju B, Murphy E, Levy LA, Hall RD, London RE (1989) A fluorescent indicator for measuring cytosolic free magnesium. American Journal of Physiology 256:C540-8

Ramaswami M, Krishnan KS, Kelly RB (1994) Intermediates in synaptic vesicle recycling revealed by optical imaging of Drosophila neuromuscular junctions. Neuron 13:363-75

Redman S (1990) Quantal analysis of synaptic potentials in neurons of the central nervous system. Physiol Rev 70:165-98

Rettig J, Heinemann C, Ashery U, Sheng ZH, Yokoyama CT, Catterall WA, Neher E (1997) Alteration of Ca2+ dependence of neurotransmitter release by disruption of Ca2+ channel/syntaxin interaction. J Neurosci 17:6647-56

Rieke F, Schwartz EA (1994) A cGMP-gated current can control exocytosis at cone synapses. Neuron 13:863-73

Rieke F, Schwartz EA (1996) Asynchronous transmitter release: control of exocytosis and endocytosis at the salamander rod synapse. J Physiol (Lond) 493:1-8

Roberts WM, Jacobs RA, Hudspeth AJ (1990a) Colocalization of ion channels involved in frequency selectivity and synaptic transmission at presynaptic active zones of hair cells. J Neurosci 10:3664-84

Robinson IM, Finnegan JM, Monck JR, Wightman RM, Fernandez JM (1995) Colocalization of calcium entry and exocytotic release sites in adrenal chromaffin cells. Proceedings of the National Academy of Sciences of the United States of America 92:2474-8

Rodieck RW (1998) The first steps in seeing. Sinauer Associates, Inc, Sunderland, Mass.

Rosenmund C, Stevens CF (1996) Definition of the readily releasable pool of vesicles at hippocampal synapses. Neuron 16:1197-207

Rothman JE (1994) Mechanisms of intracellular protein transport. Nature 372:55-63

Rothman JE, Orci L (1992) Molecular dissection of the secretory pathway. Nature 355:409-15

Rouze NC, Schwartz EA (1998) Continuous and transient vesicle cycling at a ribbon synapse. Journal of Neuroscience 18:8614-24

Ryan TA, Reuter H, Wendland B, Schweizer FE, Tsien RW, Smith SJ (1993) The kinetics of synaptic vesicle recycling measured at single presynaptic boutons. Neuron 11:713-24

Ryan TA, Smith SJ (1995) Vesicle pool mobilization during action potential firing at hippocampal synapses. Neuron 14:983-9

Sakaba T, Tachibana M, Matsui K, Minami N (1997) Two components of transmitter release in retinal bipolar cells: exocytosis and mobilization of synaptic vesicles. Neurosci Res 27:357-70

Schacher SM, Holtzman E, Hood DC (1974) Uptake of horseradish peroxidase by frog photoreceptor synapses in the dark and the light. Nature 249:261-3

Schaeffer SF, Raviola E (1978) Membrane recycling in the cone cell endings of the turtle retina. Journal of Cell Biology 79:802-25

Schneggenburger R, Meyer AC, Neher E (1999) Released fraction and total size of a pool of immediately available transmitter quanta at a calyx synapse. Neuron 23:399-409

Schweizer FE, Dresbach T, DeBello WM, O'Connor V, Augustine GJ, Betz H (1998) Regulation of neurotransmitter release kinetics by NSF. Science 279:1203-6

Seagar M, Leveque C, Charvin N, Marqueze B, Martin-Moutot N, Boudier JA, Boudier JL, Shoji-Kasai Y, Sato K, Takahashi M (1999) Interactions between proteins implicated in exocytosis and voltage-gated calcium channels. Philosophical Transactions of the Royal Society of London – Series B: Biological Sciences 354:289-97

Seward EP, Chernevskaya NI, Nowycky MC (1995) Exocytosis in peptidergic nerve terminals exhibits two calcium- sensitive phases during pulsatile calcium entry. J Neurosci 15:3390-9

Sheng ZH, Rettig J, Takahashi M, Catterall WA (1994) Identification of a syntaxin-binding site on N-type calcium channels. Neuron 13:1303-13

Siegel JH, Brownell WE (1986) Synaptic and Golgi membrane recycling in cochlear hair cells. Journal of Neurocytology 15:311-28

Simon SM, Llinas RR (1985) Compartmentalization of the submembrane calcium activity during calcium influx and its significance in transmitter release. Biophys J 48:485-98

Smith C, Moser T, Xu T, Neher E (1998) Cytosolic Ca2+ acts by two separate pathways to modulate the supply of release-competent vesicles in chromaffin cells. Neuron 20:1243-53

Smith C, Neher E (1997) Multiple forms of endocytosis in bovine adrenal chromaffin cells. J Cell Biol 139:885-94

Smith CB, Betz WJ (1996) Simultaneous independent measurement of endocytosis and exocytosis. Nature 380:531-4

Spruce AE, Breckenridge LJ, Lee AK, Almers W (1990) Properties of the fusion pore that forms during exocytosis of a mast cell secretory vesicle. Neuron 4:643-54

Stanley EF, Mirotznik RR (1997) Cleavage of syntaxin prevents G-protein regulation of presynaptic calcium channels. Nature 385:340-3

Stern MD (1992) Buffering of calcium in the vicinity of a channel pore. Cell Calcium 13:183-92

Stevens CF (1993) Quantal release of neurotransmitter and long-term potentiation. Cell 72 Suppl:55-63

Stevens CF, Tsujimoto T (1995) Estimates for the pool size of releasable quanta at a single central synapse and for the time required to refill the pool. Proc Natl Acad Sci USA 92:846-9

Stevens CF, Wang Y (1995) Facilitation and depression at single central synapses. Neuron 14:795-802

Stevens CF, Wesseling JF (1998) Activity-dependent modulation of the rate at which synaptic vesicles become available to undergo exocytosis. Neuron 21:415-24

Stevens CF, Wesseling JF (1999) Identification of a novel process limiting the rate of synaptic vesicle cycling at hippocampal synapses. Neuron 24:1017-28

Steyer JA, Horstmann H, Almers W (1997) Transport, docking and exocytosis of single secretory granules in live chromaffin cells. Nature 388:474-8

Sudhof TC (1995) The synaptic vesicle cycle: a cascade of protein-protein interactions. Nature 375:645-53

Sudhof TC, De Camilli P, Niemann H, Jahn R (1993) Membrane fusion machinery: insights from synaptic proteins. Cell 75:1-4

Tachibana M, Okada T (1991) Release of endogenous excitatory amino acids from ON-type bipolar cells isolated from the goldfish retina. J Neurosci 11:2199-208

Tachibana M, Okada T, Arimura T, Kobayashi K, Piccolino M (1993) Dihydropyridine-sensitive calcium current mediates neurotransmitter release from bipolar cells of the goldfish retina. J Neurosci 13:2898-909

Takeuchi A (1958) The long-lasting depression in neuromuscular transmission of frog. Jap J Physiol 8:102-113

Tang CM, Margulis M, Shi QY, Fielding A (1994) Saturation of postsynaptic glutamate receptors after quantal release of transmitter. Neuron 13:1385-93

Thomas P, Lee AK, Wong JG, Almers W (1994) A triggered mechanism retrieves membrane in seconds after Ca2+-stimulated exocytosis in single pituitary cells. J Cell Biology 124:667-675

Thomas P, Surprenant A, Almers W (1990) Cytosolic Ca2+, exocytosis, and the endocytosis in single melanotrophs of the rat pituitary. Neuron 5:723-33

Thomas P, Wong JG, Almers W (1993a) Millisecond studies of secretion in single rat pituitary cells stimulated by flash photolysis of caged Ca2+. Embo J 12:303-6

Thomas P, Wong JG, Lee AK, Almers W (1993b) A low affinity Ca2+ receptor controls the final steps in peptide secretion from pituitary melanotrophs. Neuron 11:93-104

Tolar LA, Pallanck L (1998) NSF function in neurotransmitter release involves rearrangement of the SNARE complex downstream of synaptic vesicle docking [In Process Citation]. J Neurosci 18:10250-6

Travis ER, Wightman RM (1998) Spatio-temporal resolution of exocytosis from individual cells. Annual Review of Biophysics & Biomolecular Structure 27:77-103

Trussell LO, Fischbach GD (1989) Glutamate receptor desensitization and its role in synaptic transmission. Neuron 3:209-18

Trussell LO, Zhang S, Raman IM (1993) Desensitization of AMPA receptors upon multiquantal neurotransmitter release. Neuron 10:1185-96

Vogel SS, Smith RM, Baibakov B, Ikebuchi Y, Lambert NA (1999) Calcium influx is required for endocytotic membrane retrieval. Proceedings of the National Academy of Sciences of the United States of America 96:5019-24

von Gersdorff H, Matthews G (1994a) Dynamics of synaptic vesicle fusion and membrane retrieval in synaptic terminals. Nature 367:735-9

von Gersdorff H, Matthews G (1994b) Inhibition of endocytosis by elevated internal calcium in a synaptic terminal. Nature 370:652-5

von Gersdorff H, Matthews G (1997) Depletion and replenishment of vesicle pools at a ribbon-type synaptic terminal. J Neurosci 17:1919-27

von Gersdorff H, Sakaba T, Berglund K, Tachibana M (1998) Submillisecond kinetics of glutamate release from a sensory synapse. Neuron 21:1177-88

von Gersdorff H, Schneggenburger R, Weis S, Neher E (1997) Presynaptic depression at a calyx synapse: the small contribution of metabotropic glutamate receptors. J Neurosci 17:8137-46

von Gersdorff H, Vardi E, Matthews G, Sterling P (1996) Evidence that vesicles on the synaptic ribbon of retinal bipolar neurons can be rapidly released. Neuron 16:1221-7

von Ruden L, Neher E (1993) A Ca-dependent early step in the release of catecholamines from adrenal chromaffin cells. Science 262:1061-5

Wang LY, Kaczmarek LK (1998) High-frequency firing helps replenish the readily releasable pool of synaptic vesicles. Nature 394:384-8

Weber T, Zemelman BV, McNew JA, Westermann B, Gmachl M, Parlati F, Sollner TH, Rothman JE (1998) SNAREpins: minimal machinery for membrane fusion. Cell 92:759-72

Weis S, Schneggenburger R, Neher E (1999) Properties of a model of Ca++-dependent vesicle pool dynamics and short term synaptic depression. Biophysical Journal 77:2418-2429

Wiedemann C, Schafer T, Burger MM, Sihra TS (1998) An essential role for a small synaptic vesicle-associated phosphatidylinositol 4-kinase in neurotransmitter release. Journal of Neuroscience 18:5594-602

Wightman RM, Jankowski JA, Kennedy RT, Kawagoe KT, Schroeder TJ, Leszczyszyn DJ, Near JA, Diliberto EJ, Jr., Viveros OH (1991) Temporally resolved catecholamine spikes correspond to single vesicle release from individual chromaffin cells. Proceedings of the National Academy of Sciences of the United States of America 88:10754-8

Wiser O, Trus M, Hernandez A, Renstrom E, Barg S, Rorsman P, Atlas D (1999) The voltage sensitive Lc-type Ca2+ channel is functionally coupled to the exocytotic machinery. Proceedings of the National Academy of Sciences of the United States of America 96:248-53

Wu LG, Betz WJ (1996) Nerve activity but not intracellular calcium determines the time course of endocytosis at the frog neuromuscular junction. Neuron 17:769-79

Wu LG, Borst JG (1999) The reduced release probability of releasable vesicles during recovery from short-term synaptic depression. Neuron 23:821-32

Xin Q, Wightman RM (1998) Simultaneous detection of catecholamine exocytosis and Ca2+ release from single bovine chromaffin cells using a dual microsensor. Analytical Chemistry 70:1677-81

Xu T, Binz T, Niemann H, Neher E (1998) Multiple kinetic components of exocytosis distinguished by neurotoxin sensitivity. Nat Neurosci 1:192-200

Yamada WM, Zucker RS (1992) Time course of transmitter release calculated from simulations of a calcium diffusion model. Biophysical Journal 61:671-82

Yang SN, Larsson O, Branstrom R, Bertorello AM, Leibiger B, Leibiger IB, Moede T, Kohler M, Meister B, Berggren PO (1999) Syntaxin 1 interacts with the L(D) subtype of voltage-gated Ca(2+) channels in pancreatic beta cells. Proceedings of the National Academy of Sciences of the United States of America 96:10164-9

Zhong H, Yokoyama CT, Scheuer T, Catterall WA (1999) Reciprocal regulation of P/Q-type Ca2+ channels by SNAP-25, syntaxin and synaptotagmin. Nature Neuroscience 2:939-41

Zhou Z, Misler S (1995a) Action potential-induced quantal secretion of catecholamines from rat adrenal chromaffin cells. J Biol Chem 270:3498-505

Zhou Z, Misler S (1995b) Amperometric detection of stimulus-induced quantal release of catecholamines from cultured superior cervical ganglion neurons. Proceedings of the National Academy of Sciences of the United States of America 92:6938-42

Zhou Z, Misler S, Chow RH (1996) Rapid fluctuations in transmitter release from single vesicles in bovine adrenal chromaffin cells. Biophysical Journal 70:1543-52

Zimmerberg J, Curran M, Cohen FS, Brodwick M (1987) Simultaneous electrical and optical measurements show that membrane fusion precedes secretory granule swelling during exocytosis of beige mouse mast cells. Proceedings of the National Academy of Sciences of the United States of America 84:1585-9

Editor-in-charge: Professor R. Jahn

Transport of Proteins Into Mitochondria

K. N. Truscott, N. Pfanner, W. Voos

Institut für Biochemie und Molekularbiologie, Universität Freiburg,
Hermann-Herder-Straße 7, D-79104 Freiburg, Germany

Contents

Abbreviations

AAC	ADP/ATP carrier
CoxIV	Cytochrome c oxidase subunit IV
pCoxIV	Presequence of cytochrome c oxidase subunit IV
$\Delta\psi$	inner membrane potential
DHFR	Dihydrofolate reductase
E. coli	*Escherichia coli*
GIP	General import pore
IM	Inner membrane
IMS	Intermembrane space
kDa	Kilodalton
MPP	Mitochondrial processing peptidase
MCC	Multiconductance channel
MSF	Mitochondrial stimulation factor
mtHsp70	Mitochondrial heat shock protein 70
NAC	Nascent polypeptide-associated complex
N. crassa	*Neurospora crassa*
NMR	Nuclear magnetic reasonance
OTC	Ornithine transcarbamylase
OM	Outer membrane
PAGE	Polyacrylamide gel electrophoresis
S. cerevisiae	*Saccharomyces cerevisiae*
Su9-DHFR	F_o-ATPase subunit 9-dihydrofolate reductase
TIMX	Translocase of the inner membrane, subunit of X kDa
TOMX	Translocase of the outer membrane, subunit of X kDa

1
Summary

Most mitochondrial proteins are nuclear-encoded and synthesised as preproteins on polysomes in the cytosol. They must be targeted to and translocated into mitochondria. Newly synthesised preproteins interact with cytosolic factors until their recognition by receptors on the surface of mitochondria. Import into or across the outer membrane is mediated by a dynamic protein complex coined the translocase of the outer membrane
(TOM). Preproteins that are imported into the matrix or inner membrane of
mitochondria require the action of one of two translocation complexes of
the inner membrane (TIMs). The import pathway of preproteins is prede-
termined by their intrinsic targeting and sorting signals. Energy input in the
form of ATP and the electrical gradient across the inner membrane is required for protein translocation into mitochondria. Newly imported proteins may require molecular chaperones for their correct folding.

2
Introduction

Mitochondria are specialised organelles found only in eukaryotic cells.
Structurally they can be divided into four distinct compartments. They possess an outer membrane which provides a barrier between the cytosol and
essential processes that occur within. A pore protein termed porin permits
the passage of metabolites across the outer membrane. An aqueous space
separates the outer membrane from a second lipid bilayer, the inner membrane, that contains the protein-pumping complexes of the respiratory
chain. The matrix is the innermost compartment of mitochondria, an aqueous environment housing notably maternally inherited genetic information.
Mitochondrial DNA codes for a few protein components of the inner membrane as well as the ribosomal and transfer RNA required for their synthesis.
However this limited amount of genetic information accounts for a very
small percentage of all mitochondrial proteins synthesised. Most mitochondrial proteins are encoded on nuclear DNA hence the need for sophisticated
protein trafficking systems. The transport of proteins is achieved via three
key mechanistic processes: specific targeting of preproteins to the mitochondrial surface; passage through lipid bilayers facilitated by proteinaceous
channels and assisted folding of newly imported proteins. Transport of preproteins into mitochondria is an energy-dependent process.

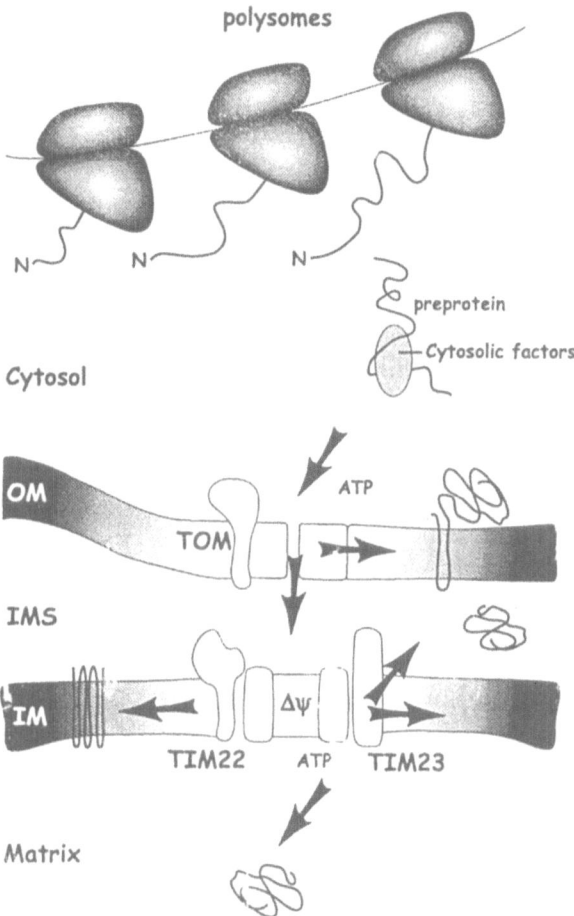

Fig. 1. Import of preproteins into mitochondrial compartments. Nuclear-encoded mitochondrial proteins are targeted to one of four compartments, the outer membrane (OM), intermembrane space (IMS), inner membrane (IM) or matrix. Preproteins are firstly synthesised on polysomes in the cytosol and then, perhaps in a cotranslational manner, interact with cytosolic proteins. Translocation into or across the outer membrane of mitochondria is facilitated by the TOM machinery, a large multisubunit complex. Import into the matrix or inner membrane and in many cases the intermembrane space requires the action of the TIM import machinery (TIM23 and TIM22 complexes) localised at the inner membrane. The point at which energy from ATP or the membrane potential ($\Delta\psi$) is require for import is illustrated. The illustration is not to scale

Following synthesis on polysomes nuclear-encoded mitochondrial pre-proteins interact with numerous translocation factors before reaching their final destination (Fig. 1). Many of these specific interactions depend on intrinsic targeting signals carried within the preprotein. Preproteins firstly interact with cytosolic molecular chaperones which presumably prevent premature folding, misfolding or aggregation prior to their recognition by receptors. Once a preprotein is localised to the surface of mitochondria through receptor binding, it is transported across the outer membrane as-sisted by components of the outer membrane translocase (TOM). Depend-ing on its final destination a preprotein may engage one of two distinct translocation complexes of the inner membrane (TIM) for import into the matrix, intermembrane space or the inner membrane. Energy for this proc-ess is derived from ATP and the electrochemical gradient across the inner membrane. Matrix-located proteins recruit molecular chaperones for their correct folding. In this article we provide an overall view of protein translo-cation into mitochondria focusing on the properties and functions of indi-vidual members of this complex pathway. Most of the discussion is directed toward protein import in the eukaryotic model organism *Saccharomyces cerevisiae* since the components and processes involved are best character-ised for this species. Given that mitochondria of higher eukaryotes contain homologues of the yeast import machinery the basic principles described here most likely apply to these organisms also.

3
Preprotein Recognition

3.1
Targeting Signals

How is a mitochondrial protein, which is synthesised in the cytosol, specifi-cally transported to its correct location in mitochondria? Nuclear-encoded mitochondrial proteins carry signal sequences or targeting sequences that are decoded by specific mitochondrial receptors. The signals may take the form of an amino-terminal extension (presequence) or an internal signal which is retained within the functional protein.

3.1.1
Presequences

Proteins targeted to the matrix of mitochondria have amino-terminal extensions relative to homologous non-mitochondrial counterparts. The length of the extension varies from one preprotein to another but may be up to 80 amino acids in length. In most cases the amino-terminal extension is cleaved off by a specific peptidase once the preprotein is imported into the mitochondrial matrix. For a few matrix-targeted preproteins such as rhodanese (Miller et al. 1991), 3-oxoacyl-CoA thiolase (Amaya et al. 1988), the β-subunit of human electron-transfer flavoprotein (Finocchiaro et al. 1993) and chaperonin 10 (Rospert et al. 1993; Ryan et al. 1994b) the targeting presequence is retained on the mature protein. Presequences are not reserved for matrix-located proteins alone as some intermembrane space and inner membrane proteins also have presequences which play a role in their targeting (Section 6).

Early experimental evidence indicating that amino-terminal extensions act as targeting signals was achieved by assessing the ability of a preprotein to import into isolated mitochondria following removal of the presequence with purified peptidases. The truncated preprotein failed to import into mitochondria unlike the full length precursor indicating that the presequence extension is necessary for import into the matrix (Gasser et al. 1982). In order to determine if the presequence alone was also sufficient for targeting, the ability of presequences to direct non-mitochondrial proteins into mitochondria following fusion of the presequence to passenger proteins was demonstrated (Horwich et al. 1985; Hurt et al. 1984; Hurt et al. 1985). In the first reports of this kind relating to mitochondrial import the cleavable presequence of human ornithine transcarbamylase (OTC) or yeast cytochrome c oxidase subunit IV were attached to the amino-terminus of the cytosolic mouse protein dihydrofolate reductase. The chimeric proteins were imported into the matrix of isolated mitochondria (Hurt et al. 1984; Horwich et al. 1985). There are however some exceptions reported where the mature portion of the preprotein is also believed to contribute to targeting (Bedwell et al. 1987; Pfanner et al. 1987b; Zara et al. 1992; Arnold et al. 1998).

3.1.2
Structure of Presequences

Close examination of mitochondrial presequences reveals an absence of primary sequence homology. In general presequences have a high content of basic, hydrophobic and hydroxylated amino acid residues with a bias against acidic amino acids (von Heijne et al. 1989). A specific primary amino acid sequence is not important for targeting as many randomly generated prese- quences which resemble natural presequences in their amino acid content can function as artificial mitochondrial presequences (Allison and Schatz 1986; Baker and Schatz 1987). Early theoretical and biophysical studies lead to the prediction that mitochondrial presequences form positively-charged amphipathic α-helices important for their targeting function (von Heijne 1986; Roise et al. 1988; Lemire et al. 1989). Indeed for many presequences separation of positive and hydrophobic amino acids can be observed on a helical wheel projection (Fig. 2). Chemically synthesised peptides which have a similar amino acid content to natural presequences, but are non- amphipathic, fail to act as functional presequences (Allison and Schatz 1986). The importance of an amphipathic α-helix for targeting function has subsequently been strengthened following the structural determination of chemically synthesised presequence peptides by two-dimensional NMR

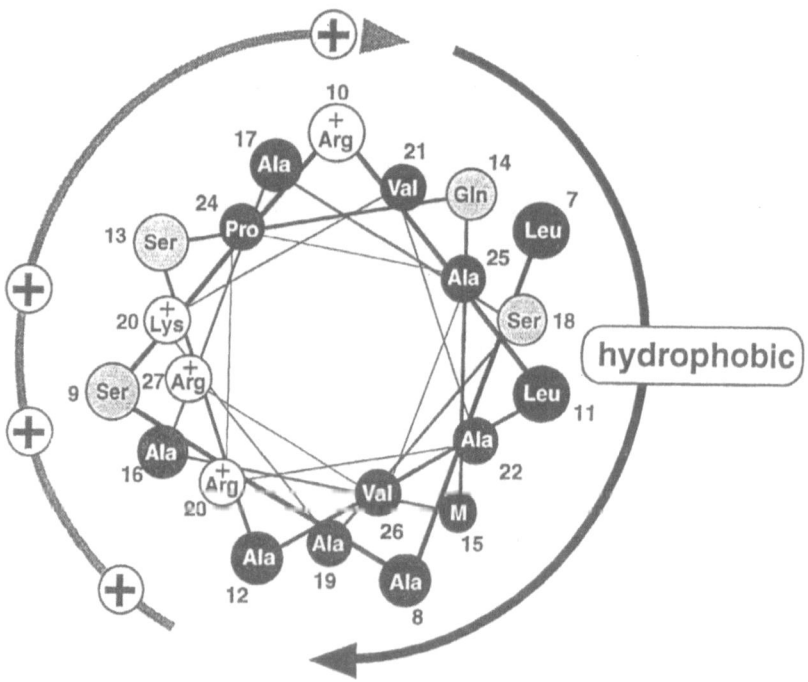

Fig. 2. An amphipathic α-helical presequence. A helical wheel projection showing the end-on view of *Neurospora crassa* F$_o$-ATPase subunit 9 presequence from amino acids 7 to 28. Positively charged residues are indicated (+)

spectroscopy. The presequence of cytochrome c oxidase subunit IV (Endo et al. 1989), the β-subunit of the F_1-ATPase (Bruch and Hoyt 1992), aldehyde dehydrogenase (Karslake et al. 1990), rhodanese (Hammen et al. 1994), 3-oxoacyl-CoA thiolase (Hammen et al. 1994) and chaperonin 10 (Jarvis et al. 1995) all form amphipathic α-helices in non-aqueous environments.

Prior to the identification and characterisation of import receptors it was suggested that the amphipathic nature of a presequence was important for a direct interaction with membranes. It is now generally believed that the formation of an amphipathic α-helix in presequences is important for their specific recognition and subsequent binding to import receptors on the surface of mitochondria. The direct role of an amphipathic α-helices in targeting preproteins to import receptors has been demonstrated for one mitochondrial preprotein. Recently, the structure of the soluble cytosolic core-domain of rat Tom20 was solved by NMR in complex with the C-terminal half of rat aldehyde dehydrogenase presequence (Abe et al. 2000). In this case the hydrophobic face of the helix lies in a hydrophobic groove formed by the receptor. Site-directed mutagenesis confirmed that hydrophobic interactions indeed mediate binding between the presequence and the receptor. The importance of an amphipathic α-helical presequence for receptor recognition is also evident from *in vitro* binding studies. Brix et al. (1997) showed that the chemically synthesised amphiphatic α-helical presequence of cytochrome c oxidase subunit IV (pCoxIV) competes with natural presequence-containing preproteins for binding to Tom20 and Tom22. A synthetic peptide (SynB2) which contains the same number of positive charged residues as pCoxIV but lacks a hydrophobic face (Allison and Schatz 1986; Roise et al. 1988) provided only slight competition for binding to import receptors. As more structural information is gathered it may soon be possible to describe a consensus presequence peptide.

3.1.3
Internal Targeting Signals

Apart from preproteins which are synthesised with non-cleavable amino-terminal signals many mitochondrial proteins are synthesised as mature proteins. The amino acid residues which contribute information required to direct the preprotein to the right mitochondrial compartment are hidden amongst structural and functional elements of the protein. This makes it very difficult to define the characteristics of internal targeting signals. Consequently there is very little known about the nature and distribution of such signals. Some information has been gained by studying the import of the ADP/ATP carrier (AAC). This protein is a member of the metabolic carrier family of the inner membrane. Members of this family are responsible for the exchange of metabolites between the intermembrane space and the matrix. Targeting information have been reported to exist in both the N- and C-terminal domains of AAC (Pfanner et al. 1987a; Smagula and Douglas 1988) but the molecular nature of these signals is not known. Using a systematic peptide-scanning method potential receptor binding regions were identified throughout the linear sequence of the phosphate carrier (Brix et al. 1999). This information indicates that a complicated targeting mechanism exists whereby efficient import depends on multiple targeting signals within these proteins. This is in contrast to presequence-containing preproteins in which the binding sequence is located almost exclusively at the N-terminus.

3.2
Import Receptors on the Surface of Mitochondria

When isolated mitochondria are exposed to a mild protease treatment their ability to bind and import preproteins is reduced substantially. Such an effect is due to the removal of membrane-bound preprotein import receptors on the surface of mitochondria. To date, four mitochondrial outer membrane proteins which exhibit properties of a preprotein receptor have been described. These are Tom20, Tom22, Tom70 and Tom37. Deletion of the genes encoding any of these four proteins specifically impairs preprotein import.

3.2.1
Topology of Tom Receptors

Tom20, Tom22 and Tom70 each possess a large soluble cytosolic domain and a single putative transmembrane anchor. The N-terminal portion of Tom22 is exposed to the cytosol with its membrane anchor positioned near the C-terminus (Kiebler et al. 1993). In addition the extreme C-terminal domain of Tom22 faces the intermembrane space and acts as a preprotein binding site. Tom20 and Tom70 on the other hand have the opposite orientation in the outer membrane. They are anchored to the lipid bilayer via their amino-termini with large receptor domains facing the cytosol (Söllner et al. 1990; Schneider et al. 1991; Schlossmann et al. 1994). The topology of Tom37 has not been clearly defined. A considerable portion of the protein is exposed to the cytosol as it is degraded by a mild protease treatment of isolated mitochondria (Gratzer et al. 1995). However, in contrast to the other three receptors, it can be extracted from the lipid bilayer with sodium carbonate (Gratzer et al. 1995; Ryan et al. 1999).

3.2.2
Tom20 is Required for Optimal Protein Translocation

Tom20 was the first preprotein receptor identified (Söllner et al. 1989). Antibodies or Fab fragments specifically directed against Tom20 inhibit the import of several but not all preprotein types (Söllner et al. 1989; Ramage et al. 1993; Goping et al. 1995; Hanson et al. 1996; Terada et al. 1997). When Tom20 is deleted from *N. crassa* using the genetic method of sheltered disruption a severe growth defect is observed (Harkness et al. 1994). In addition the import of most preproteins into Tom20 deficient mitochondria is inhibited but to varying degrees. Deletion of Tom20 from *S. cerevisiae* yields variable preprotein import inhibition *in vivo* (Ramage et al. 1993; Moczko et al. 1994). However *in vitro* import of presequence containing preproteins into *tom20Δ* mitochondria is strongly impaired (Moczko et al 1994). Interpretation of these findings is complicated by the fact that *tom20Δ* mitochondria have reduced levels of the vital translocation component Tom22 (Harkness et al. 1994; Lithgow et al. 1994) as its biogenesis is strongly dependent on receptors (Keil and Pfanner 1993). The import defects observed in Tom20 deletion strains could therefore be partly due to the reduced levels of Tom22. Lithgow et al. (1994) observed that a *tom20Δ* strain of *S. cerevisiae* adapts to the loss of Tom20 by regaining normal levels of Tom22. Overexpression of Tom22 suppresses the respiratory growth defect of yeast cells lacking Tom20 however the protein import defect *in vitro* is not fully rescued indicating that Tom20 is required for optimal protein translocation (Hönlinger et al. 1995).

3.2.3
The Receptor Function of Tom22

Tom22 is a multifunctional protein acting as a preprotein receptor in the cytosol, providing a preprotein binding site in the intermembrane space as well as playing a key role in the organisation of the TOM translocation machinery (van Wilpe et al. 1999). Cross-linking experiments indicate that Tom22 comes into direct contact with preproteins in transit across the outer membrane (Hönlinger et al. 1995). This interaction is thought to occur following initial binding of preproteins to the receptors Tom20 or Tom70 as antibodies directed against the cytosolic domain of Tom22 do not prevent receptor binding but inhibit preprotein insertion into the outer membrane (Kiebler et al. 1993). It is generally believed that Tom22 mediates the transfer of preproteins from either receptor Tom20 or Tom70 to the general import pore (GIP). Several reports, using standard procedures for gene deletion, indicated that Tom22 was essential for cell viability (Lithgow et al. 1994; Hönlinger et al. 1995; Nargang et al. 1995) but recently a yeast strain was generated in which Tom22 was deleted and yet the cells survived, although at a strongly reduced growth rate (van Wilpe et al. 1999). Mitochondria isolated from tom22Δ import different types of preproteins but with greatly reduced efficiency compared to wild-type mitochondria. This gross import defect can be attributed to a loss of receptor function but is also probably related to defects arising from the loss of structural organisation and dynamics of the outer membrane translocation machinery.

3.2.4
Tom70, the Receptor for Preproteins With Internal Targeting Signals

The import of preproteins with internal targeting sequences such as AAC are only marginally inhibited when the function of Tom20 is abolished by either biochemical or genetic means. Although Tom20 has some affinity for these proteins it is not responsible for their efficient import (Steger et al. 1990). It is clear that Tom70 is the major receptor for preproteins with internal targeting signals such as members of the carrier family. In N. crassa antibodies raised against this outer membrane protein specifically inhibit the import of AAC (Söllner et al. 1990; Schlossmann et al. 1994). Both genetic and biochemical studies reveal that S. cerevisiae Tom70 plays a role in the import of carrier proteins as well as F_1-ATPase β-subunit and cytochrome c_1 (Hines et al. 1990). The import of cytochrome c_1, a presequence containing preprotein, is also impaired by deletion of Tom20. This information combined indicates that Tom20 and Tom70 have partially overlapping specificity for preproteins and therefore it is not possible to assign mitochondrial preprotein families entirely to distinct receptor import pathways.

3.2.5
Receptor-Preprotein Binding *in vitro*

A direct assessment of the relative affinities of purified Tom receptor domains for individual preproteins has helped to clarify their respective roles in preprotein recognition. Supporting biochemical and genetic evidence Schlossmann et al. (1994) showed that the soluble receptor domain of S. cerevisiae Tom70 could directly and selectively interact with AAC and the phosphate carrier with high affinity. In addition, the precursor of cytochrome c_1 also bound to Tom70 but less tightly. Similar studies were carried out using purified recombinant receptor domains of S. cerevisiae Tom20, Tom70 and Tom22 (Brix et al. 1997). Each receptor displays bias toward particular preprotein types. Consistent with the findings of Schlossmann et al. (1994) Tom70 binds with greatest affinity to preproteins with internal targeting information. This binding is not disrupted by the addition of synthetic presequence peptides indicating that the receptor exclusively recognises internal targeting signals. Tom20 has significant affinity for both presequence containing preproteins such as Su9-DHFR (a chimeric protein consisting of the presequence of Fo-ATPase subunit 9 fused to entire dihydrofolate reductase) and preproteins with internal targeting signals. Tom22 on the other hand binds preferentially to presequence containing preproteins. This interaction appears to be specific as presequence peptides compete for binding. The ability of each receptor domain to bind preproteins independently of other translocation components indicates that they can function as receptors in their own right. However, there is considerable evidence indicating that *in vivo* Tom20 and Tom22 cooperate resulting in efficient recognition and translocation of preproteins (Mayer et al. 1995; van Wilpe et al. 1999; Pfanner 2000).

3.2.6
The Molecular Nature of Preprotein Recognition

What is the molecular nature of preprotein-receptor interactions? As described previously presequences form positively-charged amphipathic α-helices. The receptor domains of Tom20 and Tom22 possess an overall net negative charge. It seems reasonable that presequences bind to Tom20 or Tom22 through electrostatic interactions. In the case of Tom22 most experimental evidence supports this notion (Bolliger et al. 1995; Brix et al. 1997). The interaction is salt sensitive and synthetic presequence peptides can compete for binding. However, the nature of preprotein binding to Tom20 is not clear. Following an assessment of preprotein binding to intact receptors on the surface of mitochondria Haucke et al. (1995) concluded that Tom20 binds to the positively charged presequence of a preprotein via electrostatic interactions. In addition deletion of the C-terminal acidic domain of Tom20 strongly inhibited *in vitro* import of a selection of preproteins providing further evidence for a charge-based affinity (Bolliger et al 1995). However, binding of preproteins to the purified receptor domain of Tom20 is stimulated by salt, indicating a hydrophobic type of interaction (Brix et al. 1997). The three dimensional structure of Tom20 in complex with a presequence peptide provides compelling evidence that the interaction is predominately hydrophobic (Abe et al. 2000). Salt bridges are not observed in the NMR structure between the receptor and presequence (Abe et al. 2000). It is possible however that other regions on the surface of Tom20 bind preproteins via ionic interactions. The binding of preproteins to Tom70 is not inhibited by salt or presequence peptides (Brix et al. 1997). It is likely that the molecular nature of this interaction is largely hydrophobic.

3.2.7
Role of Tom37 and Tom72

With only one detailed report describing the characteristics and function of Tom37 it is not possible to evaluate its role in protein import. It has been proposed to act with Tom70 in the recognition of preproteins with internal targeting sequences since a simultaneous deletion of both Tom37 and Tom70 is lethal (Gratzer et al. 1995). Tom37 and Tom70 comigrate on a detergent-containing sucrose gradient (Gratzer et al. 1995) however they are found as separate complexes following electrophoresis under native conditions (Ryan et al. 1999). Moreover, a deletion of Tom37 did not impair binding of preproteins to Tom70 (Ryan et al. 1999). It is thus still open if Tom37 plays a direct role in protein import. A homologue of Tom70 called Tom72 has also been described. As deletion of this protein does not effect cell growth or mitochondrial function it seems likely that its function is redundant (Bömer et al. 1996; Schlossmann et al. 1996).

3.3
Cytosolic Import Factors

As described above the fundamental mechanism underlying the early steps of protein import is the specific binding of the preprotein, facilitated by its intrinsic targeting information, to import receptors on the surface of mitochondria. In the cell the execution of this process can be assisted by cytosolic proteins. It is thought that cytosolic import factors may act either to ensure that targeting regions of the preprotein are suitably exposed for recognition by the receptors, actively aid the delivery of the preprotein to the mitochondrial surface, prevent misfolding and aggregation or maintain the preprotein in a loose or unfolded conformation. The degree with which cytosolic factors co-operate or overlap in function and their specificity of targeting *in vivo* has not been characterised.

Stimulation of mitochondrial import *in vitro* by cytosolic factors of reticulocyte lysate, yeast and liver was demonstrated some years ago but the factors responsible for the measured activity in these experiments were not characterised further (Argan et al. 1983; Ono and Tuboi 1988; 1990; Ohta and Schatz 1984; Miura et al. 1983). It is known however that cytosolic molecular chaperones belonging to the Hsp70 family assist protein import into mitochondria as well as other organelles such as the endoplasmic reticulum and nucleus (Chirico et al. 1988; Deshaies et al. 1988; Murakami et al. 1988). A role in import is therefore considered to be a consequence of their general chaperone function rather than relating to a specific targeting function.

Cytosolic proteins which seem to preferentially interact with mitochondrial preproteins are the mammalian proteins referred to as targeting factor (Ono and Tuboi 1988, 1990), presequence binding factor (Murakami and Mori, 1990) and mitochondrial stimulation factor (MSF) (Hachiya et al. 1993). MSF is the only one of these three factors in which a functional interaction with characterised mitochondrial receptors has been demonstrated (Hachiya et al. 1995; Komiya et al. 1996; Komiya et al. 1997). MSF, a heterodimeric protein, was affinity purified from rat liver cytosol through a specific interaction with the presequence of yeast cytochrome oxidase (pCoxIV). MSF stimulates the *in vitro* import of mitochondrial preproteins such as adrenodoxin, the presequence of superoxide dismutase, porin, and a yeast CoxIV precursor (Hachiya et al. 1993). The protein also possesses ATPase activity which is specifically stimulated by preproteins targeted to mitochondria (Hachiya et al. 1994). Heterologous *in vitro* studies between MSF and yeast mitochondria revealed that MSF bound to adrenodoxin interacts with the receptor Tom70 (Hachiya et al. 1995). Following release of MSF, a reaction dependent on the hydrolysis of ATP, the preprotein is transferred to the receptors Tom20 and Tom22 followed by import across the outer membrane. Mihara and colleagues expanded upon these studies by showing that the type of chaperone bound, either MSF or Hsp70, determines the receptor import pathway taken by a preprotein (Komiya et al. 1996; Komiya et al. 1997). For example *in vitro* studies indicate that the specific binding of a preprotein to the cytosolic domain of Tom20 only occurs in the presence of Hsp70 (Komiya et al. 1997). Upon binding of the preprotein to Tom20 Hsp70 is released in an ATP-independent manner allowing translocation to proceed through the outer membrane. However, if MSF was also bound to the preprotein in addition to Hsp70 then it would first interact with the Tom70 receptor complex.

Using a homologous yeast *in vitro* assay system Fünfschilling and Rospert (1999) recently showed that the cytosolic protein nascent polypeptide-associated complex (NAC) stimulates the import of yeast mitochondrial proteins *in vitro*. They propose that ribosome associated NAC facilitates recognition of newly synthesised proteins by the import machinery. *In vivo* deletion of the α-subunit of NAC in yeast inhibits the biogenesis of artificial fusion proteins in mitochondria however it does not significantly alter the overall composition of mitochondrial proteins (George et al. 1998).

4
Translocase of the Outer Membrane (TOM)

Transport of preproteins through or into the outer membrane of mitochondria in most cases depends on a high molecular weight multisubunit protein complex referred to as the TOM complex (Fig. 3). In *S. cerevisiae* there are five tightly-associated integral membrane components of this complex Tom40, Tom22, Tom5, Tom6 and Tom7 each of which performs a distinct function in protein translocation. The association of the outer membrane receptors Tom20 and Tom70 with the TOM complex is either weak or dynamic in nature but together all components are necessary for efficient pro-

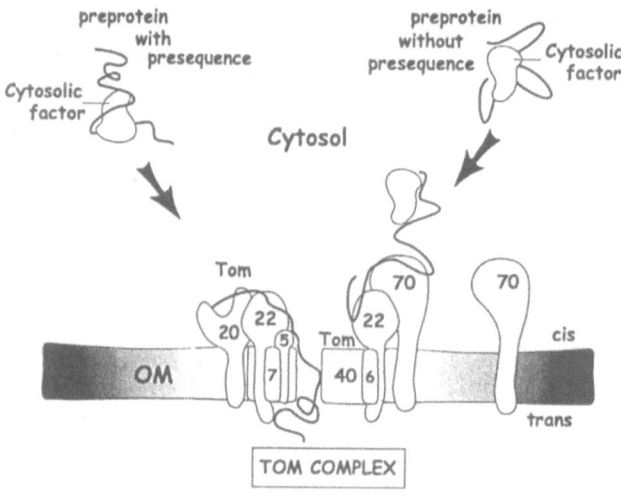

Fig. 3. Components of the TOM complex in *S. cerevisiae*. Seven different integral membrane proteins comprise the TOM complex. Tom20 acts as a receptor mainly for preproteins with presequences but can also recognise preproteins with internal signals. *In vivo* it is most likely largely associated with the TOM complex via an interaction with the cytosolic domain of Tom22. Tom70 is the main receptor for preproteins with internal targeting signals. It is only loosely associated with the TOM complex. Following receptor binding preproteins are sequentially transfered to the cytosolic receptor domains of Tom22 and Tom5 which directs them to the general import pore (GIP). Tom40 forms the translocation pore through which preproteins of all types pass. Tom6, Tom7 and the membrane spanning domain of Tom22 influence complex stability and mobility. The intermembrane space domain of Tom22 provides a *trans* binding site for the efficient translocation of presequence preproteins

tein translocation and comprise the translocase of the outer membrane (Fig. 3). In the following we describe what is currently known about the composition and function of individual components of the TOM complex.

4.1
Tom40 Forms a Cation-Selective Channel

The most critical component of the TOM complex is Tom40 as it forms a hydrophilic channel in the lipid bilayer which is permeable to preproteins. Tom40 was first identified as a component of the outer membrane preprotein translocase machinery through a specific photocross-link to an artificial preprotein blocked in outer membrane import sites (Vestweber et al. 1989). However, direct evidence that Tom40 could form a translocation pore was achieved by studying its properties following reconstitution into a planar lipid bilayer. It was revealed that Tom40 alone can form a cation-selective pore sufficiently large (~22 Å) to allow the passage of preproteins through the membrane in an unfolded conformation (Hill et al. 1998). This important function is reflected through *in vivo* studies which indicate that there is an absolute requirement for Tom40 as it is essential for cellular viability under all growth conditions examined (Baker et al. 1990; Kassenbrock et al. 1993). The fact that it is the only truly essential translocation component of the outer membrane translocation machinery suggests that indeed it alone forms a translocation channel and that other Tom components can neither substitute for its function nor form a vital component of the pore forming scaffold.

4.2
Tom22 and Tom5 Link Preprotein Recognition
and Outer Membrane Translocation

The other translocation components of the TOM complex facilitate either the passage of preproteins from receptors to the translocation channel, often referred to as the general import pore (GIP), or play a role in the structure and dynamics of the complex itself. As described (section 3.2.3) Tom22 acts as a preprotein receptor downstream in the import pathway of both Tom20 and Tom70. The cytosolic domain of Tom22 harbours binding sites for the soluble domains of both Tom20 and Tom70 (van Wilpe et al. 1999). Thus the association of Tom20 and Tom70 to the GIP complex is mediated by Tom22 and facilitates the transfer of preproteins from Tom20 and Tom70 to the GIP via its receptor domain. Prior to preprotein insertion into the GIP another translocation component, Tom5, exerts its influence (Dietmeier et al. 1997).

Tom5 is in close proximity to Tom40. Yeast deleted of Tom5 are inhibited in import of all preprotein types. Inhibition occurs at a post receptor binding stage. It has therefore been proposed that the cytosolic domain of Tom5 mediates the transfer of preproteins from receptors to the GIP via charge based interactions. Although Tom5 is tightly associated with the GIP complex via a membrane anchor it is not required for the structural integrity of the TOM complex (Dekker et al. 1998).

4.3
Distinct Roles for Tom6 and Tom7

Unlike Tom5 the other two small translocation components of the TOM complex do not come into direct contact with preproteins but they rather influence complex stability. When Tom6 is deleted from *S. cerevisiae* Tom40 migrates as a 100 kDa species containing Tom5 and Tom7 but not Tom22 (Dekker et al. 1998). However, Tom6 is not a permanent structural element necessary for the interaction between Tom22 and Tom40 as it can be removed from the TOM complex with the detergent Triton X100 without disturbing the association between these two proteins (Dekker et al. 1998). Therefore, it seems that Tom6 is an assembly factor which facilitates the association of Tom22 and Tom40 but is not required for the maintenance of this interaction. The consequence of a lack of Tom6 and hence fully assembled TOM complexes is a delay in transfer of preproteins from receptors to the GIP (Alconada et al. 1995). Deletion of Tom7 on the other hand does not destabilise the 400K TOM complex (Dekker et al. 1998). In fact a lack of Tom7 enables Tom20 and Tom22 to bind more tightly to Tom40. This change in TOM complex dynamics, due to the absence of Tom7, only slightly impairs the import of preproteins destined for the inner membrane or matrix of mitochondria (Hönlinger et al. 1996). However, the import of porin into the outer membrane of mitochondria is significantly reduced in *tom7Δ* mitochondria. Porin is imported into the outer membrane of mitochondria via the TOM translocation machinery showing a strong dependence on most translocation components for import (Krimmer, Pfanner, unpublished). Thus, Tom7 may destabilise the GIP complex so that the rate of lateral movement out of the translocation machinery and into the outer membrane is increased ensuring efficient import.

4.4
Organisation of the TOM Machinery

The structural organisation of the TOM complex has been investigated extensively using blue native PAGE (Dekker et al. 1998). The estimated molecular weight of the TOM complex is approximately 400 kDa. However, when the membrane spanning domain of Tom22 is deleted from yeast the TOM complex dissociates into a basic translocation unit of 100 kDa which is still able to import preproteins (van Wilpe et al. 1999). It consists of Tom40, Tom5, Tom6 and Tom7. The large size shift, in the order of 300 kDa, is not due to the removal of multiple Tom22 subunits but is rather a consequence of the loss of structural complexity. It has been proposed that each translocation pore is formed by a dimer of Tom40 and that an estimated six molecules of Tom40 are present in each TOM complex (van Wilpe et al. 1999; Dekker et al. 1998). In addition electron microscopy of the TOM complex isolated from *N. crassa* reveals that up to three pores are present (Künkele et al. 1998). Together this information indicates that three basic translocation units, containing the pore forming dimer of Tom40 and Tom5, 6 and 7, associate through the single membrane anchors of Tom22 subunits to form the 400 kDa TOM complex. The advantage of having up to three translocation pores in one complex is not known. It may be a question of efficiency or perhaps it allows the coordinated control of channel gating as the open probability of the translocation pore is reduced in the presence of Tom22 (van Wilpe et al. 1999).

In *S. cerevisiae* the majority of Tom20 can only be detected as a stable member of the TOM complex on blue native PAGE following mild detergent solubilisation. Under the same conditions Tom70 migrates as a separate smaller complex (Meisinger, Ryan, Pfanner unpublished). However, the cytosolic domains of Tom70 and Tom22 interact *in vitro* indicating that Tom70 can associate directly with the TOM complex via Tom22 (van Wilpe et al. 1999). The Tom complex isolated from *N. crassa* via affinity purification and gel filtration contains all of the known *N. crassa* Tom components, Tom20, Tom70, Tom40, Tom22, Tom6 and Tom7 (Künkele et al. 1998). However, regardless of these differences in the structural dynamics of Tom components the known receptors link initial preprotein recognition to translocation across the GIP via Tom22. Thus all preproteins converge at the GIP. Subsequent entry into the GIP is then facilitated by the cytosolic domain of Tom5.

4.5
Separation of Import Pathways at the General Import Pore

At some point the import pathways of different preprotein types diverge at
or near to the GIP as they are sorted to different compartments. As mentioned previously the negatively charged C-terminal domain of Tim22 pro-
trudes into the intermembrane space (Lithgow et al. 1994; Bolliger et al.
1995; Hönlinger et al. 1995; Nakai and Endo 1995). Due to the confinement
of Tom22 in the TOM complex its intermembrane space domain is closely
opposed to the exit site of the import pore. This domain provides a *trans*
binding site for preproteins with N-terminal targeting signals improving
their rate of translocation (Moczko et al. 1997). However, deletion of this
domain does not inhibit the import rates of porin or the noncleavable preprotein AAC (Moczko et al. 1997). Thus, the import pathways for different
preproteins diverge at the latest as soon as they reach the intermembrane
space. For preproteins destined for the outer membrane exit from the GIP
may occur at an earlier stage.

4.6
Factors Driving Import Across the Outer Membrane

While the contribution of each translocation component to protein import
has been described what combination of factors promotes the movement of
a preprotein through the outer membrane via the general import pore? For
presequence containing preproteins translocation may be promoted by Tom
and Tim subunits via electrostatic interactions (acid chain hypothesis). It
has been reported that the affinity of a preprotein for each translocation
component increases sequentially along the import route (Komiya et al.
1998). Using cross-linkers to probe the spatial arrangement of Tom40, Rapaport et al. (1998) showed that binding of preproteins to the GIP complex
altered the conformation of this protein. They propose that preprotein
binding induced conformational changes in Tom40 facilitate the movement
of preproteins across the membrane. This type of movement may be applicable to all preprotein types. However, an external energy source that drives
translocation into or across the outer membrane has not been found. Translocational of presequence preproteins across the outer membrane is coupled
to their translocation across the inner membrane. Thus, the ATP-dependent
mechanisms that draw preproteins across the inner membrane also sustain
translocation of preproteins across the outer membrane. The movement of
preproteins across the outer membrane may also be facilitated by partial

by partial folding of preprotein domains in the intermembrane space or at the inner membrane.

5
The General Translocase of the Inner Membrane

Once a preprotein has traversed the outer membrane with the aid of the TOM complex several import pathways may be followed depending on the intrinsic targeting signals of the preprotein. At least two distinct protein translocation complexes of the inner membrane are responsible for trafficking of preproteins to internal mitochondrial compartments. For matrix-targeted preproteins which possess a presequence import is mediated by the TIM23 complex. Some preproteins have sorting signals in addition to the presequence which direct them, also via the TIM23 complex, to the inner membrane or intermembrane space. Inner membrane proteins that lack a presequence depend on the large TIM22 complex for their import (Section 7). Import of preproteins via either the TIM23 or TIM22 complexes is strictly dependent on the inner membrane potential.

5.1
Composition of the TIM23 Translocase

The complexity of the TIM23 inner membrane translocase with respect to the type and number of translocation components does not match that of the TOM complex. Three Tim subunits make up the lipid-associated TIM23 translocase (Table 1). These are Tim23, Tim17 and Tim44. Screening of yeast mutants for partial defects in mitochondrial protein import led to the identification of the genes encoding these previously unidentified mitochondrial inner membrane proteins (Maarse et al. 1992; Dekker et al. 1993; Emtage and Jensen 1993; Maarse et al. 1994). Further evidence that these proteins formed part of the mitochondrial inner membrane import machinery was achieved through biochemical means when Tim17, Tim23 and Tim44 were cross-linked to translocating preproteins (Scherer et al. 1992; Blom et al. 1993; Ryan and Jensen 1993; Kübrich et al. 1994). In addition, specific antibodies directed against Tim23 and Tim44 blocked import of preproteins into mitoplasts (Emtage and Jensen 1993; Horst et al. 1993). Also found to play a role in protein import was a fourth gene that encoded the known molecular chaperone mitochondrial Hsp70 (mtHsp70). MtHsp70 is a soluble matrix protein which interacts with the TIM23 translocase by binding to Tim44 and Tim17 (Kronidou et al. 1994; Rassow et al. 1994; Börner et al. 1997). How-

Table 1. Components of the ¯TIM23 and TIM22 complexes

Translocase	Component	Inner membrane association	Proposed role
	Tim23	Integral	Preprotein binding, forms a pore for preprotein translocation, mitochondrial morphology
	Tim17	Integral	Forms a pore for preprotein translocation
TIM23	Tim44	Peripheral	Membrane anchor for mtHsp70
	mtHsp70 (Ssc-1)	Indirect via Tim44	Translocation motor
	Mge1	Soluble in matrix	Nucleotide exchange factor for mtHsp70
	Tim22	Integral	Forms a pore for preprotein translocation
	Tim54	Integral	Unknown, possibly preprotein binding or assembly factor
	Tim18	Integral	Unknown, possibly an assembly factor
TIM22	Tim12	Peripheral	Preprotein acceptor
	Tim9	Peripheral & soluble in IMS	Carrier preprotein transfer factor
	Tim10	Peripheral & soluble in IMS	Carrier preprotein transfer factor
	Tim8	Soluble in IMS	Preprotein transfer factor
	Tim13	Soluble in IMS	Preprotein transfer factor

IMS (intermembrane space).

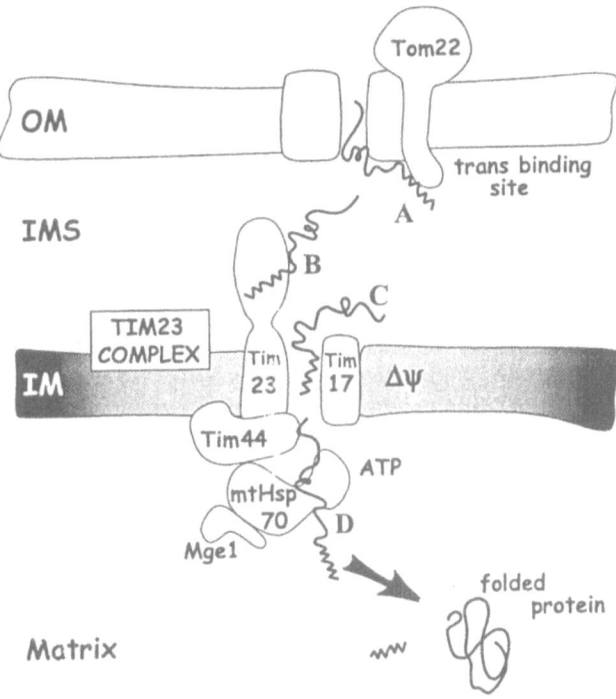

Fig. 4. Import of matrix targeted preproteins across the inner membrane of mitochondria. The *trans* binding site of Tom22 which extends into the intermembrane space enhances the translocation efficiency of presequence-containing preproteins through the outer membrane TOM complex (**A**). The preprotein is then transferred to the inner membrane translocase perhaps due to the presence of a high affinity binding site on the intermembrane space domain of Tim23 (**B**). This event may increase the open probability of the import pore so that the preprotein can enter the translocation channel. The membrane potential ($\Delta\psi$) provides an electrophoretic force in which the positively charged presequence is attracted to the net negative charge in the matrix thereby drawing it into this compartment (**C**). This force is not responsible for the translocation of the mature portion of the preprotein. This is the role of the translocation motor which consists of Tim44, mtHsp70 and Mge1p (**D**). The molecular chaperone Hsp70 is positioned at the import site through a specific interaction with Tim44. As the unfolded preprotein emerges through the import pore it binds to mtHsp70. This interaction provides unidirectional movement of the preprotein into the matrix. But the action of mtHsp70 may also be more active that this especially for preproteins that have tightly folded domains on the outside of mitochondria. In this case conformational changes in mtHsp70 provide a pulling force on the preprotein drawing it into the matrix. The action of mtHsp70 is driven by its ability to obtain energy from the hydrolysis of ATP. The cochaperone Mge1p regulates the ATPase activity of Hsp70. The preprotein may be processed to the mature form. It is then free to fold into its active conformation

ever, only a small portion of total matrix localised mtHsp70 associates with the inner-membrane translocase (Rassow et al. 1994; Horst et al. 1997). The translocation function of mtHsp70 is regulated by its cochaperone Mge1p. The membrane associated population of mtHsp70 acts as an unfoldase for unfolding of preproteins on the cytosolic side of the mitochondria and a translocase assisting the import of presequence containing preproteins into the matrix (Schneider et al. 1994; von Ahsen et al. 1995; Voos et al. 1996; Horst et al. 1997).

All three TIM23 translocation components, Tim23, Tim17, Tim44 and associated molecular chaperones mtHsp70 and Mge1p are essential for the viability of yeast under all growth conditions (Craig et al. 1989; Maarse et al. 1992; Emtage and Jensen 1993; Bolliger et al. 1994; Ikeda et al. 1994; Laloraya et al. 1994). Although the basic characteristics of the TIM23 translocase have been determined it is not clear how each component cooperates to facilitate protein import. However a simplified model for the import of matrix targeted preproteins across the inner membrane is illustrated in Fig. 4.

5.2
Features and Functions of TIM23 Translocation Components

5.2.1
Tim23 and Tim17

The basic functions of Tim23 and especially Tim17 are not known for certain even though they were identified some years ago. They share common structural features and associate in mitochondria to form the core components of the TIM23 translocase. Tim17 and Tim23 behave as integral membrane proteins with each protein containing four predicted membrane spanning domains (Dekker et al. 1993; Emtage and Jensen 1993; Kübrich et al. 1994; Maarse et al. 1994). These two proteins also contain similarities at the amino acid sequence level as Tim17 is 46% homologous to the C-terminal membrane domain of Tim23 (Maarse et al. 1994; Ryan et al. 1994a). However, Tim23 is more complex than Tim17 as it carries an approximately 100 amino acid hydrophilic N-terminal extension which performs an essential function in yeast (Ryan et al. 1998). Despite the similarities between Tim23 and Tim17 they are not able to functionally substitute for each other (Maarse et al. 1994; Ryan et al. 1994a). Each protein must therefore perform a specialised function. Both proteins closely associate with translocating preproteins and for this reason it has been proposed that they form the inner membrane translocation pore for presequence-containing preproteins. The translocation pore may be formed by both Tim23 and Tim17 given that their properties are very similar. Alternatively the inner membrane may be composed of two distinct but essential translocation pores in the TIM23 complex one formed by Tim23 and the other formed by Tim17. This latter notion is inspired by the recently determined properties of OEP16 an amino acid transporter in the outer envelope of chloroplast. OEP16 is an integral membrane protein with some homology to Tim23 and Tim17 (Rassow et al. 1999). Reconstitution of either pure *bona fide* or recombinant OEP16 into planar lipid bilayers reveals an ability to form an amino acid-selective channel without additional protein components (Pohlmeyer et al. 1997; Steinkamp et al. 2000). Lohret et al. (1997) studied the relationship between Tim23 and the multiconductance channel (MCC) of the inner membrane which is specifically blocked by presequence peptides. They showed that antibodies specifically directed against Tim23 inhibited MCC activity and in addition the properties of the conductance were altered in inner membrane vesicles which harboured a mutation in Tim23. The relationship between the inner membrane multiconductance channel and the TIM23 complex is yet to be determined. Perhaps the reconstitution of purified Tim translocation components may be the only way to clearly define the proteins responsible for forming the TIM23 protein translocation channel.

One of the most intriguing questions relating mitochondria protein import is how hundreds of preproteins are imported into the matrix of mitochondria without the free exchange of ions across this membrane. There must be selective transport as an electrochemical gradient is maintained across the inner membrane of mitochondria. Bauer et al. (1996) provided circumstantial evidence that the hydrophilic intermembrane space domain of Tim23 influences the gating of the translocation channel. The ability of the hydrophilic domain to form dimers was demonstrated. Dimer formation of Tim23 *in vivo* is dependent on the presence of a membrane potential across the inner membrane but could be disrupted when a preprotein was arrested across both mitochondrial membranes. Thus, in the absence of preproteins the hydrophilic domain of Tim23 forms a dimer which may, via an unknown mechanism, reduce the conductance of the translocation channel. What is known for certain is that the Tim23 hydrophilic domain is essential for function. Expression of Tim23 lacking the complete intermembrane space domain cannot complement a disruption of the chromosomal TIM23 gene (Bauer et al. 1996; Ryan et al. 1998).

Perhaps the essential function of the Tim23 intermembrane space domain relates to its proposed role as a preprotein receptor in this compartment. The negative charges of the Tim23 hydrophilic domain could provide a high affinity binding site for presequence-containing preproteins. This domain thus constitutes the final member of the acid-chain pathway prior to the membrane potential-directed preprotein insertion into the inner membrane. However, to date, there are only a few examples demonstrating the direct binding of preproteins to this domain (Bauer et al. 1996; Komiya et al. 1998).

Following detergent solubilisation of mitochondria both Tim17 and Tim23 migrate as a stable 90 kDa complex following separation by blue native PAGE electrophoresis. Binding of Tim23 to Tim17 occurs via its hydrophobic membrane domain (Ryan et al. 1998). This complex shifts to a stable 130 kDa species following accumulation of a preprotein across the inner membrane. When the stability of the 90K complex is disrupted by a point mutation in the first putative transmembrane domain of Tim23, import of cleavable preproteins into the matrix is inhibited (Dekker et al. 1997). The 90K complex is therefore the major inner membrane import site for presequence-containing preproteins. Tim subcomplexes of ~140K and 240K which contain Tim23 and Tim17 have also been identified (Dekker et al. 1997). The composition and function of these additional complexes is not known. They do however seem to constitute functional entities as they are incorporated into a 600K supercomplex when a preprotein spans both the outer and inner membrane translocases linking them together (Dekker et al. 1997).

Preproteins that cross the outer and inner membranes of the mitochondria do so at specific sites called translocation contact sites (Schleyer and Neupert 1985). At these points the outer and inner membranes are in close proximity but do not form a single stable channel for the passage of preproteins (Hwang et al. 1989; Rassow and Pfanner 1991; Segui-Real et al. 1993). In fact, the TIM23 preprotein translocase of the inner membrane is capable of importing preproteins in the absence of the outer membrane (Ohba and Schatz 1987; Hwang et al. 1989). However, recent re-examination of the membrane topology of Tim23 reveals that it has, in addition to its inner membrane integrated segment, an extreme N-terminal segment that spans the outer membrane (Donzeau et al. 2000). Thus, Tim23 provides a physical link between the inner and outer membrane at translocation sites. The first 20 amino acids protrude into the cytosol while the following 30 amino acids form a membrane spanning domain proposed to be in a ß-sheet formation. The first 50 amino acids of Tim23 are only essential for viability at elevated temperatures. A direct association of the N-terminal domain of Tim23 with the outer membrane TOM complex was not detected.

5.2.2
Tim44 the Anchor Protein

Fractionation experiments indicate that Tim44 is a peripheral membrane protein residing on the matrix side of the mitochondrial inner membrane (Scherer et al. 1992; Blom et al. 1993; Rassow et al. 1994). Its membrane association may be due to a combination of both protein and lipid interactions as the purified recombinant protein can interact directly with liposomes consisting of phosphatidylcholine and cardiolipin (Weiss et al. 1999). There are however conflicting views with respect to its topology at the inner membrane since a portion of Tim44 is believed to be exposed on the outer surface of the inner membrane (Maarse et al. 1992; Scherer et al. 1992; Kanamori et al. 1997). However, hydrophathy profiles of both yeast and mammalian Tim44 predict a largely hydrophilic protein lacking membrane spanning domains. In the case of mammalian Tim44 it displays a double distribution following fraction. It is found as a soluble form in the matrix as well as being peripherally associated with the matrix side of the inner membrane (Ishihara and Mihara 1998). Although Tim44 is closely opposed to the site of translocation it does not seem to form part of the translocation pore as functional inactivation of a temperature-sensitive mutant of Tim44 does not alter the insertion or passage of preproteins though the TIM23 translocase (Bömer et al. 1998). Co-immunoprecipitation experiments performed under stringent conditions lead to the identification of a Tim23-Tim44 complex (Bömer et al. 1997). The complex transiently interacts in a functional manner with mtHsp70 via an association through Tim44. Crosslinking experiments showed that Tim44 is also in direct contact with the incoming precursor polypeptide (Blom et al. 1993). The cooperative function of Tim44 and mtHsp70 as a translocation motor is discussed further in section 8.2.4 with respect to the energetics of mitochondrial import.

6
Sorting of Preproteins to the Intermembrane Space or Inner Membrane

There are several pathways that direct preproteins to either the intermembrane space or inner membrane. The simplest route to the intermembrane space is one in which only the TOM import machinery is required. For example cytochrome heme lyases and some components of the TIM22 complex follow this pathway. However, other preproteins with amino-terminal presequences engage the TIM23 complex but are then subsequently translocated to the intermembrane space or inner membrane. Examples are cytochrome b_2 which is a soluble intermembrane space protein and cytochrome c_1 which is anchored to the inner membrane close to its C-terminus with a hydrophilic heme binding domain protruding into the intermembrane space (Reid et al. 1982; Guiard 1985). Both preproteins contain cleavable bipartite presequences. The first part of the bipartite sequence directs the preprotein to the TIM23 translocase while the second hydrophobic sequence acts as a sorting signal. The protein components involved in this sorting process have not been identified. The model of "conservative sorting" states that the protein is translocated across both the outer and inner membrane into the matrix then reexported to the inner membrane or intermembrane space (Hartl et al. 1987; Koll et al. 1992; Gruhler et al. 1995) reminiscent of the bacterial protein export pathway. In contrast the "stop transfer" model indicates that the presequence is translocated across the inner membrane through the TIM23 complex. However, the sorting signal causes the arrest of the preprotein at the inner membrane so that the mature portion of the preprotein is only translocated across the outer membrane directly into the intermembrane space (Blobel et al. 1980; Kaput et al. 1982; Hurt and Van Loon 1986). Gärtner et al. (1995) proposed a variation on the stop transfer model for the sorting of cytochrome b_2 that takes into consideration aspects of both models. The import of these proteins may be even more complex as cytochrome c_1 contains in addition to its bipartite presequence an internal C-terminal targeting signal which directs translocation of the preprotein from the intermembrane space into the inner membrane (Arnold et al. 1998). The current view on the sorting of cytochromes c_1 and b_2 is that their mature parts do not enter the matrix but follow a stop transfer like model. Proteins such as Rieske Fe/S protein and nuclear-encoded Fo-ATPase subunit 9 are probably sorted by a conservative mechanism (Hartl et al. 1986; van Loon and Schatz 1987; Mahlke et al. 1990; Rojo et al. 1995).

7
The TIM22 Complex is Required for Import of Inner Membrane Proteins

Nuclear-encoded mitochondrial inner membrane proteins that are not directed by a presequence to the TIM23 translocase depend on the TIM22 translocase for their import. Proteins utilising this pathway are either members of the carrier protein family or other polytopic proteins such as integral inner membrane translocation components themselves. Internal targeting signals of an unknown nature direct them to this translocase. Preproteins with internal targeting signals bind to a specialised receptor, Tom70, but utilise the GIP for translocation across the outer membrane. The pathway of metabolic carrier proteins separates in the intermembrane space. For example members of the carrier family do not require the Tom22 *trans* binding site for their efficient import through the GIP. Instead soluble intermembrane space translocation components belonging to the TIM22 translocase assist the transfer of polytopic inner membrane proteins from the GIP to the TIM22 membrane complex where insertion into the lipid bilayer takes place.

7.1
Composition of the TIM22 Translocase

The TIM22 translocase is made up of soluble intermembrane space oligomers and membrane-associated components (Table 1). The soluble intermembrane space components comprise the related proteins Tim9, Tim10, Tim8 and Tim13. Two 70 kDa hetero-oligomers are formed by these proteins a Tim9–Tim10 complex and a Tim8–Tim13 complex respectively. The TIM22 membrane complex is made up of Tim22, Tim54, Tim18, Tim12 and a small quantity of Tim9 and Tim10. Following separation of detergent solubilised mitochondria by blue native PAGE, the TIM22 membrane associated complex migrates as a distinct species of approximately 300 kDa (Koehler et al. 1998a; Koehler et al. 1999; Koehler et al. 2000; Kerscher et al. 2000).

Tim22, from which the complex gains its name, was the first component of the TIM22 complex identified (Sirrenberg et al. 1996). It was initially identified by data base searching since it has significant sequence homology with the membrane-associated portions of Tim17 and Tim23. The predicted structure of Tim22 resembles that of Tim23 and Tim17 as it possesses multiple hydrophobic stretches in its primary sequence which may constitute transmembrane domains. Indeed Tim22 is an integral membrane protein as determined by its resistance to extraction from the inner membrane with high pH treatment. Fractionation experiments demonstrate its localisation in the inner membrane however its topology has not been reported. It is said to be accessible to protease when the outer membrane of mitochondria is selectively disrupted. As its migration on gel filtration (following solubilisation of mitochondria in digitonin) was distinct from that of the TIM23 translocase, it was deemed a component of a new translocase. Like its homologous counterparts Tim22 is essential for the viability of *S. cerevisiae*. Given its similarity to Tim23 and Tim17 and other amino acid transporters it most likely performs a similar function having evolved from a common ancestor (Rassow et al. 1999).

The largest component of the TIM22 translocase is Tim54. It is an integral protein of the inner membrane and exposes a large portion of its C-terminal domain to the intermembrane space. It was discovered in a yeast two-hybrid screen as a potential Mmm1p interacting protein (Kerscher et al. 1997). Mmm1p is an outer membrane protein involved in the maintenance of mitochondrial morphology. The significance of this interaction has not been established. However, via biochemical and genetic means the association of Tim54 with Tim22 was established. Immunoprecipitation of Tim54 following solubilisation of mitochondria in digitonin also lead to the complete coimmunoprecipitation of Tim22. TIM23 translocase components Tim23, Tim17, Tim44 or mtHsp70 were not precipitated. Thus, Tim22 and Tim54 belong to the same complex which is distinct from the TIM23 complex. A specific genetic interaction between Tim54 and Tim22 was detected as overexpression of Tim22 but not overexpression of Tim23 or Tim17 could suppress the temperature sensitive *tim54-1* mutant. Tim54 is also essential for the viability of yeast.

Tim18 is the most recently characterised component of the TIM22 translocase. This non-essential protein was identified by two independent methods. First, genetically, as a high copy suppressor of the temperature sensitive *tim54-1* mutant and second, biochemically, as a interacting partner of members of the TIM22 complex (Kerscher et al. 2000; Koehler et al. 2000). Basic characterisation revealed that it is an integral membrane protein of the inner membrane with three putative membrane spanning domains. It is synthesised with a presequence which is cleaved off when the preprotein is imported into mitochondria. Several genetic properties confirm that it is a member of the TIM22 translocase. Deletion of *tim18* is synthetically lethal with temperature-sensitive mutations in Tim54, Tim9 and Tim10. In addition overexpression of Tim22 suppresses the growth defect associated with deleting *tim18* (Kerscher et al. 2000). Tim18 was also detected in the 300 kDa TIM22 membrane complex following analysis by blue native page electrophoresis. The complex shifts to an approximately 200–250 kDa species following deletion of Tim18 confirming that this protein is a genuine member of this translocase.

The remaining known TIM22 translocation components are all members of the same family. They are the small homologous cysteine rich components referred to as Tim10, Tim9, Tim12, Tim8 and Tim13. Although Tim10 and Tim12 were first discovered as suppressors of mitochondrial RNA splicing defects they were soon linked to a role in the import of carrier proteins (Koehler et al. 1998b; Sirrenberg et al. 1998). Tim9 was discovered by two independent approaches as a spontaneous extragenic suppressor that restored growth to a temperature-sensitive mutation in Tim10 and also via a specific interaction with both Tim10 and Tim12 (Koehler et al. 1998b; Adam et al. 1999). All three subunits are essential for the viability of *S. cerevisiae* (Jarosch et al. 1996; Koehler et al. 1998a; Adam et al. 1999). Tim9 and Tim10 form a heter-oligomer in the intermembrane space of mitochondria most likely consisting of three subunits of each (Koehler et al. 1998a; Adam et al. 1999). Small quantities of Tim9 and Tim10 are also peripherally associated with the TIM22 membrane complex. Tim12, unlike Tim9 and Tim10, does not exist in a soluble form in the intermembrane space under normal conditions. It is tightly opposed to the inner membrane forming part of the 300 kDa TIM22 complex. It can be detected as a component of the 300 kDa translocase via coimmunoprecipitation experiments and blue native PAGE electrophoresis (Koehler et al. 1998b; Sirrenberg et al. 1998). A genetic link between Tim12 and Tim22 was detected as Tim22 was identified as a suppressor of a temperature-sensitive mutation in Tim12 (Koehler et al. 1998b). Tim8 and Tim13 are non-essential components of the TIM22 translocase (Koehler et al. 1999). Deletion of Tim8 causes loss of Tim13 and vice versa (Koehler et al. 1999). In addition deletion of Tim8 is synthetically lethal with a temperature-sensitive mutation in Tim10 but not Tim12 (Koehler et al. 1999). They also associate together with a small quantity of Tim9 to form a 70 kDa hetero-oligomer which is only detected in the intermembrane space (Koehler et al. 1999). Each small Tim subunit contains four cysteines arranged in two CX_3C motifs. At least for Tim10 and Tim12 this motif constitutes a zinc binding site as in the absence of metal ions the proteins become sensitive to digestion by proteases and fail to interact with translocating preproteins (Sirrenberg et al. 1998).

7.2
Import of Polytopic Inner Membrane Proteins

All the biochemical and genetic data collected so far indicate that defects in components of the TIM22 complex directly inhibit the import of polytopic inner membrane preproteins with internal targeting signals (Sirrenberg et al. 1996; Kerscher et al. 1997; Koehler et al. 1998a; Koehler et al. 1998b; Adam et al. 1999; Koehler et al. 1999; Leuenberger et al. 1999). Matrix targeted preproteins appear to be largely unaffected by defects in the TIM22 complex except perhaps in the cases in which the defect disrupts the biogenesis of Tim23 or Tim17. In what way does each TIM22 translocation component contribute to the import or polytopic proteins? This can best be explained by describing the sequential import steps of a carrier protein. The import of ATP/ADP carrier (AAC) can be monitored in stages by manipulating the availability of cytosolic ATP or the membrane potential across the inner membrane (Pfanner and Neupert 1987; Ryan et al. 1999). By monitoring the import of AAC in mitochondria functionally depleted of various TIM22 components it has been possible to pinpoint the stage at which they are acting in the import pathway (Fig. 5).

When Tim9 or Tim 10 are functionally depleted in mitochondria, AAC fails to reach stage 3 import (Koehler et al. 1998a; Koehler et al. 1998b; Sirrenberg et al. 1998; Adam et al. 1999). The bulk of the preprotein does not protrude into the intermembrane space but is exposed on the outer surface of mitochondria. Thus the Tim9–Tim10 complex in some way assists the transfer of preproteins across the outer membrane. As both subunits can be cross-linked to translocating preproteins they most likely play a direct role in the import of these proteins (Koehler et al. 1998b; Sirrenberg et al. 1998; Ryan et al. 1999). Tim12 can also be cross-linked to a translocating preprotein but its functional depletion influences a different stage of import compared to its homologous counterparts (Koehler et al. 1998b; Sirrenberg et al. 1998). Although AAC reaches stage 3 inactivation of Tim12 prevents the membrane potential-dependent insertion of the protein into the inner membrane. This is the same stage at which Tim22 influences the import of AAC (Sirrenberg et al. 1996). Its functional inactivation prevents stage 4 of AAC import, that is the insertion of the preprotein into the lipid bilayer. Cross-linking experiments demonstrated that Tim22 interacts with a translocating preprotein but not when it is arrested at stage three (Sirrenberg et al. 1996). Given that Tim22 is homologous to Tim23 and Tim17 and it interacts directly with preproteins, post stage 3, then it is most likely involved in the formation of a pore for the insertion of preproteins into the lipid bilayer. A temperature sensitive mutation in Tim54 prevents the import of AAC and Tim23 into the inner membrane (Kerscher et al. 1997). A direct role of Tim54 in the import of carrier preproteins has not been shown by cross-linking or immunoprecipitation. It has been proposed that Tim54 may stabilise the TIM22 complex or act as an inner membrane receptor for prepro-

Fig 5. Sequential import steps of a carrier preprotein. (a). Stage I and II of carrier protein import. Following synthesis in the cytosol the preprotein binds to cytosolic factors of an unknown nature (Stage I). It is then transferred to the receptor Tom70 (Stage 2). (b). Stage III of carrier import. In an ATP dependent manner AAC inserts into the GIP to a site in which most of the protein is on the *trans* side of the outer membrane largely inaccessible to protease digestion but still associated with the outer membrane translocase (Stage III). It can be arrested at this stage in the absence of a membrane potential across the inner membrane. (c). Stage IV and V of carrier import. In the presence of a membrane potential AAC inserts into the inner membrane (Stage 4) then dimerises to its active form (Stage 5)

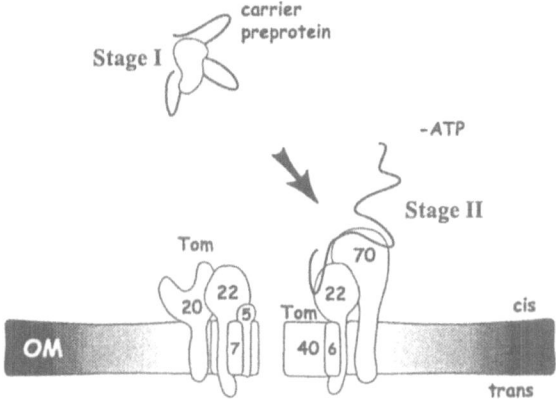

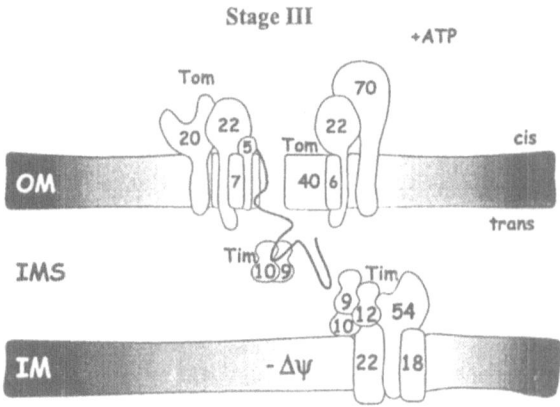

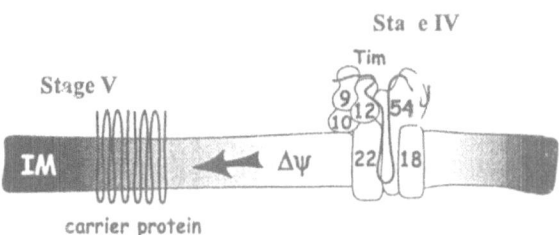

teins. Likewise the function of Tim18 is not known. It may not interact with translocating preproteins as it could not be cross-linked to imported Tim23 (Koehler et al. 2000). There are however some inconsistencies reported with respect to the phenotype of *tim18Δ* yeast. Koehler et al. (2000) reported that deletion of Tim18 decreases the growth rate of yeast cells and slows the import rate of several preproteins such as Tim23, Tim54 and matrix-targeted proteins such as Su9-DHFR. The import of Tim17 and AAC were however unaltered. On the other hand, Kerscher et al. (2000) reported that the same deletion produces a cold sensitive phenotype on glucose and that the *in vitro* import of several proteins into mitochondria including the phosphate carrier and CoxIV were not altered. The role of Tim18 for the import of inner membrane proteins requires some clarification.

7.3
The Biogenesis of Tim Translocation Components

Components of both TIM translocation machineries do not follow a single common pathway for their import. As far as we know these differences can be attributed to the nature of their targeting information. For example Tim44 is membrane-associated matrix protein which is synthesised with a cleavable presequence. It is therefore reasonable that it is imported via the TIM23 complex. Tim54, although a member of the TIM22 carrier translocase which requires Tom70 for its recognition in the outer membrane, contains a noncleavable presequence-like region at its N-terminus which is required for its translocation via the TIM23 translocase (Kurz et al. 1999). Tim18 is also synthesised with a cleavable presequence although its import pathway has not been determined. On the other hand Tims 22, 23, 17, 12, 10 , 9, 8 and 13 all contain internal targeting information. At least for Tim17 and Tim23 this information is distributed over the preprotein (Davis et al. 1998; Káldi et al. 1998). A hint that some of the Tim translocation components are imported via a TIM23-independent pathway came from the observation that a temperature-sensitive defect in Tim23 did not inhibit the import of polytopic preproteins such as Tim23 or Tim17 (Dekker et al. 1997). The same effect was observed when the TIM23 import channel was specifically blocked by arresting a preprotein across this translocase (Dekker et al. 1997). Kerscher et al. (1997) were the first to show that Tim23 was imported via the newly discovered TIM22 translocase as import of Tim23 into the inner membrane of mitochondria with a temperature sensitive mutation in Tim54 was significantly reduced. The same effect was observed when mitochondria were depleted of Tim22 (Káldi et al. 1998). Evidence that Tim22 is also imported via this pathway was obtained when functional depletion of both

Tim10 and Tim12 drastically reduced its import (Sirrenberg et al. 1996; Koehler et al. 1998b). In addition a temperature-sensitive mutation in Tim22 also inhibits the *in vitro* import of itself as well as Tim23 (Leuenberger et al. 1999). Interestingly, Tim22 does not require the receptor Tom70 for its recognition but rather Tom20 which is preferentially used by presequence containing preproteins rather than preproteins with internal targeting signals (Kurz et al. 1999). The import pathway for Tim17 is not clear. In one case it was reported that functional depletion of Tim23 or Tim22 from mitochondria reduced the import of Tim17 (Káldi et al. 1998). In contrast Leuenberger et al. (1999) find that Tim17 import is insensitive to partial inactivation of Tim22 or Tim23 suggesting that another as yet unidentified translocase may be responsible for the import of this protein.

Until recently the function of the non-essential components Tim8 and Tim13 was not known. By accessing the import of a range of integral membrane proteins into mitochondria lacking both Tim8 and Tim13 it was discovered that the import of Tim23 is drastically reduced (Leuenberger et al. 1999). The 70 kDa Tim8–13 complex could be specifically cross-linked to Tim23 indicating that the complex plays a direct role in its import. In contrast translocation of Tim17, Tim54 and Tim22 require the Tim9–Tim10 complex for their import even though each of these proteins may use different translocation complexes for their insertion into the inner membrane. This means that the Tim9–Tim10 complex is able to deliver preproteins to either of the separate import complexes TIM23 or TIM22. Perhaps in undisrupted mitochondria these two translocases are more closely associated than is realised. Finally the import of Tims 9, 10, 12, 8 and 13 is unique as only Tom5 is required for their recognition and hence subsequent translocation through the outer membrane at the GIP and into the intermembrane space (Kurz et al. 1999).

7.4
Relationship Between Protein Import and a Human Disease State

Mohr-Tranebjaerg syndrome is a rare neurodegenerative disorder which presents itself in early childhood. Patients with the syndrome all have deletions in the same gene referred to as DNF-1 (Jin et al. 1996). The product of the normal gene is an 11 kDa protein designated DDP1 which until recently was of unknown function. Using database searching the protein was found to be a homologue of the recently characterised yeast translocation components Tims 9, 10, 12, 13 and 8 (Koehler et al. 1999). It is most similar to the non-essential component Tim8. As these proteins are involved in the import of carrier proteins it has been suggested that the disease state is a consequence of a defect in mitochondrial oxidative phosphorylation (Wallace and Murdock 1999). Such details are yet to be determined. DDP1 is however a mitochondrial protein which is found in the intermembrane space like its homologous yeast counterparts (Koehler et al. 1999).

8
Driving Forces for Protein Translocation

8.1
Mitochondrial Membrane Potential

The first studies on the mechanism of mitochondrial protein import revealed that the membrane potential across the inner membrane is imperative for the translocation process (Schleyer et al. 1982; Pfanner and Neupert 1985). If the membrane potential is abolished either by inhibition of the respiratory chain or addition of ionophores, preproteins can bind specifically to the mitochondrial membranes but further translocation across the membranes is blocked completely. In general, it was observed that all preproteins that contain an amino-terminal presequence exhibit membrane potential-dependent import. Additionally, many proteins that insert into the inner membrane as multiple membrane-spanning proteins, such as members of the metabolite carrier family, require a membrane potential for complete import (Schleyer et al. 1982). As discussed previously the exact nature of these internal targeting sequences is not known. Thus a membrane potential is required for the import of preproteins that utilises either TIM23 or TIM22 translocation pathways.

The overall positive charge of mitochondrial targeting motifs together with the orientation of the membrane potential (negative inside and positive outside), lead to the hypothesis that the translocation of targeting sequences is driven by an electrophoretic effect generated by the electrical gradient across the inner membrane. This model was confirmed by studies demonstrating that indeed only the electrical component of the total proton-motive force across the inner membrane is required for import (Pfanner and Neupert 1985). Furthermore, the magnitude of the membrane potential differentially influences the import of mitochondrial preproteins. This effect is due to differences in presequence composition and especially relates to the number of positively charged residues in this targeting region (Martin et al. 1991). Another role for the membrane potential was proposed after analysing the behaviour of Tim23. The membrane potential promotes dimerisation of this protein. In the presence of a matrix targeting signal, the Tim23 dimer dissociates and presumably opens the translocation channel for the incoming polypeptide chain (Bauer et al. 1996).

8.2
ATP-Dependent Translocation Motor

Only the initial step of inner membrane protein translocation, the movement of the presequence, seems to be driven by the membrane potential. Transport of the remaining preprotein chain can occur in the absence of the electrical gradient (Pfanner et al. 1987c). However, hydrolysis of nucleoside triphosphates is generally required for complete translocation of preproteins (Pfanner and Neupert 1986; Eilers et al. 1987). A dependence on ATP/GTP was demonstrated for almost all mitochondrial precursors irrespective of their submitochondrial localisation. However, one has to distinguish between a nucleotide hydrolysis requirement occurring outside as opposed to inside mitochondria. For all imported proteins ATP outside mitochondria is required for the maintenance of import competence and preprotein recognition by the outer membrane receptors (Pfanner et al. 1987c). It is assumed that cytosolic factors like MSF or Hsp70 utilise ATP to bind and stabilise mitochondrial precursor proteins in an unfolded and import competent conformation (Wachter et al. 1994). However, as far as we know, energy derived from nucleotide hydrolysis outside mitochondria is involved in the mechanism of maintaining import competence but does not contribute significantly to the movement of precursor polypeptide chains across the mitochondrial membranes. For the import of outer membrane proteins and also most intermembrane space proteins this seems to be the only nucleotide requirement. Insertion of membrane proteins into the lipid bilayer itself is probably a spontaneous process driven by the difference in entropy between free and membrane-inserted forms. Translocation across the outer membrane is driven by sequential binding of the presequence to components of the outer membrane import machinery ("Acid-chain hypothesis") (Hönlinger et al. 1995; Schatz 1997). This mechanism would also be employed for the translocation of the amino-terminal moieties of matrix-targeted precursors.

8.2.1
Requirement for Matrix ATP

By manipulating the nucleotide levels on both sides of mitochondrial membranes it was shown that completion of protein translocation is dependent on nucleoside triphosphates in the matrix (Hwang and Schatz 1989; Pfanner et al. 1990; Cyr et al. 1993). This energy requirement is absolute for precursor proteins destined for the matrix (Beasley et al. 1992). The only exception are integral membrane proteins that insert into the inner membrane without the need for matrix ATP (Pfanner et al. 1987a; Wachter et al. 1992). In the absence of matrix nucleotides import can proceed up to the membrane potential-dependent step. At this stage the amino-terminal end of the precursor polypeptide reaches into the matrix and can be processed by the matrix processing peptidase. The carboxy-terminal part of the precursor remains on the outside of mitochondria, essentially generating a membrane-spanning translocation intermediate.

8.2.2
Involvement of mtHsp70

The major question concerning energy utilisation for import was which enzyme uses ATP hydrolysis to drive protein translocation? Indeed, the only clearly identifiable ATPase involved in mitochondrial protein import turned out to be a molecular chaperone, a member of the 70 kDa heat shock protein family, termed mtHsp70 or Ssc1p in yeast (Rassow et al. 1997). Hsp70s consist of three domains: an N-terminal ATPase domain (residues ~1–385), a peptide binding domain (residues ~385-600) and an extreme C-terminal domain (residues ~600–650) (Ha et al. 1999). Hsp70s bind reversibly to mainly hydrophobic patches in polypeptides which become exposed during cellular events such as protein synthesis, protein transport or protein denaturation. The action of Hsp70 essentially helps to prevent misfolding or aggregation of an unfolded polypeptide chain. Binding and release of polypeptide substrates by Hsp70s is regulated by a reaction cycle coupled to binding and hydrolysis of ATP. Most mechanistic details of the Hsp70 ATPase reaction cycle have been established for the major Hsp70 of *Escherichia coli*, DnaK (summarised in Buchberger et al. 1999). DnaK binds to hydrophobic amino acid stretches in polypeptides that are unstructured (Gragerov et al. 1994; Rüdiger et al. 1997; Landry et al. 1992; Zhu et al. 1996). Such regions are predicted to occur frequently in protein sequences (Rüdiger et al. 1997). In the case of the mitochondrial member of the Hsp70 family its molecular function is extended to the facilitation of preprotein movement across membranes. Assuming that mtHsp70 recognises a similar binding motif to DnaK then a preprotein is likely to expose multiple mtHsp70 binding sites during its translocation.

Several facts point to mtHsp70 as a central component of the mitochondrial import machinery. First, it belongs to a small group of mitochondrial proteins, that are essential for viability of yeast (Craig et al. 1989). Other members of this group, for example Tom40 or Hsp60, are involved in key steps of mitochondrial biogenesis, having functions that cannot be substituted by other proteins. Second, preproteins in transit physically interact with mtHsp70. This could be demonstrated by both crosslinking of translocation intermediates and by coimmunoprecipitation (Scherer et al. 1990; Gambill et al. 1993). The crosslinking data demonstrated that mtHsp70 is involved at a very early step of inner membrane polypeptide translocation, directly after the membrane potential-dependent step. Third, extensive analysis of temperature-sensitive alleles of *SSC1* in yeast showed, that in mitochondria that carry a functionally defective mtHsp70, preprotein import is arrested (Kang et al. 1990; Gambill et al. 1993). The precursor proteins are again accumulated as translocation intermediates spanning both membranes as in the case when matrix ATP is depleted. Additional experiments showed a complete correlation between the import defects of a mutant mtHsp70 unable to interact with ATP and the import defects generated by ATP depletion. This data collectively lead to the conclusion that mtHsp70 is the protein responsible for the matrix ATP-dependent import step (Kang et al. 1990; Gambill et al. 1993).

8.2.3
Activation of mtHsp70 by the Cochaperone Mge1p

Further evidence for a role of mtHsp70 as the ATP consuming protein import motor was provided by the identification of a typical cochaperone of Hsp70s in the mitochondrial matrix, termed Mge1p in yeast. A homologue of this protein in *E. coli*, GrpE, regulates the activity of the bacterial Hsp70, DnaK. Similar to its prokaryotic homologue, Mge1p forms a very stable complex with mtHsp70 in the absence of ATP (Bolliger et al. 1994; Nakai et al. 1994; Voos et al. 1994). In the presence of ATP and even nonhydrolysable analogs, the mtHsp70-Mge1p complex is quickly dissociated. In the case of GrpE, interaction with DnaK increases the ATPase activity substantially. Mge1p shows a similar activity in mitochondria. It is thought to act as a nucleotide exchange factor, facilitating the release of ADP and thereby increasing the rate of hydrolysis (Dekker and Pfanner 1997; Miao et al. 1997).

Interestingly, *MGE1* is also an essential gene in yeast. Mitochondria containing a mutant Mge1p show defects in protein import *in vivo* and *in vitro* (Laloraya et al. 1994; Laloraya et al. 1995), demonstrating the importance of the mtHsp70 ATPase activity for protein translocation. Since an interaction of Mge1p with preproteins could only be detected when they were accumulated as translocation intermediates, it is supposed, that the essential action of Mge1p is focused on the initial entry of preprotein segments into the matrix, the very same step where the crucial function of mtHsp70 for polypeptide translocation is required (Voos et al. 1994; Schneider et al. 1996).

8.2.4
Tim44-mtHsp70 Interaction

The identification of Tim44 (section 5.2.2) eventually provided the basis for a detailed understanding of the molecular function of mtHsp70 in driving the translocation process. Genetic analysis first hinted at a functional connection between Tim44 and mtHsp70. Temperature-sensitive mutants of mtHsp70 are synthetically lethal with mutations in *TIM44*. In addition the phenotype of Tim44 mutants can be suppressed by overexpression of mtHsp70. Biochemical experiments demonstrated a direct physical interaction between Tim44 and mtHsp70 (Kronidou et al. 1994; Rassow et al. 1994). Similar to the interaction of mtHsp70 and Mge1p, binding of mtHsp70 to Tim44 seems to occur only at a certain stage of the Hsp70-ATPase reaction cycle. The complex contains mtHsp70 that has predominantly bound ADP and is dissociated in the presence of ATP (von Ahsen et al. 1995; Voos et al. 1996).

Tim44 and mtHsp70 form a 1:1 complex at the import site although greater than 90% of mtHsp70 remains free in the matrix of yeast (Rassow et al. 1994; Krimmer et al. 2000). The nature of the interaction between Tim44 and mtHsp70 is not known. Tim44 does however possess a region (amino acids 185–202) which is partially related to a small portion of the highly conserved J-domain of Hsp40s and related proteins (Rassow et al. 1994). J-domains are required for a functional interaction with Hsp70s. The J-related segment of Tim44 is essential for function and its absence significantly reduces complex formation with mtHsp70 (Merlin et al. 1999). Tim44 directly interacts with the ATPase domain of mtHsp70 (Krimmer et al. 2000). Its affinity for the ATPase domain alone is not as great as that for the complete Hsp70 molecule. The addition of the peptide binding domain strengthens the interaction between Tim44 and the ATPase domain of mtHsp70 but not to its normal level. Thus all three functional domains of mtHsp70 influence its binding to Tim44. The interaction of DnaJ with DnaK also occurs via the ATPase domain but is positively influenced by the presence of the peptide binding domain (Gässler et al. 1998; Suh et al. 1998). As such, Tim44 may act in an analogous manner to DnaJ assisting the localisation of preproteins to the peptide binding site of Hsp70. It can be envisaged that Mdj1 and Tim44, both binding partners of mtHsp70, compete for the same binding site. This may explain why Mdj1 is not detected as part of the translocation motor. It is not known if Tim44 is capable of stimulating the ATPase activity of mtHsp70. Apart from its interaction with Tim44, mtHsp70 can also associate with the inner membrane translocase complex directly. This interaction is mediated by Tim17 and independent of the presence of Tim44 (Bömer et al. 1997). The functional implications of this observation are not clear, but it could indicate a regulatory role of mtHsp70.

8.2.5
Translocation Mechanisms – Proposed Models

How could the interaction of mtHsp70 with the polypeptide in transit result in its movement in the translocation pore? A basic model, dubbed the "Brownian ratchet model" states that the actual movement of the polypeptide chain during translocation is generated by random Brownian motion. Binding of the precursor chain to mtHsp70 inside would restrict the backward movement. Further, movement would expose additional binding sites for mtHsp70 inside, eventually resulting in complete translocation into the matrix (Simon et al. 1992; Ungermann et al. 1994; Gaume et al. 1998). This model was refined taking into account the function of Tim44 and Mge1p. Tim44 would direct the otherwise soluble mtHsp70 to the membraneous import site, making binding of the precursor much more efficient. ATP hydrolysis and indirectly Mge1p function would be required for recycling mtHsp70 to its most active state with respect to translocation.

However, this model does not take into account some additional properties of the import reaction. First, apart from the mere movement of the polypeptide chain in the translocation channel, the precursor protein has to be unfolded during import. The polypeptide chain in transit is most probably in an extended conformation. A detailed analysis of protein import demonstrated that unfolding is mainly accomplished by the action of mtHsp70 (Gambill et al. 1993; Voos et al. 1993; Matouschek et al. 1997). The polypeptide chain is indeed unravelled from the amino-terminal end during import (Huang et al. 1999). Second, any stronger interaction of the polypeptide in transit with components of the channel complexes would restrict the random movement due to Brownian motion and stall the translocation process (Chauwin et al. 1998). Indeed, there is experimental evidence, that membrane spanning translocation intermediates are quite stable even in the absence of bound mtHsp70 and Tim44 (Dekker et al. 1997).

Both effects together would require a more active role of mtHsp70 during import as formulated in a model named "active motor" or "pulling" model (Voos et al. 1996). Here, the following three properties of the mtHsp70 machinery would combine to generate an inward directed force on the polypeptide in transit, generating forward movement through the channel and also unfolding on the outside. i) MtHsp70 stably binds the incoming unfolded preprotein chain. ii) MtHsp70 is anchored to the inner membrane by its interaction with Tim44, providing the leverage for force generation. iii) A conformational change due to binding and/or hydrolysis of ATP generates the inward directed force resulting in movement of the polypeptide chain. Then, mtHsp70 would dissociate from Tim44 and the preprotein. The cycle would start over again resulting in preprotein translocation. The cycling of mtHsp70 action is catalysed by Mge1p.

Experimental evidence for the "pulling" model is again provided by the analysis of mtHsp70 and Tim44 mutants (Merlin et al. 1999; Voisine et al. 1999). A temperature-sensitive mutant of mtHsp70 in yeast mitochondria, Ssc1-2, shows reduced interaction with Tim44 but increased binding to imported preprotein. The mutant cells are not viable at elevated temperatures and the mitochondria are unable to efficiently import folded preproteins due to a lack of generation of inward directed force. Second site mutations in Ssc1-2 that restore both binding to Tim44 and the generation of the import force, are viable and fully import competent (Voisine et al. 1999). However, the suppressor mutants still show increased interaction to precursor polypeptides in transit. It was concluded, that binding to the preprotein and import efficiency do not correlate and therefore the Brownian ratchet model alone is not sufficient to describe the function of mtHsp70 during import. On the other hand, mutations in Tim44 that reduce the efficiency of mtHsp70 binding do indeed mainly affect the unfolding of preproteins with conformational restrictions during import (Merlin et al. 1999). It is most likely that a combination of both trapping and pulling are used during protein translocation.

8.3
A Possible Role for GTP

Although a completely consistent concept has not been developed so far, there is recent experimental evidence indicating that hydrolysis of GTP might contribute to the overall efficiency of preprotein translocation (Sepuri et al. 1998a). In these experiments import of mitochondrial precursors was stimulated by addition of GTP on the outside of the inner membrane. Non-hydrolysable analogs were not functional, demonstrating the involvement of nucleotide hydrolysis. It was concluded, that in addition to the ATP-dependent "pull" in the matrix, a GTP-utilising "push" might contribute to the import reaction (Sepuri et al. 1998b). However, a GTPase responsible for this effect has not been identified so far. One has to keep in mind, that the involvement of GTP could be connected with the possibility of co-translational import into mitochondria. There is some data showing ribosomes associated with mitochondria in a GTP-dependent manner (Crowley and Payne 1998). A cotranslational mechanism can therefore not be excluded completely, even if most available evidence demonstrates a predominantly posttranslational import of preproteins in the case of mitochondria (Crowley et al. 1994).

9
Processing and Folding of Newly Imported Matrix Proteins

In most cases the presequence which directs a preprotein into the matrix of mitochondria is cleaved off by a processing peptidase once it reaches this compartment. The protein responsible is called matrix processing peptidase (MPP) which consists of two subunits of approximately 55 and 52 kDa termed α-MPP and β-MPP respectively. In yeast the proteins associate to form a dimer (Geli et al. 1990). But some preproteins are processed twice first by MPP and then by a second peptidase called mitochondrial intermediate peptidase (Isaya et al. 1994; Kalousek et al. 1992). During or subsequent to proteolytic processing the mature protein may fold into its active form. This event may occur spontaneously or only with the help of molecular chaperones. Chaperones assist protein folding essential by forming a reversible association with polypeptides in an unfolded state thereby preventing unproductive pathways that end in aggregation. The matrix of mitochondria is very similar to that of eubacterial cytosol with respect to its repertoire of molecular chaperones. There are representatives from the major chaperone families such as the Hsp100s, Hsp70s and their cochaperones, the chaperonins and folding catalysts the peptidyl-prolyl *cis-trans* isomerases (PPIases). A comprehensive discussion of protein folding in the matrix of mitochondria is beyond the scope of this review but has been discussed in detail elsewhere (Neupert, 1997; Dekker and Pfanner 1999).

10
Perspectives

There are many major and minor properties of the mitochondria protein import reaction that have not been established to date. For example it is not clear to what extent translocation into mitochondria occurs in a posttranslational or cotranslational manner. If it is predominately posttranslational then to what degree do most preproteins fold before they are translocated?

Do hydrophobic membrane proteins form water soluble conformations during their translocation through aqueous compartments. While the basic features of a presequence can be described we do not know the nature of internal targeting signals or in either case the molecular nature of receptor recognition. The structures of translocation components which are essentially lacking so far will help immensely to unravel the mechanisms of mitochondrial protein import.

Only discovered a couple of years ago was a completely new translocase of the mitochondrial inner membrane, the TIM22 complex. Sequencing of the yeast genome facilitated its detection and also provided a link between a human disease and mitochondrial biogenesis. Undoubtedly there are still more translocation components that have so far evaded detection. As the components responsible for the sorting of preprotein to the inner membrane and intermembrane space are not known it is tempting to speculate that these factors also await discovery. But for some of the translocation components that have been described in this review all that we know about them is (1) their existence, (2) to which translocation complex they belong and (3) their basic characteristics. Their specific contribution to preprotein translocation is yet to be determined.

Perhaps the most difficult problem to tackle is the bioenergetics of protein translocation. For example what drives preprotein translocation across or into the outer membrane of mitochondria. Is the acid chain hypothesis fundamentally correct? Currently of great interest is how a molecular chaperone that assists protein folding in many different cellular compartments has evolved to play an essential function in protein translocation. Under truly physiological conditions does mitochondrial Hsp70 actively pull preproteins into the mitochondrial matrix? The solutions to these and so many other questions require the freedom of fundamental and innovative research that has so successfully brought us to this stage.

Acknowledgments. Work of the authors' laboratory was supported by the Deutsche Forschungsgemeinschaft, the Sonderforschungsbereich 388 and the Fonds der Chemischen Industrie and by a research fellowship (to K.N.T.) from the Alexander von Humboldt Foundation.

References

Abe Y, Shodia T, Muto T, Mihara K, Torii H, Nishikawa S, Endo T, Kohda D (2000) Structural basis of presequence recognition by the mitochondrial protein import receptor Tom20. Cell 100:551-560

Adam A, Endres M, Sirrenberg C, Lottspeich F, Neupert W, Brunner M (1999) Tim9, a new component of the TIM22•54 translocase in mitochondria. EMBO J 18:313-319

Alconada A, Kübrich M, Moczko M, Hönlinger A, Pfanner N (1995) The mitochondrial receptor complex: the small subunit Mom8b/Isp6 supports association of receptors with the general insertion pore and transfer of preproteins. Mol Cell Biol 15:6196-6205

Allison DS, Schatz G (1986) Artificial mitochondrial presequences. Proc Natl Acad Sci USA 83:9011-9015

Amaya Y, Arakawa H, Takiguchi M, Ebina Y, Yokota S, Mori M (1988) A noncleavable signal for mitochondrial import of 3-oxoacyl-CoA thiolase. J Biol Chem 263:14463-14470

Argan C, Lusty CJ, Shore G (1983) Membrane and cytosolic components affecting transport of the precursor of ornithine carbamyltransferase into mitochondria. J Biol Chem 258:6667-6670.

Arnold I, Fölsch H, Neupert W, Stuart RA (1998) Two distinct and independent mitochondrial targeting signals function in the sorting of an inner membrane protein, cytochrome c_1. J Biol Chem 273:1469-1476

Baker A, Schatz G (1987) Sequences from a prokaryotic genome or the mouse dihydrofolate reductase gene can restore the import of a truncated precursor protein into yeast mitochondria. Proc Natl Acad Sci USA 84:3117-3121

Baker KP, Schaniel A, Vestweber D, Schatz G (1990) A yeast mitochondrial outer membrane protein essential for protein import and cell viability. Nature 348:605-609

Bauer MF, Sirrenberg C, Neupert W, Brunner M (1996) Role of Tim23 as voltage sensor and presequence receptor in protein import into mitochondria. Cell 87:33-41

Beasley EM, Wachter C, Schatz G (1992) Putting energy into mitochondrial protein import. Curr Opin Cell Biol 4:646-651

Bedwell DM, Klionsky DJ, Emr SD (1987) The yeast F1-ATPase beta subunit precursor contains functionally redundant mitochondrial protein import information. Mol Cell Biol 7:4038-4047

Blobel G (1980) Intracellular protein topogenesis. Proc Natl Acad Sci USA 77:1496-1500

Blom J, Kübrich M, Rassow J, Voos W, Dekker PJ, Maarse AC, Meijer M, Pfanner N (1993) The essential yeast protein MIM44 (encoded by *MPI1*) is involved in an early step of preprotein translocation across the mitochondrial inner membrane. Mol Cell Biol 13:7364-7371

Bolliger L, Deloche O, Glick BS, Georgopoulos C, Jenö P, Kronidou N, Horst M, Morishima N, Schatz G (1994) A mitochondrial homologue of bacterial GrpE interacts with mitochondrial hsp70 and is essential for viability. EMBO J 13:1998-2006

Bolliger L, Junne T, Schatz G, Lithgow T (1995) Acidic receptor domains on both sides of the outer membrane mediate translocation of precursor proteins into yeast mitochondria. EMBO J 14:6318-6326

Bömer U, Pfanner N, Dietmeier K (1996) Identification of a third yeast mitochondrial Tom protein with tetratrico peptide repeats. FEBS Lett 382:153-158

Bömer U, Meijer M, Maarse AC, Hönlinger A, Dekker PJ, Pfanner N, Rassow J (1997) Multiple interactions of components mediating preprotein translocation across the inner mitochondrial membrane. EMBO J 16:2205-2216

Bömer U, Maarse AC, Martin F, Geissler A, Merlin A, Schönfisch B, Meijer M, Pfanner N, Rassow J (1998) Separation of structural and dynamic functions of the mitochondrial translocase: Tim44 is crucial for the inner membrane import sites in translocation of tightly folded domains, but not of loosely folded preproteins. EMBO J 17:4226-4237

Brix J, Dietmeier K, Pfanner N (1997) Differential recognition of preproteins by the purified cytosolic domains of the mitochondrial import receptors Tom20, Tom22, and Tom70. J Biol Chem 272:20730-20735

Brix J, Rüdiger S, Bukau B, Schneider-Mergener J, Pfanner N (1999) Distribution of binding sequences for the mitochondrial import receptors Tom20, Tom22, and Tom70 in a presequence-carrying preprotein and a non-cleavable preprotein. J Biol Chem 274:16522-16530

Bruch MD, Hoyt DW (1992) Conformational analysis of a mitochondrial presequence derived from the F1-ATPase beta-subunit by CD and NMR spectroscopy. Biochim Biophys Acta 1159:81-93

Buchberger A, Reinstein J, Bukau B (1999) in Molecular Chaperones and Folding Catalysts. Regulation, Cellular Function and Mechanism (Bukau B, ed) pp, 609-635 Harwood Academic Publishers, Amsterdam

Chauwin JF, Oster G, Glick BS (1998) Strong precursor-pore interactions constrain models for mitochondrial protein import. Biophys J 74:1732-1743

Chirico W, Waters MG, Blobel G (1988) 70K heat shock related proteins stimulate protein translocation into microsomes. Nature 332:805-810

Craig EA, Kramer J, Shilling J, Werner-Washburne M, Holmes S, Kosic-Smithers J, Nicolet CM (1989) SSC1, an essential member of the yeast HSP70 multigene family, encodes a mitochondrial protein. Mol Cell Biol 9:3000-3008

Crowley KS, Liao S, Worrell VE, Reinhart GD, Johnson AE (1994) Secretory proteins move through the endoplasmic reticulum membrane via an aqueous, gated pore. Cell 78:461-471

Crowley KS, Payne RM (1998) Ribosome binding to mitochondria is regulated by GTP and the transit peptide. J Biol Chem 273:17278-17285

Cyr DM, Stuart RA, Neupert W (1993) A matrix ATP requirement for presequence translocation across the inner membrane of mitochondria. J Biol Chem 268:23751-23754

Davis AJ, Ryan KR, Jensen RE (1998) Tim23p contains separate and distinct signals for targeting to mitochondria and insertion into the inner membrane. Mol Biol Cell 9:2577-93

Dekker PJT, Keil P, Rassow J, Maarse AC, Pfanner N, Meijer M (1993) Identification of MIM23, a putative component of the protein import machinery of the mitochondrial inner membrane. FEBS Lett 330:66-70

Dekker PJT, Pfanner N (1997) Role of mitochondrial GrpE and phosphate in the ATPase cycle of matrix Hsp70. J Mol Biol 270:321-327

Dekker PJT, Martin F, Maarse AC, Bömer U, Müller H, Guiard B, Meijer M, Rassow J, Pfanner N (1997) The Tim core complex defines the number of mitochondrial translocation contact sites and can hold arrested preproteins in the absence of matrix Hsp70-Tim44. EMBO J 16:5408-5419

Dekker PJT, Ryan MT, Brix J, Müller H, Hönlinger A, Pfanner N (1998) Preprotein translocase of the outer mitochondrial membrane: molecular dissection and assembly of the general import pore complex. Mol Cell Biol 18:6515-6524

Dekker PJT, Pfanner N (1999) in Molecular Chaperones and Folding Catalysts. Regulation, Cellular Function and Mechanism (Bukau B, ed) pp 235-262, Harwood Academic Publishers, Amsterdam

Deshaies R, Koch B, Werner-Washburne M, Craig EA, Schekman R (1988) A subfamily of stress proteins facilitates translocation of secretory and mitochondrial precursor polypeptides. Nature 332:800-805

Dietmeier K, Hönlinger A, Börner U, Dekker PJ, Eckerskorn C, Lottspeich F, Kübrich M, Pfanner N (1997) Tom5 functionally links mitochondrial preprotein receptors to the general import pore. Nature 388:195-200

Donzeau M, Kaldi K, Adam A, Paschen S, Wanner G, Guiard B, Bauer MF, Neupert W, Brunner M (2000) Tim23 links the inner and outer mitochondrial membranes. Cell 101:401-412

Eilers M, Oppliger W, Schatz G (1987) Both ATP and an energized inner membrane are required to import a purified precursor protein into mitochondria. EMBO J 6:1073-1077

Emtage JL, Jensen RE (1993) MAS6 encodes an essential inner membrane component of the yeast mitochondrial protein import pathway. J Cell Biol 122:1003-1012

Endo T, Shimada I, Roise D, Inagaki F (1989) N-terminal half of a mitochondrial presequence peptide takes a helical conformation when bound to dodecylphosphocholine micelles: a proton nuclear magnetic resonance study. J Biochem 106:396-400

Finocchiaro G, Colombo I, Garavaglia B, Gellera C, Valdameri G, Garbuglio N, Didonato S (1993) cDNA cloning and mitochondrial import of the beta-subunit of the human electron-transfer flavoprotein. Eur J Biochem 213:1003-1008

Fünfschilling U, Rospert S (1999) Nascent Polypeptide-associated Complex stimulates protein import into yeast mitochondria. Mol Biol Cell 10:3289-3299

Gambill BD, Voos W, Kang PJ, Miao B, Langer T, Craig EA, Pfanner N (1993) A dual role for mitochondrial heat shock protein 70 in membrane translocation of preproteins. J Cell Biol 123:109-117

Gärtner F, Börner U, Guiard B, Pfanner N (1995) The sorting signal of cytochrome b_2 promotes early divergence from the general mitochondrial import pathway and restricts the unfoldase activity of matrix Hsp70. EMBO J 14:6043-6057

Gasser SM, Daum G, Schatz G (1982) Import of proteins into mitochondria. Energy-dependent uptake of precursors by isolated mitochondria. J Biol Chem 257:13034-13041

Gässler CS, Buchberger A, Laufen T, Mayer MP, Schröder H, Valencia A, Bukau B (1998) Mutations in the DnaK chaperone affecting interaction with the DnaJ co-chaperone. Proc Natl Acad Sci USA 95:15229-15234

Gaume B, Klaus C, Ungermann C, Guiard B, Neupert W, Brunner M (1998) Unfolding of preproteins upon import into mitochondria. EMBO J 17:6497-6507

Geli V, Yang MJ, Suda K, Lustig A, Schatz G (1990) The MAS-encoded processing protease of yeast mitochondria. Overproduction and characterization of its two nonidentical subunits. J Biol Chem 265:19216-19222

George R, Beddoe T, Landl K, Lithgow T (1998) The yeast nascent polypeptide-associated complex initiates protein targeting to mitochondria in vivo. Proc Natl Acad Sci USA 95:2296-2301

Goping IS, Millar DG, Shore GC (1995) Identification of the human mitochondrial protein import receptor, huMas20p. Complementation of Δmas20 in yeast. FEBS Lett 373:45-50

Gragerov A, Zeng L, Zhao X, Burkholder W, Gottesman ME (1994) Specificity of DnaK-peptide binding. J Mol Biol 235:848-854

Gratzer S, Lithgow T, Bauer RE, Lamping E, Paltauf F, Kohlwein SD, Haucke V, Junne T, Schatz G, Horst M (1995) Mas37p, a novel receptor subunit for protein import into mitochondria. J Cell Biol 129:25-34

Gruhler A, Ono H, Guiard B, Neupert W, Stuart RA (1995) A novel intermediate on the import pathway of cytochrome b2 into mitochondria: evidence for conservative sorting. EMBO J 14:1349-1359

Guiard B (1985) Structure, expression and regulation of a nuclear gene encoding a mitochondrial protein: the yeast L(+)-lactate cytochrome c oxidoreductase (cytochrome b_2). EMBO J 4:3265-3272

Ha J-H, Johnson ER, Mckay DB, Sousa MC, Takeda S, Wilbanks SM (1999) in Molecular Chaperones and Folding Catalysts. Regulation, Cellular Function and Mechanism (Bukau B, ed) pp 573-607, Harwood Academic Publishers, Amsterdam

Hachiya N, Alam R, Sakasegawa Y, Sakaguchi M, Mihara K, Omura T (1993) A mitochondrial import factor purified from rat liver cytosol is an ATP-dependent conformational modulator for precursor proteins. EMBO J 12:1579-1586

Hachiya N, Komiya T, Alam R, Iwahashi J, Sakaguchi M, Omura T, Mihara K (1994) MSF, a novel cytoplasmic chaperone which functions in precursor targeting to mitochondria. EMBO J 13:5146-5154

Hachiya N, Mihara K, Suda K, Horst M, Schatz G, Lithgow T (1995) Reconstitution of the initial steps of mitochondrial protein import. Nature 376:705-709

Hammen PK, Gorenstein DG, Weiner H (1994) Structure of the signal sequences for two mitochondrial matrix proteins that are not proteolytically processed upon import. Biochem 33:8610-8617

Hanson B, Nuttal S, Hoogenraad N (1996) A receptor for the import of proteins into human mitochondria. Eur J Biochem 235:750-753

Harkness TA, Nargang FE, van der Klei I, Neupert W, Lill R (1994) A crucial role of the mitochondrial protein import receptor MOM19 for the biogenesis of mitochondria. J Cell Biol 124:637-648

Hartl F-U, Schmidt B, Wachter E, Weiss H, Neupert W (1986) Transport into mitochondria and intramitochondrial sorting of the Fe/S protein of ubiquinol-cytochrome c reductase. Cell 47:939-951

Hartl F-U, Ostermann J, Guiard B, Neupert W (1987) Successive translocation into and out of the mitochondrial matrix: targeting of proteins to the intermembrane space by a bipartite signal peptide. Cell 51:1027-1037

Haucke V, Lithgow T, Rospert S, Hahne K, Schatz G (1995) The yeast mitochondrial protein import receptor Mas20p binds precursor proteins through electrostatic interaction with the positively charged presequence. J Biol Chem 270:5565-5570

Hill K, Model K, Ryan MT, Dietmeier K, Martin F, Wagner R, Pfanner N (1998) Tom40 forms the hydrophilic channel of the mitochondrial import pore for preproteins. Nature 395:516-521

Hines V, Brandt A, Griffith G, Horstmann H, Brütsch H, Schatz G (1990) Protein import into yeast mitochondria is accelerated by the outer membrane protein MAS70. EMBO J 9:3191-3200.

Hönlinger A, Kübrich M, Moczko M, Gärtner F, Mallet L, Bussereau F, Eckerskorn C, Lottspeich F, Dietmeier K, Jacquet M, Pfanner N (1995) The mitochondrial receptor complex: Mom22 is essential for cell viability and directly interacts with preproteins. Mol Cell Biol 15:3382-3389

Hönlinger A, Bömer U, Alconada A, Eckerskorn C, Lottspeich F, Dietmeier K, Pfanner N (1996) Tom7 modulates the dynamics of the mitochondrial outer membrane translocase and plays a pathway-related role in protein import. EMBO J 15:2125-2137

Horst M, Jenö P, Kronidou NG, Bolliger L, Oppliger W, Scherer P, Manning-Krieg U, Jascur T, Schatz G (1993) Protein import into yeast mitochondria: the inner membrane import site protein ISP45 is the MPI1 gene product. EMBO J 12:3035-3041

Horst M, Oppliger W, Rospert S, Schonfeld HJ, Schatz G, Azem A (1997) Sequential action of two Hsp70 complexes during protein import into mitochondria. EMBO J 16:1842-1849

Horwich AL, Kalousek F, Mellman I Rosenberg LE (1985) A leader peptide is sufficient to direct mitochondrial import of a chimeric protein. EMBO J 4:1129-1135

Huang S, Ratliff KS, Schwartz MP, Spenner JM, Matouschek A (1999) Mitochondria unfold precursor proteins by unravelling them from their N-termini. Nat Struct Biol 6:1132-1138

Hurt EC, Pesold-Hurt B, Schatz G (1984) The cleavable prepiece of an imported mitochondrial protein is sufficient to direct cytosolic dihydrofolate reductase into the mitochondrial matrix. FEBS Lett 178:306-310

Hurt EC, Persold-Hurt B, Suda K, Oppliger W, Schatz G (1985) The first twelve amino acids (less than half of the pre-sequence) of an imported mitochondrial protein can direct mouse cytosolic dihydrofolate reductase into the yeast mitochondrial matrix. EMBO J 4: 2061-2068

Hurt EC, Van Loon APMG (1986) How proteins find mitochondria and intramitochondrial compartments. Trends Biochem Sci 11:204-207

Hwang ST, Schatz G (1989) Translocation of proteins across the mitochondrial inner membrane, but not into the outer membrane, requires nucleoside triphosphates in the matrix. Proc Natl Acad Sci USA 86:8432-8436

Hwang S, Jascur T, Vestweber D, Pon L, Schatz G (1989) Disrupted yeast mitochondria can import precursor proteins directly through their inner membrane. J Cell Biol 109:487-493

Ikeda E, Yoshida S, Mitsuzawa H, Uno I, Toh-e A (1994) YGE1 is a yeast homologue of Escherichia coli grpE and is required for maintenance of mitochondrial functions. FEBS Lett 339:265-268

Isaya G, Miklos D, Rollins RA (1994) MIP1, a new yeast gene homologous to the rat mitochondrial intermediate peptidase gene, is required for oxidative metabolism in Saccharomyces cerevisiae. Mol Cell Biol 14:5603-5616

Ishihara N, Mihara K (1998) Identification of the protein import components of the rat mitochondrial inner membrane, rTIM17, rTIM23, and rTIM44. J Biochem 123:722-732

Jarosch E, Tuller G, Daum G, Waldherr M, Voskova A, Schweyen RJ (1996) Mrs5p, an essential protein of the mitochondrial intermembrane space, affects protein import into yeast mitochondria. J Biol Chem 271:17219-17225

Jarvis JA, Ryan MT, Hoogenraad NJ, Craik DJ, Høj PB (1995) Solution structure of the acetylated and noncleavable mitochondrial targeting signal of rat chaperonin 10. J Biol Chem 270:1323-1331

Jin H, May M, Tranebjaerg L, Kendall E, Fontan G, Jackson J, Subramony SH, Arena F, Lubs H, Smith S, Stevenson R, Schwartz C, Vetrie D (1996) A novel X-linked gene, DDP, shows mutations in families with deafness (DFN-1), dystonia, mental deficiency and blindness. Nat Genet 14:177-180

Káldi K, Bauer MF, Sirrenberg C, Neupert W, Brunner M (1998) Biogenesis of Tim23 and Tim17, integral components of the TIM machinery for matrix-targeted preproteins. EMBO J 17:1569-1576

Kalousek F, Isaya G, Rosenberg LE (1992) Rat liver mitochondrial intermediate peptidase (MIP): purification and initial characterization. EMBO J 11:2803-2809

Kanamori T, Nishikawa S, Shin I, Schultz PG, Endo T (1997) Probing the environment along the protein import pathways in yeast mitochondria by site-specific photocrosslinking. Proc Natl Acad Sci USA 94:485-490

Kang PJ, Ostermann J, Shilling J, Neupert W, Craig EA, Pfanner N (1990) Requirement for hsp70 in the mitochondrial matrix for translocation and folding of precursor proteins. Nature 348:137-143

Kaput J, Goltz S, Blobel G (1982) Nucleotide sequence of the yeast nuclear gene for cytochrome c peroxidase precursor. Functional implications of the presequence for protein transport into mitochondria. J Biol Chem 257:15054-15058

Karslake C, Piotto ME, Pak YK, Weiner H, Gorenstein DG (1990) 2D NMR and structural model for a mitochondrial signal peptide bound to a micelle. Biochem 29: 9872-9878

Kassenbrock CK, Cao W, Douglas MG (1993) Genetic and biochemical characterization of ISP6, a small mitochondrial outer membrane protein associated with the protein translocation complex. EMBO J 12:3023-3034

Keil P, Pfanner N (1993) Insertion of MOM22 into the mitochondrial outer membrane strictly depends on surface receptors. FEBS Lett 321:197-200

Kerscher O, Holder J, Srinivasan M, Leung RS, Jensen RE (1997) The Tim54p-Tim22p complex mediates insertion of proteins into the mitochondrial inner membrane. J Cell Biol 139:1663-1675

Kerscher O, Sepuri NB, Jensen RE (2000) Tim18p is a new component of the Tim54p-Tim22p translocon in the mitochondrial inner membrane. Mol Biol Cell 11:103-116

Kiebler M, Keil P, Schneider H, van der Klei IJ, Pfanner N, Neupert W (1993) The mitochondrial receptor complex: a central role of MOM22 in mediating preprotein transfer from receptors to the general insertion pore. Cell 74:483-492

Koehler CM, Merchant S, Oppliger W, Schmid K, Jarosch E, Dolfini L, Junne T, Schatz G, Tokatlidis K (1998a) Tim9p, an essential partner subunit of Tim10p for the import of mitochondrial carrier proteins. EMBO J 17:6477-6486

Koehler CM, Jarosch E, Tokatlidis K, Schmid K, Schweyen RJ, Schatz G (1998b) Import of mitochondrial carriers mediated by essential proteins of the intermembrane space. Science 279:369-373

Koehler CM, Leuenberger D, Merchant S, Renold A, Junne T, Schatz G (1999) Human deafness dystonia is a mitochondrial disease. Proc Natl Acad Sci USA 96:2141-2146

Koehler CM, Murphy MP, Bally NA, Leuenberger D, Oppliger W, Dolfini L, Junne T, Schatz G, Or E (2000) Tim18p, a new subunit of the TIM22 complex that mediates insertion of imported proteins into the yeast mitochondrial inner membrane. Mol Cell Biol 20:1187-1193

Koll H, Guiard B, Rassow J, Ostermann J, Horwich AL, Neupert W, Hartl F-U (1992) Antifolding activity of hsp60 couples protein import into the mitochondrial matrix with export to the intermembrane space. Cell 68:1163-1175

Komiya T, Sakaguchi M, Mihara K (1996) Cytoplasmic chaperones determine the targeting pathway of precursor proteins to mitochondria. EMBO J 15:399-407

Komiya T, Rospert S, Schatz G, Mihara K (1997) Binding of mitochondrial precursor proteins to the cytoplasmic domains of the import receptors Tom70 and Tom20 is determined by cytoplasmic chaperones. EMBO J 16:4267-4275

Komiya T, Rospert S, Koehler C, Looser R, Schatz G, Mihara K (1998) Interaction of mitochondrial targeting signals with acidic receptor domains along the protein import pathway: evidence for the 'acid chain' hypothesis. EMBO J 17:3886-3898

Krimmer T, Rassow J, Kunau W-H, Voos W, Pfanner N (2000) Mitochondrial protein import motor: the ATPase domain of matrix Hsp70 is crucial for binding to Tim44, while the peptide binding domain and the carboxy-terminal segment play a stimulatory role. Mol Cell Biol 20:5879-5887

Kronidou NG, Oppliger W, Bolliger L, Hannavy K, Glick BS, Schatz G, Horst M (1994) Dynamic interaction between Isp45 and mitochondrial Hsp70 in the protein import system of the yeast mitochondrial inner membrane. Proc Natl Acad Sci USA 91:12818-12822

Kübrich M, Keil P, Rassow J, Dekker PJ, Blom J, Meijer M, Pfanner N (1994) The polytopic mitochondrial inner membrane proteins MIM17 and MIM23 operate at the same preprotein import site. FEBS Lett 349:222-228

Künkele KP, Heins S, Dembowski M, Nargang FE, Benz R, Thieffry M, Walz J, Lill R, Nussberger S, Neupert W (1998) The preprotein translocation channel of the outer membrane of mitochondria. Cell 93:1009-1019

Kurz M, Martin H, Rassow J, Pfanner N, Ryan MT (1999) Biogenesis of Tim proteins of the mitochondrial carrier import pathway: differential targeting mechanisms and crossing-over with the main import pathway. Mol Biol Cell 10:2461-2474

Laloraya S, Gambill BD, Craig EA (1994) A role for a eukaryotic GrpE-related protein Mge1p, in protein translocation. Proc Natl Acad Sci USA 91:6481-6485

Laloraya S, Dekker PJT, Voos W, Craig EA, Pfanner N (1995) Mitochondrial GrpE modulates the function of matrix Hsp70 in translocation and maturation of preproteins. Mol Cell Biol 15:7098-7105

Landry SJ, Jordan R, McMacken R, Gierasch LM (1992) Different conformations for the same polypeptide bound to chaperones DnaK and GroEL. Nature 355:455-457

Lemire BD, Fankhauser C, Baker A, Schatz G (1989) The mitochondrial targeting function of randomly generated peptide sequences correlates with predicted helical amphiphilicity. J Biol Chem 264:20206-20215

Leuenberger D, Bally NA, Schatz G, Koehler CM (1999) Different import pathways through the mitochondrial intermembrane space for inner membrane proteins. EMBO J 18:4816-4822

Lithgow T, Junne T, Suda K, Gratzer S, Schatz G (1994) The mitochondrial outer membrane protein Mas22p is essential for protein import and viability of yeast. Proc Natl Acad Sci USA 91:11973-11977

Lohret TA, Jensen RE, Kinnally KW (1997) Tim23, a protein import component of the mitochondrial inner membrane, is required for normal activity of the multiple conductance channel, MCC. J Cell Biol 137:377-386

Maarse AC, Blom J, Grivell LA, Meijer M (1992) MPI1, an essential gene encoding a mitochondrial membrane protein, is possibly involved in protein import into yeast mitochondria. EMBO J 11:3619-3628

Maarse AC, Blom J, Keil P, Pfanner N, Meijer M (1994) Identification of the essential yeast protein MIM17, an integral mitochondrial inner membrane protein involved in protein import. FEBS Lett 349:215-221

Mahlke K, Pfanner N, Martin J, Horwich AL, Hartl FU, Neupert W (1990) Sorting pathways of mitochondrial inner membrane proteins. Eur J Biochem 192:551-555

Martin J, Mahlke K, Pfanner N (1991) Role of an energized inner membrane in mitochondrial protein import. J Biol Chem 266:18051-18057

Matouschek A, Azem A, Ratliff K, Glick BS, Schmid K, Schatz G (1997) Active unfolding of precursor proteins during mitochondrial protein import. EMBO J
16:6727-6736

Mayer A, Nargang FE, Neupert W, Lill R (1995) MOM22 is a receptor for mitochondrial targeting sequences and cooperates with MOM19. EMBO J 14:4204-4211

Merlin A, Voos W, Maarse AC, Meijer M, Pfanner N, Rassow J (1999) The J-related
segment of Tim44 is essential for cell viability: a mutant Tim44 remains in the
mitochondrial import site, but inefficiently recruits mtHsp70 and impairs protein translocation. J Cell Biol 145:961-972

Miao B, Davis JE, Craig EA (1997) Mge1 functions as a nucleotide release factor for
Ssc1, a mitochondrial Hsp70 of *Saccharomyces cerevisiae*. J Mol Biol 265:541-552

Miller DM, Delgado R, Chirgwin JM, Hardies SC, Horowitz PM (1991) Expression of
cloned bovine adrenal rhodanese. J Biol Chem 266:4686-4691

Miura S, Mori M, Tatibana M (1983) Transport of ornithine carbamoyltransferase
precursor into mitochondria. J Biol Chem 258:6671-6674

Moczko M, Ehmann B, Gärtner F, Hönlinger A, Schäfer E, Pfanner N (1994) Deletion
of the receptor MOM19 strongly impairs import of cleavable preproteins into
Saccharomyces cerevisiae mitochondria. J Biol Chem 269:9045-9051

Moczko M, Bömer U, Kübrich M, Zufall N, Hönlinger A, Pfanner N (1997) The intermembrane space domain of mitochondrial Tom22 functions as a *trans* binding site for preproteins with N-terminal targeting sequences. Mol Cell Biol
17:6574-6584

Murakami H, Pain D, Blobel G (1988) 70-kD heat shock-related protein is one of at
least two distinct cytosolic factors stimulating protein import into mitochondria.
J Cell Biol 107:2051-2057

Murakami K, Mori M (1990) Purified presequence binding factor (PBF) forms an
import-competent complex with a purified mitochondrial precursor protein.
EMBO J 9:3201-3208

Miura S, Mori M, Tatibana M (1983) Transport of ornithine carbamoyltransferase
precursor into mitochondria. J Biol Chem 258:6671-6674.

Nakai M, Kato Y, Toh-e A, Endo T (1994) Yge1p, a eukaryotic GrpE-homologue, is
localized in the mitochondrial matrix and interacts with mitochondrial hsp70.
Biochem Biophys Res Commun 200:435-442

Nakai M, Endo T (1995) Identification of yeast *MAS17* encoding the functional
counterpart of the mitochondrial receptor complex protein MOM22 of *Neurospora crassa*. FEBS Lett 357:202-206

Nargang FE, Künkele KP, Mayer A, Ritzel RG, Neupert W, Lill R (1995) 'Sheltered
disruption' of *Neurospora crassa* MOM22, an essential component of the mitochondrial protein import complex. EMBO J 14:1099-1108

Neupert W (1997) Protein import into mitochondria. Annu Rev Biochem 66:863-917

Ohta S, Schatz G (1984) A purified precursor polypeptide requires a cytosolic protein fraction for import into mitochondria. EMBO J 3:651-657

Ohba M, Schatz G (1987) Disruption of the outer membrane restores protein import
to trypsin-treated yeast mitochondria. EMBO J 6:2117-2122

Ono H, Tubio S (1988) The cytosolic factor required for import of precursors of
mitochondrial proteins into mitochondria. J Biol Chem 263:3188-3193

Ono H, Tuboi S (1990) Purification and identification of a cytosolic factor required for import of precursors of mitochondrial proteins into mitochondria. Arch Biochem Biophys 280:299–304

Pfanner N (2000) Recognizing mitochondrial presequences. Curr Biol 10:412-415

Pfanner N, Neupert W (1985) Transport of proteins into mitochondria: a potassium diffusion potential is able to drive the import of the ADP/ATP carrier. EMBO J 4:2819-2825

Pfanner N, Neupert W (1986) Transport of F_1-ATPase subunit β into mitochondria depends on both a membrane potential and nucleoside triphosphates. FEBS Lett 209:152-156

Pfanner N, Neupert W (1987) Distinct steps in the import of the ADP/ATP carrier into mitochondria. J Biol Chem 262:7528-7536

Pfanner N, Hoeben P, Tropschug M, Neupert W (1987a) The carboxy terminal two-thirds of the ADP/ATP carrier polypeptide contains sufficient information to direct translocation into mitochondria. J Biol Chem 262:14851-14854

Pfanner N, Müller HK, Harmey MA, Neupert W (1987b) Mitochondrial protein import: involvement of the mature part of a cleavable precursor protein in the binding to receptor sites. EMBO J 6:3449-3454

Pfanner N, Tropschug M, Neupert W (1987c) Mitochondrial protein import: nucleoside triphosphates are involved in conferring import-competence to precursors. Cell 49:815-823

Pfanner N, Rassow J, Guiard B, Söllner T, Hartl FU, Neupert W (1990) Energy requirements for unfolding and membrane translocation of precursor proteins during import into mitochondria. J Biol Chem 265:16324-16329

Pohlmeyer K, Soll J, Steinkamp T, Hinnah S, Wagner R (1997) Isolation and characterization of an amino acid-selective channel protein present in the chloroplastic outer envelope membrane. Proc Natl Acad Sci USA 94:9504-9509

Ramage L, Junne T, Hahne K, Lithgow T, Schatz G (1993) Functional cooperation of mitochondrial protein import receptors in yeast. EMBO J 12:4115-4123

Rapaport D, Mayer A, Neupert W, Lill R (1998) *Cis* and *trans* sites of the TOM complex of mitochondria in unfolding and initial translocation of preproteins. J Biol Chem 273:8806-8813

Rassow J, Pfanner N (1991) Mitochondrial preproteins en route from the outer membrane to the inner membrane are exposed to the intermembrane space. FEBS Lett 293:85-88

Rassow J, Maarse AC, Krainer E, Kübrich M, Müller H, Meijer M, Craig EA, Pfanner N (1994) Mitochondrial protein import: biochemical and genetic evidence for interaction of matrix hsp70 and the inner membrane protein MIM44. J Cell Biol 127:1547-1556

Rassow J, von Ahsen O, Bömer U, Pfanner N (1997) Molecular chaperones: towards a characterization of the heat-shock protein 70 family. Trends Cell Biol 7:129-133

Rassow J, Dekker PJT, van Wilpe S, Meijer M, Soll J (1999) The preprotein translocase of the mitochondrial inner membrane: function and evolution. J Mol Biol 286:105-120

Reid GA, Yonetani T, Schatz G (1982) Import of proteins into mitochondria. Import and maturation of the mitochondrial intermembrane space enzymes cytochrome b_2 and cytochrome c peroxidase in intact yeast cells. J Biol Chem 257:13068-13074

Roise D, Schatz G (1988) Mitochondrial presequences. J Biol Chem 263:4509-4511

Rojo EE, Stuart RA, Neupert W (1995) Conservative sorting of Fo-ATPase subunit 9: export from matrix requires delta pH across inner membrane and matrix ATP. EMBO J 14:3445-3451

Rospert S, Junne T, Glick BS, Schatz G (1993) Cloning and disruption of the gene encoding yeast mitochondrial chaperonin 10, the homologue of *E. coli* groES. FEBS Lett 335:358-360

Rüdiger S, Germeroth L, Schneider-Mergener J, Bukau B (1997) Substrate specificity of the DnaK chaperone determined by screening cellulose-bound peptide libraries. EMBO J 16:1501-1507

Ryan KR, Jensen RE (1993) Mas6p can be cross-linked to an arrested precursor and interacts with other proteins during mitochondrial protein import. J Biol Chem 268:23743-23746

Ryan KR, Menold MM, Garrett S, Jensen RE (1994a) *SMS1*, a high-copy suppressor of the yeast *mas6* mutant, encodes an essential inner membrane protein required for mitochondrial protein import. Mol Biol Cell 5:529-538

Ryan MT, Hoogenraad NJ, Høj PB (1994b) Isolation of a cDNA clone specifying rat chaperonin 10, a stress-inducible mitochondrial matrix protein synthesised without a cleavable presequence. FEBS Lett 337:152-156

Ryan KR, Leung RS, Jensen RE (1998) Characterization of the mitochondrial inner membrane translocase complex: the Tim23p hydrophobic domain interacts with Tim17p but not with other Tim23p molecules. Mol Cell Biol 18:178-187

Ryan MT, Müller H, Pfanner N (1999) Functional staging of ADP/ATP carrier translocation across the outer mitochondrial membrane. J Biol Chem 274:20619-20627

Schatz G (1997) Just follow the acid chain. Nature 388:121-122

Scherer PE, Krieg UC, Hwang ST, Vestweber D, Schatz G (1990) A precursor protein partly translocated into yeast mitochondria is bound to a 70 kD mitochondrial stress protein. EMBO J 9:4315-4322

Scherer PE, Manning-Krieg UC, Jenö P, Schatz G, Horst M (1992) Identification of a 45-kDa protein import site of the yeast mitochondrial inner membrane. Proc Natl Acad Sci USA 89:11930-11934

Schleyer M, Neupert W (1985) Transport of proteins into mitochondria: Translocational intermediates spanning contact sites between outer and inner membranes. Cell 43:339-350

Schleyer M, Schmidt B, Neupert W (1982) Requirement of a membrane potential for the postranslational transfer of proteins into mitochondria. Eur J Biochem 125:109-116

Schlossmann J, Dietmeier K, Pfanner N, Neupert W (1994) Specific recognition of mitochondrial preproteins by the cytosolic domain of the import receptor MOM72. J Biol Chem 269:11893-11901

Schlossmann J, Lill R, Neupert W, Court DA (1996) Tom71, a novel homologue of the mitochondrial preprotein receptor Tom70. J Biol Chem 271:17890-17895

Schneider H, Söllner T, Dietmeier K, Eckerskorn C, Lottspeich F, Trülzsch B, Neupert W, Pfanner N (1991) Targeting of the master receptor MOM19 to mitochondria. Science 254:1659-1662

Schneider H-C, Berthold J, Bauer MF, Dietmeier K, Guiard B, Brunner M, Neupert W (1994) Mitochondrial Hsp70/MIM44 complex facilitates protein import. Nature 371:768-774

Schneider HC, Westermann B, Neupert W, Brunner M (1996) The nucleotide exchange factor MGE exerts a key function in the ATP-dependent cycle of mt-

Hsp70-Tim44 interaction driving mitochondrial protein import. EMBO J 15:5796-5803

Segui-Real B, Kispal G, Lill R, Neupert W (1993) Functional independence of the protein translocation machineries in mitochondrial outer and inner membranes: Passage of preproteins through the intermembrane space. EMBO J 12:2211-2218

Sepuri NB, Schulke N, Pain D (1998) GTP hydrolysis is essential for protein import into the mitochondrial matrix. J Biol Chem 273:1420-1424

Sepuri NBV, Gordon DM, Pain D (1998) A GTP-dependent "Push" is generally required for efficient protein translocation across the mitochondrial inner membrane into the matrix. J Biol Chem 273:20941-20950

Simon MS, Peskin CS, Oster GF (1992) What drives the translocation of proteins? Proc Natl Acad Sci USA 89:3770-3774

Sirrenberg C, Bauer MF, Guiard B, Neupert W, Brunner M (1996) Import of carrier proteins into the mitochondrial inner membrane mediated by Tim22. Nature 384:582-585

Sirrenberg C, Endres M, Fölsch H, Stuart RA, Neupert W, Brunner M (1998) Carrier protein import into mitochondria mediated by the intermembrane proteins Tim10/Mrs11 and Tim12/Mrs5. Nature 391:912-915

Smagula C, Douglas MG (1988) Mitochondrial import of the ADP/ATP carrier protein in *Saccharomyces cerevisiae*. Sequences required for receptor binding and membrane translocation. J Biol Chem 263:6783-6790

Söllner T, Griffiths G, Pfaller R, Pfanner N, Neupert W (1989) MOM19, an import receptor for mitochondrial precursor proteins. Cell 59:1061-1070

Söllner T, Pfaller R, Griffiths G, Pfanner N, Neupert W (1990) A mitochondrial import receptor for the ATP/ADP carrier. Cell 62:107-115

Steger HF, Söllner T, Kiebler M, Dietmeier KA, Pfaller R, Trülzsch KS, Tropschug M, Neupert W, Pfanner N (1990) Import of the ADP/ATP carrier into mitochondria: two receptors act in parallel. J Cell Biol 111:2353-2363

Steinkamp T, Hill K, Hinnah SC, Wagner R, Rohl T, Pohlmeyer K, Soll J (2000) Identification of the pore-forming region of the outer chloroplast envelope protein OEP16. J Biol Chem 275:11758-11764

Suh WC, Burkholder WF, Lu CZ, Zhao X, Gottesman ME, Gross CA (1998) Interaction of the Hsp70 molecular chaperone, DnaK, with its cochaperone DnaJ. Proc Natl Acad Sci USA 95:15223-15228

Terada K, Kanazawa M, Yano M, Hanson B, Hoogenraad N, Mori M, (1997) Participation of the import receptor Tom20 in protein import into mammalian mitochondria: analyses *in vitro* and in cultured cells. FEBS Lett 403:309-312

Ungermann C, Neupert W, Cyr DM (1994) The role of Hsp70 in conferring unidirectionality on protein translocation into mitochondria. Science 266:1250-1253

van Loon AP, Schatz G (1987) Transport of proteins to the mitochondrial intermembrane space: the 'sorting' domain of the cytochrome c_1 presequence is a stop-transfer sequence specific for the mitochondrial inner membrane. EMBO J 6:2441-2448

van Wilpe S, Ryan MT, Hill K, Maarse AC, Meisinger C, Brix J, Dekker PJ, Moczko M, Wagner R, Meijer M, Guiard B, Hönlinger A, Pfanner N (1999) Tom22 is a multifunctional organiser of the mitochondrial preprotein translocase. Nature 401:485-489

Vestweber D, Brunner J, Baker A, Schatz G (1989) A 42K outer-membrane protein is a component of the yeast mitochondrial protein import site. Nature 341:205-209

Voisine C, Craig EA, Zufall N, von Ahsen O, Pfanner N, Voos W (1999) The protein import motor of mitochondria: unfolding and trapping of preproteins are distinct and separable functions of matrix Hsp70. Cell 97:1-20

von Ahsen O, Voos W, Henninger H, Pfanner N (1995) The mitochondrial protein import machinery. Role of ATP in dissociation of the Hsp70-Mim44 complex. J Biol Chem 270:29848-29853

von Heijne G (1986) Mitochondrial targeting sequences may form amphipathic helices. EMBO J 5:1335-1342

von Heijne G, Steppuhn J, Herrmann RG (1989) Domain structure of mitochondrial and chloroplast targeting peptides. Eur J Biochem 180:535-545

Voos W, Gambill BD, Guiard B, Pfanner N, Craig EA (1993) Presequence and mature part of preproteins strongly influence the dependence of mitochondrial protein import on heat shock protein 70 in the matrix. J Cell Biol 123:119-126

Voos W, Gambill BD, Laloraya S, Ang D, Craig EA, Pfanner N (1994) Mitochondrial GrpE is present in a complex with hsp70 and preproteins in transit across membranes. Mol Cell Biol 14:6627-6634

Voos W, von Ahsen O, Müller H, Guiard B, Rassow J, Pfanner N (1996) Differential requirement for the mitochondrial Hsp70-Tim44 complex in unfolding and translocation of preproteins. EMBO J 15:2668-2677

Wachter C, Schatz G, Glick BS (1992) Role of ATP in the intramitochondrial sorting of cytochrome c_1 and the adenine nucleotide transporter. EMBO J 11:4787-4794

Wachter C, Schatz G, Glick BS (1994) Protein import into mitochondria: the requirement for external ATP is precursor-specific whereas intramitochondrial ATP is universally needed for translocation into the matrix. Mol Biol Cell 5:465-474

Wallace DC, Murdock DG (1999) Mitochondria and dystonia: the movement disorder connection? Proc Natl Acad Sci USA 96:1817-1819

Weiss C, Oppliger W, Vergères G, Demel R, Jeno P, Horst M, de Kruijff B, Schatz G, Azem A (1999) Domain structure and lipid interaction of recombinant yeast Tim44. Proc Natl Acad Sci USA 96:8890-8894

Zara V, Palmieri F, Mahlke K, Pfanner N (1992) The cleavable presequence is not essential for import and assembly of the phosphate carrier of mammalian mitochondria but enhances the specificity and efficiency of import. J Biol Chem 267:12077-12081

Zhu X, Zhao X, Burkholder WF, Gragerov A, Ogata CM, Gottesman ME, Hendrickson WA (1996) Structural analysis of substrate binding by the molecular chaperone DnaK. Science 272:1606-16014

Editor-in-charge: Professor N. Pfanner

Input-Output Functions
of Mammalian Motoneurons

R. K. Powers and M. D. Binder

Department of Physiology & Biophysics, University of Washington
School of Medicine, Box 357290, Seattle, Washington 98195-7290, USA

Contents

1
Introduction

Alpha motoneurons have been the most extensively studied cells in the mammalian central nervous system (CNS). The reasons for their favor among neurobiologists include the ease and certainty with which they can be identified and their large somata that facilitate stable intracellular recordings. Many of their presynaptic input fibers can be selectively activated, such as primary muscle spindle afferents, which make monosynaptic excitatory connections with motoneurons (reviewed in Henneman and Mendell 1981). Further, motoneurons innervate muscle cells that can be characterized both mechanically and histochemically (reviewed in Burke 1981), making motoneurons the only class of CNS cells for which variations in electrophysiological and anatomical properties have been linked to variations in the properties of their target cells. The one-to-one relation between the occurrence of action potentials in a motoneuron and in the muscle fibers they innervate makes it possible to record the discharge behavior of single motoneurons using intramuscular electrodes, as well as the activity of populations of motoneurons using electrodes placed over the surface of the muscle. Consequently, motoneurons are the only CNS neurons whose individual discharge behavior can be recorded in awake human subjects during the execution of normal movements.

In light of the unique features of the motoneuron listed above, it is not surprising that many of our notions regarding the integrative properties of neurons have been based on our knowledge of the anatomy and physiology of motoneurons. By the mid 1980's, motoneurons had become the model cell for understanding a number of key aspects of neuronal function, including: (1) the transfer of synaptic current from dendritic synapses to the soma (reviewed in Jack, Noble, and Tsien 1975; Rall 1977; Redman 1976): (2) the relation between cellular properties, synaptic input distribution and excitability (reviewed in Burke 1981, Henneman and Mendell 1981; Binder et al. 1996; Reckling et al. 2000), (3) the relation between synaptic or injected current and firing rate (reviewed in Schwindt and Crill 1984; Kernell 1983, 1984) and (4) the relationship between the size and shape of a postsynaptic potential and its effects on firing probability (Fetz and Gustafsson1983: Ashby and Zilm 1982a,b; Gustafsson and McCrea 1989; Cope et al. 1987).

The development of techniques for recording from neurons in brain slices *in vitro* made it possible to obtain stable intracellular recordings from the somata of many other neurons in the CNS. The initial *in vitro* studies in motoneurons were largely confined to neonatal cells (cf., Fulton and Walton 1986; Harada and Takahashi 1983; Walton and Fulton 1986), and attention was increasingly focused on other CNS cells, particularly pyramidal cells in the neocortex and hippocampus and Purkinje cells in the cerebellum (Llinas 1988). The dendritic trees of the latter two cell types are confined to specific neuronal laminae that are relatively free of cell bodies, making it possible to systematically record from the dendrites of these cells (cf. Mel 1994). The additional development of techniques for dendritic patch clamping and high-speed fluorescence imaging have provided direct measurements of the electrophysiological properties of the dendrites of a variety of CNS neurons, revealing the presence of a wide range of voltage- and calcium-sensitive channels in dendrites (for recent reviews, see Johnston et al. 1996; Mel 1994; Midtgaard 1994, 1996). With the exception of a recent study on cultured rat spinal motoneurons (Larkum et al. 1996), there have been no comparable studies of motoneuron dendrites.

The historical importance of motoneuron studies in the development of our views of neuronal function together with the shift in experimental focus to other cells of the nervous system has led to the persistence of the notion that the mechanisms of input-output transduction in motoneurons are relatively simple compared to those of other neurons. However, a variety of evidence obtained over the past fifteen to twenty years suggests that the electrophysiological properties of motoneurons are every bit as complex as those of other neurons. As a result, the continuing study of the input-output properties of motoneurons is likely to contribute not only to our understanding of motor control but also will provide insights into the biophysical mechanisms of neuronal information processing elsewhere in nervous system.

In the following review, we consider the relationship between the biophysical properties of motoneurons and the means by which they transduce their synaptic inputs into modulations in firing rate. We will first present the classical view of the functional properties of motoneurons, based primarily on the results of experimental and theoretical studies from the early 1950's to the early 1980's. This will be followed by a consideration of how our understanding of motoneuron function has evolved as a result of more recent studies of the electrophysiological properties of motoneurons. Finally, we will consider three specific aspects of input-output transduction that are not yet well understood: (1) the effects of concurrently activated synapses and of voltage-and calcium-dependent dendritic conductances on the transfer of synaptic current to the soma; (2) the mechanisms underlying the decline in firing rate following the onset of a constant stimulus (spike frequency adaptation), (3) and the relation between the magnitude and time course of synaptic potentials and their effects on firing probability.

2
Basic Features of Motoneuron Input-Output Functions

2.1
Variations in Motoneuron Excitability

A number of early studies of single motor unit activity in animals and humans revealed that as the strength of muscle contraction is increased motor units are brought into activity (recruited) in a relatively orderly fashion (rev. in Binder et al. 1996; Cope and Sokoloff 1999). In 1957, Henneman reported that the recruitment order of motoneurons was correlated with the size of the action potential recorded extracellularly from small ventral root filaments (Henneman 1957). Under these conditions, it was assumed that extracellular spike amplitude was a function of axon diameter, which in turn was thought to be correlated with cell body size. Thus, Henneman proposed that motoneurons were recruited in order of increasing cell body size.

After the formulation of this "size principle" (Henneman et al. 1965), there were a number of studies that addressed the contribution that the anatomical properties of motoneurons make to their susceptibility to recruitment. It was confirmed that soma size is positively correlated with axonal conduction velocity (Cullheim 1978; Kernell and Zwaagstra 1981), which is in turn proportional to axon diameter (Cullheim 1978). Total dendritic surface area, either directly measured or estimated from dendritic stem diameter was also found to be correlated with conduction velocity (Barrett and Crill 1974b; Burke et al. 1982; Kernell and Zwaagstra 1981).

The formulation of the size principle stimulated a great deal of experimental and theoretical work on the relationship between the recruitment order of a motoneuron and the mechanical properties of its muscle unit. The extent to which the consistency of recruitment order depends upon the type of movement performed or the source of afferent input used to activate motoneurons has also been extensively investigated (rev. in Binder et al. 1996; Binder and Mendell 1990; Burke 1981; Cope and Clark 1995; Henneman and Mendell 1981). Burke and colleagues demonstrated that motor units could be subdivided into distinct subtypes based on the histochemical and mechanical properties of their muscle units (slow twitch (S), fast-twitch, fatigue-resistant (FR) and fast-twitch fatigable (FF); Burke et al. 1974), and that motoneurons are generally recruited in order of type: S, FR, FF (Burke 1981). However, there has been considerable controversy over the precision of size-related recruitment and whether or not significant exceptions can occur (Binder et al 1996; Cope and Clark 1996; Cope and Sokoloff 1999). As the present review is concerned primarily with the input-output properties of motoneurons, we will focus only on the biophysical mechanisms of re-cruitment rather than on its precision and functional importance.

2.2
Motoneuron Firing Rate Modulation

Increases in synaptic drive to motoneurons beyond their recruitment threshold lead to repetitive discharge. Motor units fire at rather low rates at their recruitment thresholds and then undergo rapid increases in firing rate as force is increased (reviewed in Burke, 1981; Henneman and Mendell 1981; Binder et al. 1996). There are differences in the reported range of unit dis-charge rates at recruitment threshold (i.e., minimal rates) during steady isometric contractions in humans, with some studies reporting that the minimal discharge rates of motor units increase systematically with re-cruitment threshold (De Luca et al. 1982; Kanosue et al. 1979) and others reporting no systematic differences (Monster and Chan 1977; Tanji and Kato 1973). Rate modulation in hindlimb motor units studied in the decerebrate cat appears to be qualitatively similar to that reported in humans. The in-crease in motor unit firing rates during the stretch reflex is clear (Cordo and Rymer 1982), as it is for most units during various isometric forces gener-ated by the crossed-extension reflex (Powers and Rymer 1988). The minimal rates of discharge tend to increase with recruitment threshold (Powers and Rymer 1988), but maximum rates have not been studied due to the inability to generate controlled maximum forces.

Under steady-state conditions, type S units reach their maximal force output at much lower firing rates than do type F (FR & FF) units (Botterman et al. 1986; Kernell et al. 1983). Firing rates in excess of 20 to 30 imp/s are energetically wasteful for type S units, but are necessary for type F units to reach their maximal forces. Thus, one would expect type S units to have the most restricted range of rate modulation (Kernell 1983). Studies of rate modulation in human motor units (Gydikov and Kosarov 1973; Kanosue et al. 1979; Monster and Chan 1977) do indeed show that the rates of low threshold units tend to saturate as higher-threshold units are recruited and increased their discharge rates. This phenomenon has been referred to as "rate limiting" (Heckman and Binder 1991b;1993a).

2.3
A Simple Model of Motoneuron Input-Output Properties

Early experimental studies of motoneurons led to the development of a relatively straightforward model of motoneuron recruitment and rate modulation. This model is based on electrophysiological measurements made at the motoneuron soma, and relates the output of the motoneuron to the amount of current reaching the soma, either as a result of synaptic current flowing in from dendritic synapses or due to current injected directly from the recording electrode. A motoneuron is recruited when its somatic membrane potential is displaced from its resting value (V_r) to the threshold value for initiating an action potential (V_{thr}). For a given current (I), the change in somatic voltage (ΔV) depends upon the effective input resistance of the cell (R_N) according to Ohm's law:

$$\Delta V = R_N * I \tag{1}$$

As described below, R_N depends on both the morphological and biophysical properties of the motoneuron. An action potential will be produced if ΔV exceeds the difference between the threshold voltage and the resting potential:

$$\Delta V > V_{thr} - V_r . \tag{2}$$

Sustained suprathreshold currents lead to repetitive discharge, and under steady-state conditions, the relation between injected or synaptic current and the discharge rate (f) is linear over a large range of currents (Kernell 1965a):

$$f = f_{min} + [f/I] * (I - I_{thresh}), \tag{3}$$

where I_{thresh} is the minimum current needed to elicit steady repetitive discharge, f_{min} is the minimum firing rate, and f/I is the slope of the frequency-current relation. For the population of motoneurons that innervates a single muscle (i.e., motoneuron pool), there are systematic variations in the values of I_{thresh} and f_{min}, which are inversely correlated with systematic variations in R_N and the duration of the post-spike afterhyperpolarization (AHP), respectively (Kernell 1983).

Figure 1 provides a schematic illustration of these basic motoneuron responses, as well as their contribution to the development of muscle tension. The schematic motoneuron models illustrate the range of variation of response properties within a motoneuron pool, from the small, low-threshold (LT) motoneurons to the large, high-threshold (HT) motoneurons. The low threshold motoneurons have higher input resistances and longer AHPs than the high threshold motoneurons. These differences are associated with a

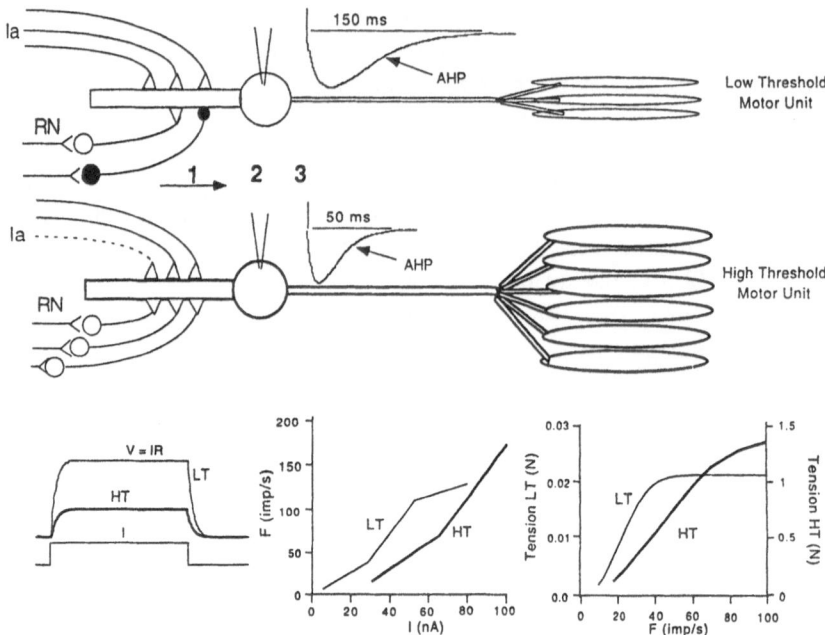

Fig. 1. Schematic representation of the transformation of synaptic input into muscle force. The upper two rows show cartoons of low (top) and high threshold motoneurons (bottom) along with their innervated muscle fibers and two different sets of presynaptic afferent connections: primary (Ia) muscle spindle afferents and spinal interneurons activated by axons originating in the red nucleus (RN). The bold numbers between the two motoneurons indicate the three stages in the synaptic transduction process: (1) the transfer of synaptic current to the soma, (2) the transformation of that current into a change in somatic voltage, and (3) the effect of the change in somatic voltage on the probability of spike initiation in the adjacent initial segment. The diagram also illustrates a number of characteristic differences between low and high threshold motoneurons, including (1) the distribution of effective synapses from different afferent sources, (2) neuronal surface area, (3) axonal diameter, (4) the duration of the postspike afterhyperpolarization (AHP), and (5) the number and diameter of innervated muscle fibers. The bottom row of panels illustrate some of the consequences of the differences in the intrinsic properties of low threshold (LT) and high threshold (HT) motoneurons, including differences in input resistance (left), current-frequency relations (middle) and tension-frequency relations (right). See text for further details

lower threshold for repetitive discharge (I_{thresh}) and a lower minimum firing rate (f_{min}). The differences in repetitive discharge properties are in turn associated with differences in the tension-frequency relations of the innervated muscle fibers: low threshold motoneurons innervate relatively fewer, smaller diameter, slow contracting muscle fibers, resulting in lower tetanic tensions and lower fusion frequencies than muscle units innervated by high threshold motoneurons (Binder et al. 1996; Burke 1981; Henneman and Mendell 1981).

Figure 1 also denotes two examples of the differences in the synaptic inputs to the low- and high-threshold motoneurons within an extensor motoneuron pool (rev. in Binder et al. 1996; Binder 2000). The first example is quantitative: direct excitatory connections from the Ia afferent fibers of muscle spindles have a greater efficacy on the low-threshold motoneurons than on high-threshold motoneurons. The second example is more qualitative: low-threshold motoneurons receive both excitatory and inhibitory oligosynaptic input from the contralateral red nucleus, whereas the high-threshold motoneurons receive predominantly excitatory input.

Under physiological conditions, displacement of the somatic membrane potential toward threshold is achieved by synaptic current originating in the dendrites. Conductance changes at dendritic synapses cause localized synaptic currents whose time course and amplitude depend on the conductance time course and the synaptic driving potential:

$$I_{syn}(t) = G_{syn}(t) * (V_m(t) - V_{eq}),$$ (4)

where $G_{syn}(t)$ is the conductance time course, $V_m(t)$ is the time course of the local membrane potential change, and V_{eq} is the synaptic reversal potential. Only a fraction of the synaptic current reaches the soma as much of it is lost to leak across the interposed membrane. Our understanding of the mechanisms of current transfer of the soma has been greatly aided by the developments of a means of representing the complex dendritic trees of motoneurons as a single equivalent cylinder (Rall 1959).

The use of this equivalent cylinder model (described below) together with a simple description of the threshold and repetitive firing properties of motoneurons permits us to represent the input-output properties of motoneurons in terms of three distinct processes: (1) the transfer of synaptic current to the soma, (2) the resultant change in somatic membrane potential and its modification by somatic conductances and (3) the initiation of an action potential in the initial segment (Fig. 1). This simple formulation relates the variations in the behavior of a population of motoneurons to variations in the values of a number of measurable parameters including R_N, V_{th} and AHP duration.

The development of the Rall model was a crucial step in advancing our understanding of the transfer of current from dendritic synapses to the soma as well as the relation between motoneuron properties and R_N. Thus, we will briefly review the original Rall model here, with an emphasis on its application to the mechanisms of synaptic current flow and motoneuron excitability. We will then review the mechanisms governing the initiation of the action potential, as well as the mechanisms underlying repetitive firing in motoneurons.

2.4
Rall's Equivalent Cylinder Model

Early anatomical studies revealed that motoneurons have extensive dendritic trees, the surface area of which is many times larger than that of the soma (reviewed in Rall 1959). Beginning in the late 1950's, Rall and his colleagues provided a compact description of the geometry and passive electrical properties of motoneurons based on the representation of the soma and dendritic trees of motoneurons as an equivalent cylinder (rev. in Rall 1977; Rall et al. 1992). They demonstrated that under certain conditions, the motoneuron can be represented as a lumped soma consisting of a resistance and capacitance in parallel attached to a uniform diameter cylinder of finite length. This representation provides a reasonable approximation of many of the electrical properties of the cell provided that a number of conditions are met, including isopotentiality of the soma, uniformity of membrane electrical properties over the entire somatodendritic surface, and several geometrical constraints on the branching and termination of dendritic trees (Rall 1959). If, at each branch point, the sum of the diameters of the daughter branches raised to the 3/2 power are equal that of the mother branch raised to the 3/2 power ($\Sigma^{3/2}_d = D^{3/2}_m$), then the entire structure can be represented as an unbranched cylinder with a diameter equal to that of the parent branch. In the equivalent cylinder representation, the surface area of the dendritic tree is preserved, so that the physical length of the equivalent cylinder representation is greater than that of the original dendritic tree. If all of the dendritic trees of a motoneuron can be represented by cylinders of equivalent electrotonic length, then they can be further collapsed into a single equivalent cylinder whose diameter is determined by the same 3/2 power rule governing the collapsing of dendritic branches.

Although the actual geometrical features of motoneurons often depart from the constraints required by the Rall model, these departures turn out to have relatively minor effects on the subthreshold behavior of the motoneuron as seen from an intrasomatic recording electrode. In contrast, nonuniform electrical properties may have more profound effects (see below).

The properties of the Rall or equivalent cylinder model can be specified by four parameters: ρ, the ratio of the input conductance of the dendritic cylinder to that of soma; L, the normalized length of the dendritic cylinder in units of space constant, (λ see below); τ_m, the membrane time constant and R_N, the input resistance of the whole cell, measured by current injected into the soma. The electrical properties of the equivalent cylinder representing the dendrites depend upon both its size and specific membrane properties. A steady voltage applied at one end of an infinite cable will decay to $1/e$ of its value at point that is one space constant (λ) away from the applied voltage. The value of λ depends upon cable properties as follows:

$$\lambda = [(R_m/R_i)(d/4)]^{1/2}, \tag{5}$$

where R_m is the resistance across a unit area of membrane (Ωcm^2), R_i is the volume resistivity of the intracellular medium (Ωcm) and d is the cylinder diameter. The normalized length of the cylinder (L) is simply the actual length divided by λ.

The time course of the voltage response of the cable model to a step of current of magnitude I, can be described in terms of the final steady-state response ($I*R_N$) and an infinite sum of exponential terms:

$$V(t) = I * R_N (1 - \Sigma\, C_i \exp(-t/\tau_i)). \tag{6}$$

In practice, only the two longest time constants (τ_0 and τ_1) can be re-solved from experimental voltage transients. The shorter time constants represent the effects of the voltage wave initiated in the soma traveling to the dendritic tips and being 'reflected' back (Holmes et al. 1992). As a result of these reflections, the entire cable becomes more uniformly polarized at later times, so that the longest time constant, τ_0, simply depends on the specific properties of the membrane. When R_m is uniform, the longest time constant $\tau_0 = \tau_m$, the time constant of an isopotential patch of membrane. Since the membrane time constant is equal to the product of the specific membrane resistance (R_m) and the specific membrane capacitance (C_m), the time constant can be used to estimate R_m, given an assumed value of C_m. Rall (1969) demonstrated that if the ratio of dendritic to somatic conductance (ρ) is fairly large, the electrotonic length of the equivalent cylinder representation of a motoneuron can be estimated from the ratio of the two longest time constants (τ_0/τ_1):

$$L \approx [\pi/(\tau_0/\tau_1 - 1)]^{1/2}. \tag{7}$$

A final useful simplification arises if the soma compartment is combined with the dendritic cable so that the entire neuron is represented as a single equivalent cylinder. In this case, the input conductance of the neuron can be expressed as:

$$G_N = (A_N/R_m)(\tanh L/L), \tag{8}$$

and the input resistance is simply the reciprocal of this expression:

$$R_N = (R_m/A_N)(L/\tanh L), \tag{9}$$

where A_N represents the surface area of the entire neuron and L is now the electrotonic length of the soma plus the dendritic cable, which is virtually identical to the length of the dendritic cable alone. By multiplying both the numerator and denominator of the above expression by C_m, remembering that $\tau_m = R_m C_m$, and rearranging terms it is possible to derive an estimate of the total surface area of the neuron based on purely physiological measurements:

$$A_N = \tau_m L/(C_m R_N \tanh L). \tag{10}$$

Thus, the surface area of the motoneuron can be estimated from measurements of τ_m and R_N, an estimate of L based on the τ_0/τ_1 ratio, and an assumed value of C_m (generally thought to be about 1 μF cm^{-2}, Hille 1992).

2.5
Application of the Rall Model to Studies of Synaptic Currents Measured at the Soma

The development of the Rall model has been crucial to our understanding of how synaptic currents in the dendrites are transferred to the soma and the types of interactions that may occur between concurrently-active synaptic inputs. The effects of activation of a set of synapses that contact different dendritic branches at equal electrotonic lengths from the soma can be modeled by placing the current injection site at a single point on an equivalent cylinder. If the synaptic current transient is brief (i.e., $\ll \tau_m$), then the time course of the synaptic potential recorded at the soma will depend upon the cable properties of the neuron and the electrotonic distance from the soma to the current injection site. Rall (1967) first applied cable theory to the analysis of composite Ia excitatory postsynaptic potentials (Ia EPSPs) in motoneurons, and this approach has since been extended to single fiber Ia EPSPs by Jack and Redman and colleagues (reviewed in Jack et al. 1975, Redman 1976). One important result of these analyses is that if τ_m, ρ and L can be estimated, then the electrotonic location of a synapse can be estimated from the rise time and half width of the PSP recorded in the soma. As illustrated in Fig. 2 (from Jack et al. 1971) PSPS produced by presumed distal synapses have longer rise times and half widths than those produced by proximal synapses. The areas enclosed by the series of overlapping polygons starting at the lower left and proceeding toward the upper right represent the features of somatically recorded PSPs expected to occur following a given conductance time course applied in the soma (lower left) or in successive dendritic compartments that are 0.2 λ in length. The fact that the

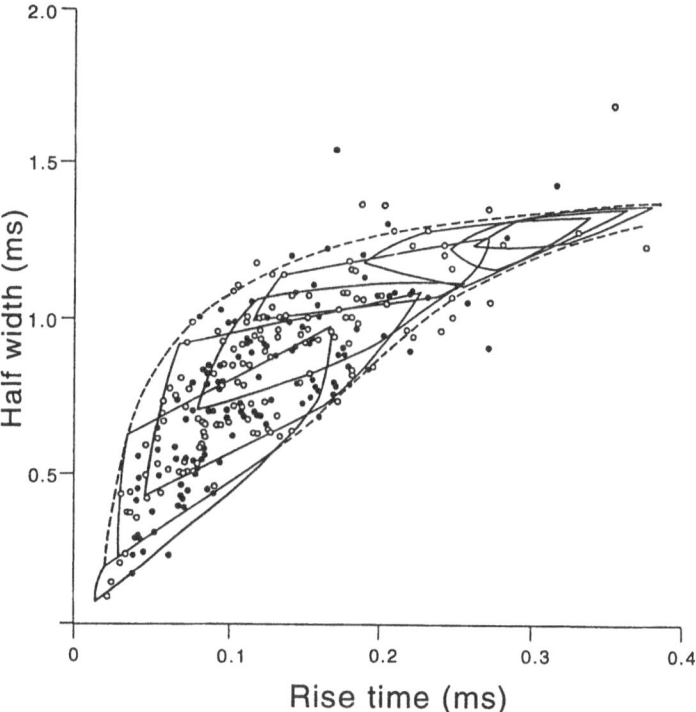

Fig. 2. Half-width versus rise-time for single fiber Ia EPSPs recorded from knee flexor (filled circles) and ankle extensor motoneurons (open circles). Both values have been normalized (divided) by the time constant of the motoneuron. The dashed lines show the theoretical boundaries for the relation between half-width and rise-time based on a synaptic current time course of $\alpha^2 T e^{-\alpha T}$, and an assumed range of values for motoneuron dendritic-to-somatic conductance ratio (4–25), electrotonic length (0.75–1.5), and current time course (α: 12–100). The set of areas bounded by continuous lines, show the theoretical boundaries for particular distances from the soma, starting at 0 for the lower left area and proceeding in steps of 0.2λ to 1.4λ for the upper right area. (Modified from Fig. 8 of Jack et al. 1971)

vast majority of the observed PSPs fall within expected boundaries supports the validity of the Rall model, at least as a first approximation. Further support comes from combined electrophysiological and anatomical measurements, which indicate that identified Ia synaptic boutons generally fall at a spatial location close to that predicted by the Rall model from the characteristics of PSPs recorded at the soma (Redman and Walmsley 1983). Although the time course of PSPs suggests identical conductance changes at all Ia synapses, the peak amplitude of single fiber Ia EPSPs should decline with increasing electrotonic distance, but in fact does not (Iansek and Redman 1973a; Jack et al. 1981), suggesting that for this system, the effective transfer of charge from distal synapses is greater than that from proximal synapses.

The equivalent cylinder representation is not appropriate for calculating the local change in voltage at the subsynaptic site because the local input impedance of a small dendritic branch is much greater than that of an equivalent cylinder compartment representing all of the dendritic branches at a given electrotonic length. If a synaptic conductance of sufficient magnitude to produce a 100 µV PSP for a somatic synapse is placed on a distal dendritic branch, the peak subsynaptic PSPs are expected to be in the range of 20–40 mV (Barrett and Crill 1974a; Rinzel and Rall 1974). This amount of depolarization should significantly reduce the driving force for synaptic current flow ($V_m(t) - V_{eq}$) and consequently reduce the amount of charge injected by distal synapses (Barrett and Crill 1974b). Simultaneous activation of many synapses will further reduce the driving force, and it has been estimated that simultaneous activation of four synapses on the same distal dendritic branch could reduce the local charge injected to about half of the amount that would be injected without a change in driving force (Barrett and Crill 1974a; Binder et al. 1996; Rose and Cushing 1999).

The amount of charge transferred to the soma from a given synaptic site will also be influenced by the effective value of R_m (R'_m) for the dendritic membrane interposed between the synaptic site and the soma. It can be seen from equation 5 that the effective space constant of a cylindrical dendritic compartment (λ) is proportional to the square root of R_m. The effective specific membrane resistivity (R'_m) of a dendritic compartment will be decreased during synaptic activity by an amount that depends upon the resting value of R_m, the conductance change produced by activating a single synapse and the net synaptic activation rate:

$$R'_m = 1/(G_m + G_{syn}), \tag{11}$$

where G_m is the resting specific membrane conductance and G_{syn} is the net conductance change during repetitive activation of a set of synapses. For large values of G_{syn} (i.e. $G_{syn} \gg G_m$), R'_m is approximately equal to the reciprocal of G_{syn}. Based on a unitary conductance change estimated for somatic Ia synapses, an average synaptic density of 20 synaptic boutons/100 μm^2 and a resting R_m value of 2000 Ω cm^2, Barrett (1975) predicted the effects of activating all of the synapses in a compartment on the effective value of R_m. The results of this analysis are indicated by the solid curve in Fig. 3, and it can be seen that a five-fold decrease in R'_m occurs when all of the synapses on a

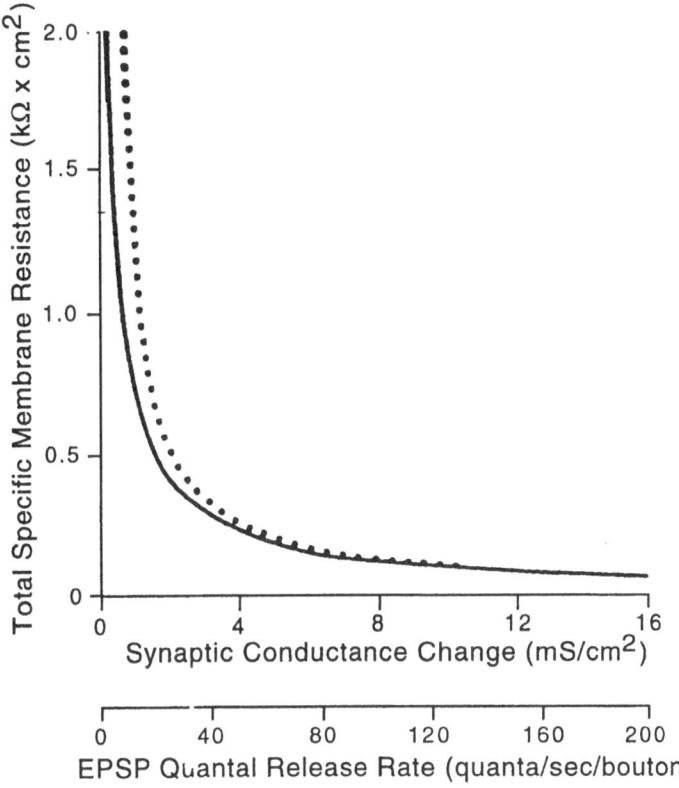

Fig. 3. Total specific membrane resistance as a function of the synaptic conductance change per unit membrane area. The synaptic conductance change is calculated as the product of the density of synaptic boutons (assumed to be 20/100 μm^2), the frequency of synaptic transmitter release (lower x-axis) and the area under a quantal conductance change (assumed to be 4 pS-s). The total specific membrane resistance is the reciprocal of the sum of the resting membrane conductance and the synaptic conductance. Solid line shows the relation for a resting specific membrane resistivity of 2000 Ωcm^2, whereas dotted line shows the relation for an infinite resting specific membrane resistivity. (Modified from Fig. 4 of Barrett 1975)

compartment are activated at about 30 Hz. Although the resting value of R_m used is probably an underestimate (see below), at moderate rates of synaptic activity the value of R'_m is not very sensitive to the resting R_m value, as indicated by the dotted curve, which is based on an infinite resting R_m.

The combined effects of changing the synaptic driving force and R_m suggest that the summed effects of activating two synaptic inputs should be markedly less than the linear sum of their individual effects. Comparisons of the sizes of somatically recorded PSPs in response to combined stimulation of different groups of afferent fibers have indeed shown the expected less-than-linear summation, although PSP amplitudes recorded on combined stimulation are generally 80% or more of the amplitude predicted from the linear sum of the individual PSPs (Burke 1967; Burke et al. 1971; Clements et al. 1986; Powers and Binder 2000; Rall et al. 1967). These relatively small departures from linearity could be due to the fact that the synaptic boutons activated in the cited experiments were rather widely distributed, so that there was relatively little interaction between adjacent synapses. Alternatively, less than linear summation could be masked by the presence of voltage-dependent inward currents in the dendrites which add to the charge transferred to the soma (Clements et al. 1986). The evidence in favor of active dendritic conductances in motoneurons is considered in more detail in section 3.3.3.

2.6
Application of the Rall Model
to Motoneuron Input Resistance Measurements

As described in equation 1, spike initiation occurs when the synaptic current reaching the soma displaces the somatic membrane voltage from its resting value (V_r) to the voltage threshold for spike initiation (V_{th}). This simple formulation implies that the recruitment order of a population of motoneurons depends upon the distribution of synaptic current (I), input resistance (R_N), and the difference between the resting potential and the voltage threshold ($V_{th} - V_r$). The value of input resistance estimated from the equivalent cylinder representation of the motoneuron (equation 9) is inversely proportional to surface area (A_N). Provided that specific membrane resistivity and electrotonic length do not vary systematically among a population of motoneurons, R_N will be strictly determined by cell surface area (size). Under such conditions, if there were also a uniform distribution of synaptic current and no significant variation in resting potential or spike threshold among a population of motoneurons, recruitment would be strictly ordered in terms of increasing cell size.

The Rall model provided the theoretical underpinning necessary to understand the relative contributions of motoneuron morphology and membrane properties to the range of input resistance values measured in motoneurons. Input resistance measured in cat lumbar motoneurons *in vivo* exhibits about a 10-fold range from about 0.4–4.0 MΩ (e.g., Gustafsson and Pinter 1984b; Zengel et al. 1985). Average R_N values are largest in type S, smaller in type FR and smallest in type FF motoneurons (Fleshman et al. 1981; Zengel et al. 1985). Input resistance also decreases with increasing conduction velocity (Kernell 1966; Kernell and Zwaagstra 1981; Barrett and Crill 1974a; Gustafsson and Pinter 1984a), so that given the correlation between conduction velocity, soma size and estimated dendritic area (see above), R_N is inversely proportional to cell size. As expressed in equation 9, R_N is equal to the ratio of specific membrane resistance (R_m) to neuron area (A_N) times a factor dependent upon the electrotonic length of the equivalent cylinder representation (L/tanhL). Estimates of L based on the Rall model have varied from 1–2 with a mean of about 1.5 (Barrett and Crill 1974b, Burke and ten Bruggencate 1971; Jack and Redman 1971; Lux et al. 1970; Nelson and Lux 1970) and have no reported covariance with R_N or motor unit type. On this basis, the ratio L/tanhL will vary between 0.48 (L=2) and 0.76 (L=1) with a mean value of about 0.6. Even if L were to vary systematically with R_N, the range of motoneuron surface areas for motoneurons within a given pool is not sufficient to explain the 10-fold range of input resistances. For a given motoneuron pool, neuronal surface area varies over at most a 4-fold range, regardless of whether A_N is estimated from electrophysiological measurements alone (Gustafsson and Pinter 1984b), from estimates of dendritic surface area based on dendritic stem diameter (Burke et al. 1982; Ulfhake and Kellerth 1984) or from complete anatomical reconstructions (Barrett and Crill 1974b)

Whereas these anatomical arguments suggest that variations in R_m contribute to variations in R_N, this contribution can be demonstrated more directly by plotting measures of R_N against estimates of A_N made in the same cells. Figure 4 (taken from Fig. 6 of Burke et al. 1982 and Fig. 15 of Gustafsson and Pinter 1984b) shows measurements of R_N in cat lumbar motoneurons plotted against estimates of A_N made from dendritic stem diameter measurements (A) or electrophysiological measurements (B). The different symbols indicate measurements from different motoneuron types (either based on direct type-identification (A) or on AHP duration (B)). Superimposed on the data are plots of R_N versus A_N derived from equation 9 using a range of values for R_m and L. Both data sets indicate that the observed variation in R_N is likely to be associated with a 3–5 fold variation in R_m and that R_m is systematically higher in type S than in type F motoneurons. These findings, together with the correlation between motoneuron size and conduction velocity or motoneuron type (see anatomy section above), suggest that systematic variation in both size and specific membrane resistance contribute to orderly motoneuron recruitment.

2.7
Mechanisms of Action Potential Initiation and Variations in Current and Voltage Thresholds Among Motoneurons

Alpha motoneurons were the first cells in the mammalian central nervous system to be studied with intracellular electrodes (Brock et al. 1951, 1952, 1953; Woodbury and Patton 1952) and by the mid-1950s there were a number of published reports on the factors influencing the generation of action potentials in motoneurons by synaptic activation, intracellular current injection and antidromic stimulation (Frank and Fuortes, 1956; Fatt 1957; Coombs et al. 1955; Coombs et al. 1957a,b; cf. Eccles 1957). These early studies also showed that action potentials are followed by a slow, potassium-meditated afterhyperpolarization (AHP) whose duration can vary from about 50 to 200 ms in different hindlimb motoneurons of the cat (cf. Kernell 1965c). Regardless of the mode of excitation, intrasomatic recordings showed that the rising phase of the action potential consists of two components, but that the separation of these two components is most distinct for antidromic action potentials. Eccles and colleagues (Brock et al. 1953) proposed that the biphasic nature of the rising phase of the action potential reflected a two-stage invasion of the motoneuron, i.e., that the first component resulted from an action potential in the initial segment that subse-

---

Fig. 4. Relation between motoneuron morphology and input resistance. A. Scatter plot of measured input resistance and total surface area of HRP-filled triceps surae motoneurons (estimated from the diameters of their stem dendrites). Different symbols represent different motor unit types (see inset). The superimposed hyper bolae are curves described by the equation 7 for input resistance: $R_N = (R_m/A_N)(L/\tanh L)$. (Modified from Fig. 6 of Burke et al. 1982) B. Scatter plot of measured input resistance and total cell capacitance for cat lumbar motoneurons with AHP durations > 100 ms (open circles) and < 55 ms (filled circles). Total capacitance was calculated from the equation $C_{cell} = \tau_m L/(R_N \tanh L)$, and can be used to estimate surface area by assuming a value for specific capacitance of 1 $\mu F/cm^2$. Superimposed hyperbolae have same meaning as in part A. Note that surface area (in A) or cell capacitance (in B) alone is not sufficient to account for the variation in cell input resistance and that variations in membrane resistance (Rm) are required to account for the observed differences in input resistance among type FF, FR and S motoneurons. (Modified from Fig. 15 of Gustafsson and Pinter 1984)

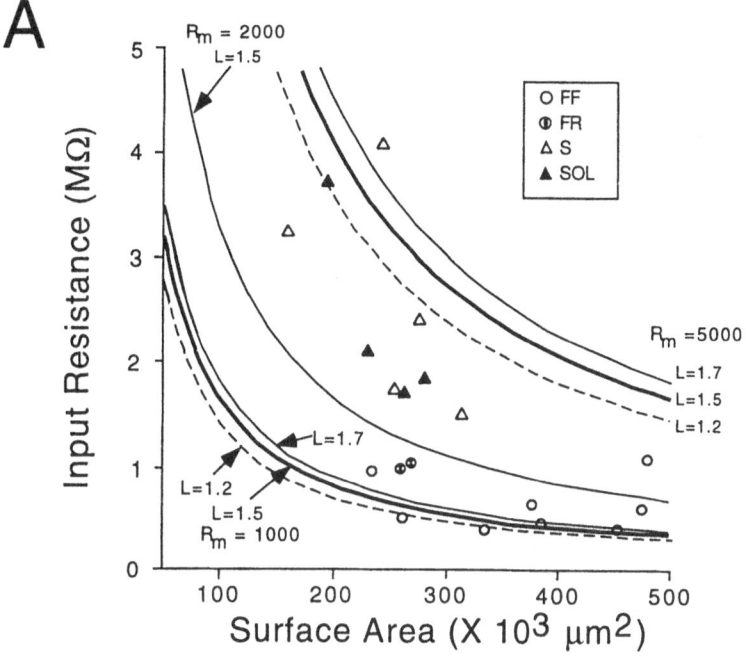

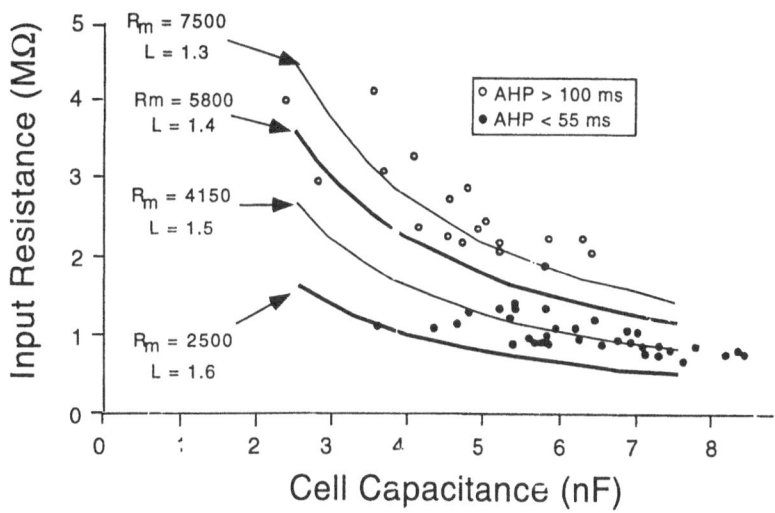

quently elicited an action potential in the soma and perhaps the proximal dendrites. The first component (A or IS) spike can be elicited in the absence of the second component (B or SD) spike by hyperpolarizing the soma or by eliciting two spikes in quick succession (Brock et al. 1953; Coombs et al, 1957b; Fuortes et al. 1957). Association of the first component with the initial segment (or at least some other location on the axon side of the soma) is supported by the fact that this first component alone can give rise to action potentials that can be recorded in the motor axon (Coombs et al. 1957a) and can collide with antidromic action potentials originating from the motor axon (Fuortes et al. 1957).

The evidence of action potential initiation in the initial segment, together with the close electrical coupling between the soma and initial segment (Richter et al. 1974) justifies the view of the soma as a summing junction that integrates current flowing in from the dendrites into a change in somatic membrane potential that will in turn determine the probability of spike initiation in the adjacent initial segment. As indicated earlier (equations 1 and 2), under steady-state conditions the amount of current needed to produce an action potential (i.e., the rheobase current or I_{rh}) depends both on the effective input resistance of the cell (R_N) and the voltage threshold for spike initiation (V_{thr}). Rheobase values exhibit about a ten-fold range across the pool of motoneurons innervating a single muscle (e.g., Fleshman et al. 1981; Gustafsson and Pinter 1984a) and are strongly correlated with input conductance (i.e., $1/R_N$), suggesting that variations in rheobase are largely the result of variations in R_N. Although there is a much smaller range of variation of voltage thresholds (V_{thr}; Gustafsson and Pinter 1984a; Pinter et al. 1983), there is some tendency for V_{thr} to be lowest in low-rheobase, high-resistance motoneurons (Gustafsson and Pinter 1984a). In addition, the value of V_{thr} is generally greater than that predicted by the product of I_{rh} and R_N, suggesting that the simple model of spike initiation expressed in equations 1 and 2 is not strictly valid. This point will be addressed below.

2.8
Frequency-Current Relations in Motoneurons

A series of studies by Granit and Kernell and colleagues in the 1960's described the response of rat and cat motoneurons to long-lasting suprathreshold injected currents and superimposed synaptic currents. Motoneurons respond to a suprathreshold injected current step with a series of action potentials whose instantaneous frequency is a function of both time and current intensity. Firing rate decreases as a function of time after the onset of the current step, a process called spike-frequency adaptation (rev in

Powers et al. 1999). The largest decrease in rate takes place in the first few interspike intervals (Kernell 1965a,b) and a near steady-state frequency is reached by 0.5–1 s after current onset (see however, below). If current injection is prolonged beyond 1 s, an even more gradual decline in frequency takes place (Granit et al. 1963; Kernell 1965a). In cat lumbar motoneurons, the relation between the 'steady-state' discharge rate (i.e., that measured at 0.5–1 s) and the magnitude of injected current (f-I relation) can be described by one or two linear segments, with the second segment (secondary range) having a slope 2–6 times steeper than that of the first segment (primary range; Kernell 1965b; see also Baldissera and Gustafsson, 1974a,b; Schwindt and Calvin, 1972, 1973; Schwindt and Crill 1982; cf. Fig. 1). Further increments in current may reveal a third segment (tertiary range) that generally has a lower slope than that of the secondary range (Schwindt 1973). The f-I relations for the initial interspike interval can also be described by two or three line segments, but the slopes of the segments are much steeper than those describing the steady-state f-I relation. The initial rate spike-frequency adaptation over the first few interspike intervals is characterized by a progressive fall in the steepness of the f-I relations until they approach the 'steady-state' relation.

Kernell (Kernell 1965c) reported that certain characteristics of a motoneuron's f-I relation were related to the time course of the AHP that follows a single spike. The minimum steady firing rate is approximately equal to the reciprocal of AHP duration (Kernell 1965c). The steady-state frequencies attained at the end of the primary and secondary ranges of firing are also correlated with the reciprocal of AHP duration (Kernell 1965c), but the primary and secondary range f-I slopes do not vary systematically with AHP duration (Kernell 1979). The relation between AHP time course and f-I characteristics can be explained on the basis of a simple threshold-crossing model in which the conductance underlying the AHP is a simple time-dependent process that reaches a peak value during the spike and then decays exponentially (Kernell 1968). Under conditions of steady-state current injection, a spike occurs whenever the depolarizing injected current overcomes the AHP current so that the membrane potential crosses a fixed voltage threshold (V_{thr}). This model not only provides a simple explanation for the correlation between the minimum steady discharge rate and the reciprocal of AHP duration, but also generates steady-state f-I curves that roughly approximate the experimental data (Kernell 1968). Modifications of this model that incorporate a more complicated time course for the AHP conductance (Baldissera and Gustafsson 1974a) are able to reproduce both first-interval and steady-state f-I relations even more closely (Baldissera and Gustafsson 1974b, c).

If it is assumed that the conductances responsible for both the action potential and the AHP are located on or near the soma, then current flowing in from dendritic synapses would be expected to have the same effect on firing rate as an equivalent amount of injected current. In most cases, repetitive activation of different synaptic inputs simply shifts f-I curves along the current axis without changing their slopes, suggesting that synaptic and injected currents do indeed act equivalently in the control of repetitive discharge (Granit et al. 1966; Kernell 1970, Schwindt and Calvin 1973a,b). This suggests that the steady-state effects of activating a given set of synapses on motoneuron firing rate can be predicted based on the amount of current reaching the soma and the motoneuron's f-I relation (equation 3). However, in some cases synaptic inputs were found to produce a change in the slope of the frequency-current (f-I) relation, indicating nonequivalence (Kernell 1966; Shapovalov 1972). Since the vast majority of synaptic contacts are on motoneuron dendrites (Brannstrom 1993), voltage-dependent dendritic conductances will influence synaptic currents more than somatically-injected currents. The presence of these conductances makes the delivery of synaptic current to the soma dependent on the mean level of dendritic depolarization. The evidence for the presence of active dendritic conductances and their effects on current delivery to the soma and on f-I relations will be considered later (section 4.1).

2.9
Synaptic Potentials Recorded in Motoneurons

The cat spinal motoneuron played a key role in the elucidation of the mechanisms underlying chemical synaptic transmission in the central nervous system (Eccles 1964). The magnitude and distribution of synaptic inputs to motoneurons from many different types of presynaptic fibers have been studied by recording synaptic potentials produced by both physiological and transient electrical stimulation (reviewed in Binder et al. 1996). Three distinct patterns of distribution of the synaptic potentials within the constituents of a motoneuron pool have been described. When measured at the motoneuron's resting potential, the synaptic potentials are either: (1) proportional to the input resistance of the motoneurons (e.g., Ia EPSPs, Ia IPSPs, Renshaw IPSPs); (2) independent of input resistance and of roughly equal magnitude in all cells (e.g. Deiter's nucleus EPSPs, spindle group II afferents); or (3) predominantly inhibitory to low threshold motoneurons and predominantly excitatory to high threshold motoneurons (rubrospinal PSPs, pyramidal tract PSPs).

The contacts between Ia afferent fibers and motoneurons constitutes the most extensively studied of all synaptic input systems (reviewed in Munson 1990). Individual Ia afferent fibers produce monosynaptic EPSPs in from 80–100% of homonymous hindlimb motoneurons and in about 60% of synergist motoneurons. The amplitude of the Ia EPSPs in these cells generally varies directly with motoneuron input resistance. A lower degree of connectivity is observed on motoneurons innervating muscles that do not act in strict mechanical synergy (Eccles et al. 1957).

Comparisons of reciprocal Ia IPSPs with homonymous Ia EPSPs generated in cat hindlimb motoneurons demonstrate a strong correlation between these two synaptic input systems (Burke et al. 1976). As is the case for Ia EPSPs, the amplitude of composite reciprocal Ia IPSPs varies systematically in type-identified motoneurons: Ia IPSPs in type S are larger than those in type FR, which in turn are larger than those in type FF motoneurons (Burke et al. 1976; Dum and Kennedy 1980).

The amplitude of recurrent IPSPs produced in cat hindlimb motoneurons by the activation of Renshaw cells also appears to scale with input resistance (Eccles et al. 1961; Granit and Renkin 1961). Further, it has been shown that the Renshaw IPSP varies systematically with motor unit type in the order FF<FR<S (Friedman et al. 1981).

Group II afferent fibers that innervate muscle spindles also make monosynaptic connections with homonymous and synergist hindlimb motoneurons (Munson et al. 1980; Stauffer et al. 1976). The group II fibers are thought to make fewer contacts with the motoneurons in that the average projection frequency and average single fiber EPSP amplitude are much lower than the comparable values for Ia afferent fibers. Munson and colleagues (Munson et al. 1982) estimated that activating all the spindle group II afferents at 100 imp/s would generate a steady-state EPSP of about 0.3 mV in a homonymous MG motoneuron. Activating the group Ia afferent fibers in a similar fashion would generate an average steady-state EPSP that is more than 8-fold larger (2.6 mV; Harrison and Taylor 1981). Moreover, Munson and colleagues (Munson et al. 1982) reported that the amplitude of the single group II fiber EPSPs did not vary systematically with motoneuron input resistance, suggesting that the high threshold motoneurons with low input resistances may receive more contacts from the group II afferent fibers than the low threshold cells do. Activation of Group II muscle afferent fibers also produces disynaptic excitatory and inhibitory inputs to cat lumbar motoneurons mediated by interneurons in the intermediate zone/ventral horn (laminae VI–VIII; reviewed in Jankowska 1992). However, neither the magnitude nor the pattern of distribution of these effects within individual motoneuron pools has been evaluated as yet.

In cats, the lateral vestibulospinal or Deiter's nucleus (DN) neurons project primarily to neck and forelimb motoneurons, but a small percentage descend to the lumbosacral area (Fukushima et al. 1979; Shinoda et al. 1988a, b). Mono- and disynaptic DN EPSPs have been recorded in cat triceps motoneurons (Grillner et al. 1970). Although the monosynaptic EPSPs were all quite small (< 2.2 mV), they displayed a wide range of amplitude, that was not correlated with the duration of the AHP. Similarly, the amplitude of monosynaptic EPSPs produced by stimulation of DN axons within the ipsilateral ventral funiculus was not related to motor unit type (Burke et al. 1976).

The rubrospinal system is one of several oligosynaptic pathways in the cat that generate qualitatively different synaptic potentials within hindlimb motoneuron pools (Burke et al. 1970; see also Hongo et al. 1969; Endo et al. 1975; Powers et al. 1993). Both excitatory and inhibitory last-order interneurons are activated by the contralateral red nucleus, leading to short-latency EPSPs and IPSPs, respectively. Low-threshold motoneurons receive predominantly inhibitory input, whereas the high-threshold motoneurons receive predominantly excitatory input.

Endo and colleagues (1975) demonstrated that stimulation of the contralateral motor cortex in cats generates predominantly excitatory PSPs in medial gastrocnemius motoneurons, whereas the PSPs evoked in soleus motoneurons are predominantly inhibitory. When they stimulated the contralateral red nucleus and recorded in the same motoneurons, the pattern of PSPs they observed was similar in almost all cases. Other input systems that generate predominantly inhibitory synaptic potentials in low threshold motoneurons and predominantly excitatory synaptic potentials in high threshold motoneurons are low threshold cutaneous afferents and group Ib afferents (Powers and Binder 1985a,b; rev. in Binder et al. 1996; Binder 2000).

2.10
Summary of Hypotheses of Motoneuron Function Circa 1980

The prevailing view of motoneuron function in the early and mid-1980s was based on a number of simplifying assumptions regarding the structure of motoneuron dendrites, the uniformity of membrane properties and the distribution of synaptic inputs to different members of a motoneuron pool. Although problems and limitations with these assumptions were recognized early on, it is only in the past 20 years that sufficient data have been accumulated to develop a more complete model of motoneuron function. Before presenting this more contemporary view, it is useful to summarize the

precedent view of the relation between the biophysical properties of motoneurons and their recruitment and rate modulation by synaptic inputs.

By the mid 1980's, the ubiquitous phenomenon of size-ordered motoneuron recruitment was reasonably well understood. The representation of the
dendritic tree as an equivalent cylinder had revealed the relationship between dendritic surface area and motoneuron input resistance (equation 9).
The higher specific membrane resistivity (R_m) of smaller motoneurons tends
to emphasize recruitment by size. If the same amount of synaptic current
were applied to every member of the motoneuron pool, a larger voltage
change would arise in smaller, high-resistance motoneurons, according to
equation 1. This equation assumes that the input resistance of the neuron is
independent of membrane potential in the range between resting potential
and voltage threshold. According to equation 2, the larger voltage change in
small motoneurons would make it more likely that their membrane potential would exceed the threshold value for spike initiation (V_{thr}), particularly
since V_{thr} tends to be lowest in small motoneurons as discussed earlier.

Increases in synaptic current above the recruitment level lead to repetitive discharge at a rate that over the 'primary' range of discharge is proportional to current magnitude (equation 3). The limits of the discharge in this
primary range as well as changes in slope at higher levels of injected or synaptic current can be related to the time course of the AHP conductance.

3
Contemporary Views of Motoneuron Function

In the following sections we will consider what revisions of the simplified
view of motoneuron recruitment and rate modulation presented above are
required to accommodate information acquired over the past 20 years.
There are new data on the physiological basis of each step in the transformation of synaptic input into motoneuron firing rate. There have been a
number of quantitative anatomical studies that have characterized the distribution of synaptic inputs across the somatodendritic surface. In addition,
there is now detailed information on the distribution of several sources of
synaptic input to the population of motoneurons within a single pool. More
accurate descriptions of dendritic morphology together with increasing
evidence for the presence of active dendritic conductances have led to new
insights into the way in which dendritic properties govern the transfer of
synaptic current to the soma. The biophysical mechanisms governing the
relation between the current delivered to the soma and the frequency of
repetitive discharges are also better understood. Finally, it is clear that each

step in this input-output transduction process is under neuromodulatory control.

In the following sections, we will review new developments in the physiological basis of motoneuron input-output properties. We will first review new information on the organization of synaptic inputs to motoneurons, then our present understanding of how intrinsic motoneuron properties control the transduction of synaptic input into motoneuron discharge rate, and finally the neuromodulatory control of motoneuron input-output properties.

3.1
Synaptic Inputs to Motoneurons: Somatodendritic Distribution

Approximately half of the total surface area of the motoneuron is covered by synaptic boutons (Brannstrom 1993; Bras et al. 1987; Conradi 1969; Conradi et al. 1979; Kellerth et al. 1979, 1983; Rose and Neuber 1991). Although both the soma and axon hillock are covered with synaptic boutons, the vast majority of the contacts made by presynaptic neurons are located on the dendrites of motoneurons, since they collectively account for 93–99% of the total surface area (Bras et al.1987; Cullheim et al. 1987b; Rose et al. 1985). The percentage of membrane area covered by synaptic boutons is not uniform across the motoneuron surface, but appears to be highest on the proximal dendrites, slightly lower on the motoneuron soma and lowest on the distal dendrites (Brannstrom 1993; Conradi 1969; Conradi et al. 1979; Kellerth et al. 1979,1983; Lagerback and Ulfhake 1987; Rose and Neuber 1991); but see (Bras et al. 1987). Given a density of roughly 10 boutons/100 μm^2 of motoneuron surface area (Brannstrom 1993, Bras et al. 1987, Conradi 1969, Conradi et al. 1979, Kellerth, Berthold, and Conradi 1979, Kellerth, Conradi, and Berthold 1983, Rose and Neuber 1991) and an estimated total surface of around 500,000 μm^2 (e.g. Cullheim et al. 1987a), individual motoneurons are likely to be contacted by approximately 50,000 synaptic boutons (Ulfhake and Cullheim 1988). If we assume that the three groups of presynaptic neurons studied to date (Ia afferent fibers, Renshaw cell and Ia-inhibitory interneurons; see below) are representative of "typical" synaptic input systems, providing an average of from 3 to 10 synaptic contacts on a motoneuron, then the output of each motoneuron can be influenced by the activity of about 10,000 presynaptic neurons.

Several distinct types of synaptic boutons have been identified on motoneurons based on their ultrastructural characteristics. The most common types, S (spherical synaptic vesicles) and F (flattened vesicles), are thought to represent excitatory and inhibitory boutons, respectively (see Brannstrom

1993 for references). The majority of synaptic boutons are inhibitory (F) (Brannstrom 1993; Conradi 1969; Conradi et al. 1979; Kellerth et al. 1979, 1983), but the relative ratio of F to S boutons appears to be highest at the soma and proximal dendrites, while F and S boutons may occur with equal or nearly equal frequency on the distal dendrites (Brannstrom 1993).

Several different classes of axon terminals have been identified on the dorsal neck motoneurons of the cat based on the shape, density, and distribution of their synaptic vesicles (Rose and Neuber 1991). The proportion of the different classes of axon terminals on the soma and proximal dendrites was similar, but there were prominent changes in the frequency of most of the classes of axon terminals on distal dendrites. Sixty percent of axon terminals on the somata and proximal dendrites contained clumps of either spherical or pleomorphic vesicles, whereas only 11% of the axon terminals on distal dendrites had these characteristics. In contrast 40% of the axon terminals on distal dendrites contained a dense collection of uniformly distributed spherical vesicles, which accounted for only 5–7% of the axon terminals on the somata and proximal dendrites. These results suggest that afferents from different sources may preferentially contact proximal or distal regions of the dendritic trees of these cells.

The limited data available comparing synaptic covering in motoneurons innervating different types of muscle units (i.e., type S, FR, FI, or FF; cf. sect. 2.1) suggest that the total synaptic covering of the soma and proximal dendrites is similar across types (Brannstrom 1993; Conradi et al. 1979, Kellerth et al. 1979,1983), although slow motoneurons have a higher proportion of F type boutons than do fast motoneurons (Brannstrom 1993). Synaptic covering of the distal dendrites is higher in slow than in fast motoneurons, and while the ratio of F to S boutons decreases somatofugally in fast motoneurons, it remains relatively constant in slow motoneurons (Brannstrom 1993).

The projections of three types of presynaptic neurons to motoneurons have been studied in detail: Ia afferent fibers, Renshaw interneurons and Ia-inhibitory interneurons. Single Ia afferent axons provide from 4–32 synaptic boutons onto a single motoneuron (Burke et al. 1988; Burke and Glenn 1996; Burke et al. 1979; Redman and Walmsley 1983b). Although Ia contacts are widely distributed across the surface of motoneurons, there are relatively fewer contacts on or close to the soma (within 200 µm) or on the most distal parts of the dendrites (greater than 1000 µm from the soma; Burke and Glenn 1996). Ultrastructural studies of Ia synaptic boutons suggest that they are all of the S-type (Conradi et al. 1983; Fyffe and Light 1984; Pierce and Mendell 1993), but that their size, vesicle content and number of active zones can vary over a wide range (Pierce and Mendell 1993). It is possible that differences in Ia bouton morphology (Pierce and Mendell 1993) could contribute to a proposed somatofugal increase in the efficacy of single Ia boutons (Jack et al. 1981).

A single Renshaw cell makes on average 3 (range: 1–9) synaptic contacts with a motoneuron, and all of the contacts are on the motoneuron dendrites within a few hundred microns of the soma (Fyffe 1991). Individual Ia inhibitory interneurons give rise to between four and eleven synaptic boutons on a motoneuron, and these boutons are located primarily on the soma (Fyffe, personal communication).

Immunohistochemical techniques have been used to describe the synaptic contacts of boutons containing different specific neurotransmitter or neuromodulatory substances. As expected from the relative proportions of type S and type F boutons, a relatively high proportion of the boutons on the soma and proximal dendrites are associated with the identified inhibitory neurotransmitters, g- aminobutyric acid (GABA) and glycine (Destombes et al. 1992; Holstege 1991; Ramirez-Leon and Ulfhake 1993), while receptors for the excitatory neurotransmitter glutamate may be concentrated more distally (Dememes and Raymond 1982, Zieglgansberger and Champagnat 1979). However, there is probably a large degree of overlap in the spatial distributions of boutons containing excitatory or inhibitory neurotransmitters, since glycine immunoreactive terminals have been recently observed on the distal dendritic branches of motoneurons (Fyffe et al. 1993) and there is also evidence to suggest that spatial distribution of the NMDA (N-methyl-D-aspartate) subtype of glutamate receptor is quite broad (Durand et al. 1987).

There is now clear evidence that motoneurons are contacted by boutons containing a number of different neuromodulatory substances, including serotonin (5-HT), substance P, thyrotropin releasing hormone (TRH) and norepinephrine (NE). The first three of these substances are highly co-localized in synaptic boutons in the ventral horn (Arvidsson et al. 1990). An average of about 1,570 5-HT containing boutons contact the dendritic surface of cat lumbar motoneurons (Alvarez et al. 1998) while the somata have an average of 52 contacts. The spatial distribution of boutons appears to be matched to the available dendritic surface membrane, resulting in a uniform density of contacts ($<1/100$ μm^2) (Alvarez et al. 1998). A predominance of dendritic contacts has also been observed for both serotonergic and noradrenergic projections to rat motoneurons (Rajaofetra et al. 1992; Saha et al. 1991).

3.2
Synaptic Inputs to Motoneurons:
Distribution Within a Motoneuron Pool

Until recently, the relative magnitude and distribution of a specific synaptic input to the constituents of a motoneuron pool have been inferred only from the analysis of the amplitude and time course of synaptic potentials (PSPs) measured at the motoneurons' resting potential (see section 2.9 and Henneman and Mendell 1981; Burke, 1981; Munson, 1990; Binder et al. 1996). PSP characteristics are influenced by cell properties as well as by the amount of synaptic current reaching the soma. Under steady-state conditions, the change in voltage recorded at the soma will be given by the product of the current reaching the soma and the motoneuron's input resistance (equation 1). Some information on the distribution of synaptic current can thus be obtained by comparing R_N and PSP measurements in the same cells. For example, for an input that is distributed uniformly across the motoneuron pool, PSP amplitudes should be positively correlated with R_N, and the range of variation of these two variables should be similar (cf. Heckman and Binder 1990). However, estimates of synaptic current obtained in this fashion are indirect, and we have argued (Binder et al. 1996) that analyses of the distribution of synaptic input should be based on direct measurement of the synaptic current at the somatic recording electrode. This measurement has been called the "effective synaptic current" (I_N; Heckman and Binder 1988; previously referred to as "effective somatic current"; Redman 1976). Effective synaptic currents can be readily measured under steady-state conditions using a modified voltage-clamp procedure (Heckman and Binder 1988; Lindsay and Binder 1991; Powers and Binder 1995). This technique not only yields the value of I_N at the motoneuron's resting potential, but further provides an estimate of its somatic voltage dependence, somatic reversal potential, and the conductance change the synaptic input produces at the soma. From these data, one can calculate the effective synaptic current at the motoneuron's threshold for repetitive discharge. This value can then be combined with the slope of the steady-state frequency-current (f/I) relation to predict the effect of the synaptic input on motoneuron firing behavior (Binder et al. 1993, 1996, 1998; Heckman and Binder 1990; Powers and Binder 1995; Powers et al. 1992; Powers et al. 1993).

Thus far, the distribution of effective synaptic currents has been measured for six different input systems to cat lumbar motoneurons (rev. in Binder 2000). Results from these studies are briefly reviewed below and summarized in Fig. 5, which illustrates the range of synaptic currents from different afferent sources measured in cat triceps surae motoneurons as a

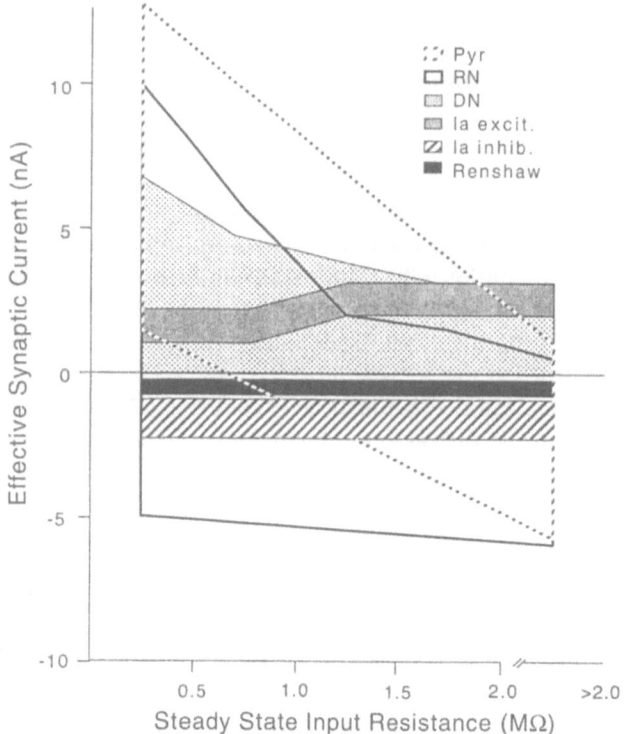

Fig. 5. Graphical representation of the magnitude and distribution of the effective synaptic currents (I_N) from six different input systems measured in cat lumbar motoneurons at rest. The stripped line represents I_N from the contralateral pyramidal tract (Binder et al. 1998); the dark, stippled band represents I_N from homonymous Ia afferent fibers (Heckman and Binder 1988); the stripped band represents I_N from Ia-inhibitory interneurons (Heckman and Binder 1991); the black band represents the I_N from Renshaw interneurons (Lindsay and Binder 1991); the thick lines outline the I_N from contralateral rubrospinal neurons (Powers et al. 1993); and the light stippled band represents I_N from ipsilateral Deiter's nucleus (Westcott et al. 1995). (Modified from Fig. 6 of Binder et al. 1998)

function of the motoneuron input resistance. The effective synaptic currents produced by three general classes of afferent inputs have been measured: (1) those that produce excitatory synaptic currents in all motoneurons (monosynaptic Ia input and input from the lateral vestibular nucleus), (2) those that produce inhibitory currents in all motoneurons (Renshaw and Ia inhibitory interneurons), and (3) those that produce a mixture of excitation and inhibition (rubrospinal and pyramidal inputs).

3.2.1
Monosynaptic Ia-Afferent Input

The observed covariance between Ia EPSPs and motoneuron input resistance has generally been assumed to result from an approximately constant synaptic current applied to cells of varying R_N values (reviewed in Heckman and Binder 1990), in keeping with the original *size principle* of recruitment (Henneman et al. 1965). However, measurements of the underlying effective synaptic currents indicate that this covariance results from systematic variance in both I_N and R_N. Within the cat medial gastrocnemius motoneuron pool, the effective synaptic currents (I_N) generated by homonymous Ia afferents display a wide range of values that covary with R_N and other motoneuron properties: I_N is about twice as large on average in motoneurons with high R_N values as in those with low R_N values (Heckman and Binder 1988; Fig 5). Adding activation of the heteronymous Ia afferent fibers from the synergist lateral gastrocnemius and soleus muscles approximately doubles the magnitude of I_N (mean value MG Ia afferents: 2 nA; mean value triceps surae Ia afferents: 4.2 nA; Westcott 1993).

3.2.2
Lateral Vestibulospinal (Deiter's Nucleus) Input

As previously discussed (section 2.9), stimulation of the ipsilateral Deiter's nucleus (DN) produces mono- and disynaptic EPSPs in cat triceps motoneurons (Grillner et al. 1970). Although the monosynaptic EPSPs are all quite small (< 2.2 mV), they display a wide range of amplitudes that are not correlated with the duration of the AHP. Similarly, the amplitudes of monosynaptic EPSPs produced by stimulation of DN axons within the ipsilateral ventral funiculus are not related to motor unit type (Burke et al. 1976).

Based on the dependence of PSP amplitude on R_N, the data on DN synaptic potentials (Grillner et al. 1970) suggest that the underlying effective synaptic currents should be inversely related to R_N. Measurements of the effective synaptic currents produced by DN stimulation in triceps surae motoneurons of the cat demonstrated that this is indeed the case (Westcott et al. 1995; Fig. 5). The DN effective synaptic currents were primarily small and depolarizing in both medial gastrocnemius (MG) and lateral gastrocnemius-soleus (LGS) motoneurons (mean = 2.5 nA). The DN input tended to be larger in putative type F (low resistance, high rheobase) motoneurons. The amplitudes of the steady-state DN synaptic potentials were similar to those previously reported for transient PSPs (Burke et al. 1976; Grillner et al. 1970). The DN PSPs showed no correlation to rheobase, resting membrane potential or either steady state or maximal input resistance of the cell. The DN synaptic input caused no significant change in motoneuron input resistance.

3.2.3
Reciprocal Ia-Inhibitory Input

Comparisons of reciprocal Ia IPSPs with homonymous Ia EPSPs generated in cat hindlimb motoneurons demonstrate a strong correlation between these two synaptic input systems (Burke et al. 1976). As is the case for Ia EPSPs, the amplitudes of composite reciprocal Ia IPSPs vary systematically in type-identified motoneurons: Ia IPSPs in type S are larger than those in type FR, which in turn are larger than those in type FF motoneurons (Burke et al. 1976; Dum and Kennedy 1980). However, there is a steeper gradient of synaptic strength across the motoneuron pool for Ia excitation than for Ia inhibition (Stein and Bertoldi 1981), and thus, one would expect that differences in the effective synaptic currents underlying Ia inhibition within a motoneuron pool would be considerably less than those found for the excitatory Ia input. This appears to be the case, as the amplitudes of the inhibitory effective synaptic currents generated in antagonist motoneurons by the activation of MG Ia afferent fibers were not correlated with the intrinsic motoneuron properties or with putative motor unit type (Heckman and Binder 1991a; Fig 5). The average value of the steady-state Ia inhibitory synaptic current in the putative type F motoneurons was 1.60 ± 0.64 nA, whereas that for putative type S motoneurons was 1.65 ± 0.71 nA.

Since the synaptic boutons of Ia inhibitory interneurons lie predominantly on the soma of motoneurons (R.E. Fyffe, personal communication), as expected, they generate substantial changes in motoneuron input resistance. As was the case for the effective synaptic currents, there was no systematic relationship between the steady-state change in conductance and the electrical properties of the motoneuron (Heckman and Binder 1991a).

3.2.4
Recurrent Inhibitory (Renshaw Cell) Input

The role of recurrent inhibition in the control of motoneuron discharge remains uncertain. One long-held hypothesis is that recurrent inhibition acts to suppress the activity of low threshold motoneurons when higher threshold motoneurons are recruited (Eccles et al. 1961; Granit and Renkin 1961; Friedman et al. 1981). Support for this hypothesis came from data demonstrating that recurrent IPSP amplitude varies systematically with

motor unit type in the order FF<FR<S (Friedman et al. 1981) and the anatomical finding that the number of end branches and "swellings" (presumed
synaptic boutons) of recurrent axon collaterals of type F motoneurons are
greater than those of type S motoneurons (Cullheim and Kellerth 1978). An
alternative hypothesis was put forth by Hultborn and colleagues (Hultborn
et al. 1979) who proposed that recurrent inhibition (and its modulation by
descending systems) provides a variable-gain regulator of the overall output
of a motoneuron pool. This hypothesis is predicated on a uniform distribution of recurrent inhibition within a motoneuron pool.

The distribution of effective synaptic currents underlying recurrent inhibition in triceps surae motoneurons does indeed appear to be uniform
(Lindsay and Binder 1991; Fig. 5). The effective synaptic currents appeared
to be entirely independent of all intrinsic motoneuron properties measured,
although the amplitudes of the steady-state recurrent IPSPs were correlated
with motoneuron input resistance, rheobase and the half-decay time of the
afterhyperpolarization ($AHP_{t1/2}$), as has been previously reported for transient recurrent IPSPs (Friedman et al. 1981; Hultborn et al. 1988). The mean
effective synaptic current produced by Renshaw cells measured at rest was
only 0.4 nA, considerably smaller than comparable values derived for other
input systems (cf. Fig. 5). Although the Renshaw input was activated in this
study by stimulating the synergist motor axons at 100 Hz, which produces
steady-state IPSPs that are about 50% of the maximum value (Lindsay,
Heckman and Binder, unpublished data), the maximal effective synaptic
current at rest is still likely to be < 1 nA on average.

Motoneurons are more likely to receive recurrent inhibition while they
are firing repetitively than when they are quiescent, and several hypotheses
of the role of recurrent inhibition in motor control emphasize the possible
effects of recurrent inhibition on firing frequency (e.g., Hultborn et al. 1979).
The mean effective synaptic current from recurrent inhibition from synergist motor axons calculated at threshold was still only 1.25 nA, and the distribution within the pool was no different from that measured at resting
potential (Lindsay and Binder 1991). These results suggest that the effect of
recurrent inhibition on the firing frequencies of motoneurons are uniform
and quite modest. Using an average value for the slope of the motoneuron
firing rate-injected current relationship (f/I curve) of 1.5 imp/s/nA (Kernell
1979), the average change in firing frequency produced by maximal recurrent inhibition (2.5 nA) should be less than 4 impulses/s. Since the slope of
the f/I curve does not appear to covary with other motor unit properties
(Kernell 1979), the change in firing frequency should not vary systematically
within a motoneuron pool. The maximum effective synaptic current calculated at threshold (4.6 nA) would only decrease firing frequency by about 7

impulses/s, a value similar to that observed in the experiments of Granit and Renkin (1961) in which whole ventral roots were stimulated to activate the Renshaw cells.

These measurements of the effective synaptic currents produced by Renshaw cell activation (Lindsay and Binder 1991) were incorporated into computer simulations aimed at assessing the effects of recurrent inhibition of motoneuron behavior (Maltenfort et al. 1998). Although the modest currents generated by recurrent inhibition were insufficient to alter the input-output gain of motoneurons in the simulation, they did appear to decorrelate the discharge of motoneurons within a motor nucleus.

3.2.5
Rubrospinal Input

The effective synaptic currents produced by stimulating the hindlimb projection area of the contralateral magnocellular red nucleus have been measured in cat triceps surae motoneurons (Powers et al. 1993). At the resting potential, the distribution of effective synaptic currents from the red nucleus was qualitatively similar to the distribution of synaptic potentials: 86% of the putative type F motoneurons received a net depolarizing effective synaptic current from the red nucleus stimulation, whereas only 38% of the putative type S units did so. However, at threshold the distribution was markedly altered. Inhibition continued to predominate in the type S cells, but among the type F cells, half received net excitatory effective synaptic currents and half received net inhibitory effective synaptic currents. Other surprising features of these data are the enormous range of effective synaptic currents observed in different cells and the fact that they are often three times larger than comparable inputs from segmental pathways (Fig. 5). Activation of red nucleus synaptic input reduced motoneuron input resistance by 40%, on average (Powers et al. 1993). The effect on input resistance was most pronounced in those motoneurons that received hyperpolarizing effective synaptic currents.

Comparison of the red nucleus input to that of Deiter's nucleus shows that the range of effective synaptic currents are similar (Fig. 5). Also for both descending systems, there are larger depolarizing currents generated in the putative type F motoneurons. However, the mean depolarizing red nucleus I_N was twice as large as that from DN, and the difference in distribution of red nucleus I_N between putative type F and S motoneurons was more substantial. These data thus indicate that the red nucleus input may provide a powerful source of synaptic drive to some high threshold motoneurons, while concurrently inhibiting low threshold cells. Thus, this input system can potentially alter the gain of the input-output function of the motoneuron pool, change the hierarchy of recruitment thresholds within the pool, and mediate rate limiting of discharge in low-threshold motoneurons (Burke et al. 1970; Burke 1981; Heckman and Binder 1990, 1991b; 1999a,b; Binder et al. 1993; Heckman 1994; Powers et al. 1993).

3.2.6
Pyramidal Tract Input

The effective synaptic currents produced by stimulating the contralateral pyramidal tract have recently been examined in cat triceps surae motoneurons (Binder et al. 1998). The magnitudes and distribution of the effective synaptic currents were quite similar to those observed earlier for the red nucleus (Powers et al. 1993). Measured at rest, the effective synaptic currents from the pyramidal tract were depolarizing in all but one motoneuron (a putative type S cell). However, the currents in the putative type F motoneurons were more than three times larger, on average, than those in the putative type S motoneurons (Fig. 5). At threshold, the distribution was markedly altered. All but one type S cell received a hyperpolarizing current from the pyramidal tract, but among the type F cells, half received net excitatory effective synaptic currents and half received net inhibitory effective synaptic currents.

3.3
Properties of Motoneuron Dendrites

During the past two decades, we have acquired a wealth of new information on the anatomy and electrophysiology of motoneuron dendrites. These new data have dictated substantial revisions in the original Rall model (section 2.4) which have had a major impact on our understanding of how synaptic currents generated in the dendrites are transmitted to the spike-generating conductances in the initial segment and axon of the motoneuron.

3.3.1
Anatomy

Prior to 1980, the most complete anatomical descriptions of motoneurons were based on reconstructions of cat hindlimb motoneurons that were filled with either Procion dye (Barrett and Crill 1974b) or tritiated glycine (Lux et al. 1970). The subsequent use of horseradish peroxidase (HRP) for intracellular labeling generated a more complete filling of fine neuronal processes, leading to a wealth of new anatomical data on the structure of alpha moto-neurons innervating a variety of muscles in both mammalian and non-mammalian species (Brown and Fyffe 1981; Clements and Redman 1989; Cullheim et al. 1987a; Kernell and Zwaagstra 1989b; Rose et al. 1985; Ulfhake and Kellerth 1981; Zwaagstra and Kernell 1980b). These new data resulted in a substantial upward revision of the maximum length of dendritic processes and of the total surface area of motoneurons. In cat motoneurons, for ex-ample, a single axon and anywhere from 6–19 dendritic trees are known to project outward from the motoneuron soma. The path length from the soma to the farthest dendritic termination can be on the order of 2 mm, and the total dendritic surface area is on the order of 500,000 μm^2 (Cullheim et al. 1987a).

Although it was apparent even from the early anatomical data that the morphology of motoneuron dendrites did not strictly satisfy the requisite constraints for collapsing the entire dendritic tree into an equivalent cylin-der (e.g., Barrett and Crill 1974b), the more recent studies have documented this discrepancy in more detail. The equivalent cylinder representation de-mands that the total surface area of all the dendritic segments at a given distance from the soma remains constant throughout the length of the tree. To further represent all the dendritic trees as one cylinder, all the trees need to be of comparable length and surface area. In fact, the dendritic surface area changes as a function of radial distance from the soma, first increasing and then decreasing with increasing somatofugal distance (Bras et al. 1987; Clements and Redman 1989; Cullheim et al. 1987a; Kernell and Zwaagstra 1989b; Rose et al. 1985). Also, the size and complexity of a dendritic tree depends upon its anatomical orientation (Cameron et al. 1983; Cullheim et al. 1987b; Egger et al. 1980; Rose 1981). The general features of motoneuron dendrites are illustrated in Fig. 6 (from Cullheim et al. 1987a). This sche-matic plot shows a single dendritic tree from a cat triceps surae motoneuron in part A. Parts B, C and D of the figure show the average relations between somatofugal distance and the number of branch points (B), the number of branch terminations (C) and membrane area (D) for a sample of type F (open symbols) and type S (filled symbols) motoneurons. At increasing distance from the soma, the number of branch points first increases then decreases sharply (Fig. 6B). A similar pattern is observed for the number of branch terminations (Fig. 6C). These factors, together with branch tapering, generate the membrane surface area distribution shown in Fig. 6D.

The size and branching complexity of individual dendritic trees are corre-lated with the diameter of the stem dendrite arising from the soma [Cameron et al. 1985; Clements and Redman 1989; Cullheim, 1978; Kernell and Zwaagstra 1989; Rose et al. 1985; Ulfhake and Kellerth 1981, 1983, 1984)

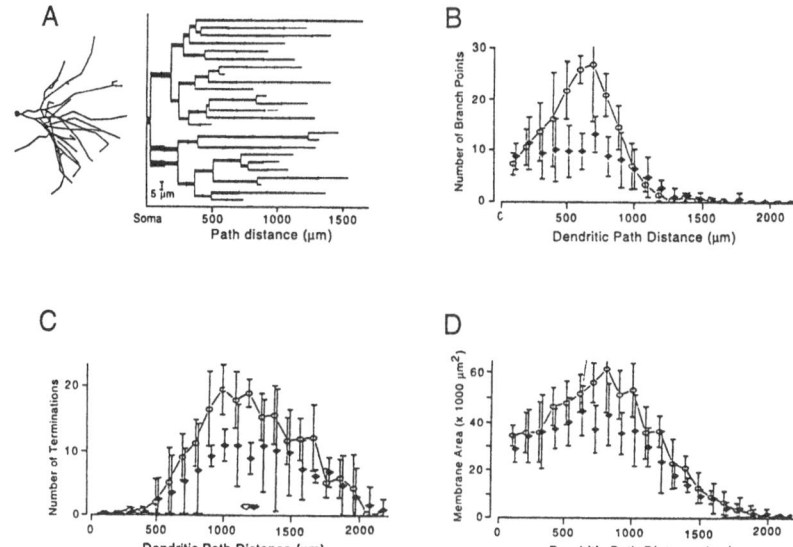

Fig. 6. Morphology of cat motoneuron dendrites. **A.** Two different representations of a dendritic tree from an HRP-filled, type FF cat triceps surae motoneuron. On the left is a two-dimensional projection drawing and on the right is a "dendrogram" of the same dendrite. **B-D.** Comparison between type F (open symbols) and type S (filled symbols) motoneuron in path distance distribution of number of branch points (**B**), number of dendritic terminations (**C**) and membrane area (**D**). (Modified from Figs. 1 and 6 of Cullheim et al. 1997)

with larger stem dendrites giving rise to larger and more complex dendritic trees than do smaller stem dendrites. Variation in the structure of the different dendritic trees of the same motoneuron is often related to the anatomical orientation of the tree. In the spinal cord, dendrites oriented along the longitudinal axis are often the largest and most complex. This rostro-caudal bias is particularly prominent in the phrenic and many cervical motoneuron pools (Cameron et al. 1985; Rose et al. 1985), although it can also be seen to a lesser extent in some lumbar motoneurons (Brown and Fyffe 1981; Cullheim et al. 1987a). The longitudinal dendritic orientation may facilitate the connectivity between pools of motoneurons (and interneurons) and primary afferent fibers, since primary afferent axons typically send off collaterals along a large longitudinal extent of the spinal cord (Brown 1981).

The variation in dendritic structure has important implications for both synaptic receptivity and for the flow of current within the neuron. The dendrites arising from the soma can exhibit a 3- to 4-fold range in their stem diameters and the various associated size-related measures (Bras et al. 1987,

1993; Cameron et al. 1985; Cullheim et al. 1987a; Kernell and Zwaagstra 1989b; Moore and Appenteng 1991) leading to marked differences in their electrotonic properties (Bras et al. 1987, 1993; Kernell and Zwaagstra 1989a; Moore and Appenteng 1991). Further, as a result of large variations in the relative change in dendritic diameter at branch points (Bras et al. 1987; Cameron et al. 1985; Clements and Redman 1989; Kernell and Zwaagstra 1989a; Rose et al. 1985) dendritic branches at the same anatomical distance from the soma can be at very different electrical distances.

Variations in the electrical properties of the different dendritic trees of a single motoneuron or even different portions of the same dendritic tree lead to a number of different pathways over which voltage transients initiated at the soma can be propagated both toward and away from the soma. This multiplicity of electrical paths may lead to significant errors in estimates of average electrotonic length based on the measurement of voltage transients in the soma (Glenn 1988). However, it has been argued that problems inherent in estimating time constants from noisy experimental data have limited the magnitude of these potential errors. In fact, the estimates of values of L based on τ_0/τ_1 may only slightly overestimate the true average L of the dendritic trees (Holmes et al. 1992; Rall et al. 1992). Although the average behavior of the dendrites can be reasonably well represented with a single cylinder, increased verisimilitude is obtained by endowing the cylinder with a variable diameter that initially increases slightly and then tapers sharply as function of distance from the soma (Clements and Redman 1989; Fleshman et al. 1988; Rall et al. 1992).

Despite the anatomical findings that the motoneuron dendrites do not strictly meet the requirements imposed by Rall's equivalent cylinder model, the original conclusions regarding the dependence of motoneuron input resistance on size and specific membrane resistivity (see above) are probably still valid (cf. Kernell and Zwaagstra 1989a). Fig. 6 B–D shows that type S motoneurons (which have high values of input resistance) have smaller, less complex dendritic trees than do type F motoneurons (which have low values of input resistance). Further, recent combined anatomical and electrophysiological studies confirm that the variations in size cannot account for the range of variation of input resistance, suggesting that differences in specific membrane resistivity are also important (Fleshman et al. 1988; Kernell and Zwaagstra 1989a). The contribution of systematic differences in active conductances to the effective input resistance and recruitment order of motoneurons will be considered later.

3.3.2
Passive Membrane Properties

A more serious problem with the original formulation of the Rall model (Rall 1959) is the assumption that R_m is uniform across the somatodendritic membrane. Recent work indicates that this is not the case, and that some of the non-uniformity is likely to arise from a leak conductance introduced by inserting a microelectrode into the motoneuron soma (somatic shunt; Durand 1984). The electrode-induced shunt conductance is variable, and its effect on the measured input conductance of a neuron will depend upon the magnitude of its 'true' (i.e., uninjured) input conductance. Since motoneurons are relatively large cells, their 'true' input conductance is also likely to be large and the relative effect of the electrode-induced shunt should be less than in smaller, low-input-conductance cells. Nonetheless, a number of recent studies suggest that the electrode induced leak conductance may make a significant contribution to the total input conductance measured in motoneurons with sharp electrodes (see below).

Measurements made on a variety of neuron types have also provided information on the biophysical basis of the resting leak conductance measured in the absence of an electrode-induced shunt. As in other neurons, the resting potential of motoneurons is determined primarily by a resting potassium conductance in combination with a relatively high internal concentration of potassium (Coombs et al. 1955; Forsythe and Redman 1988). Part of the potassium conductance active at the resting potential is apparently distinct from other potassium conductances in that it is reduced by external barium ions and is insensitive to several blocking agents that affect other potassium channels (Bayliss et al. 1992; Nistri et al. 1990). The magnitude of this barium-sensitive leak conductance (G_{Krest}) is independent of voltage over a large range (Nistri et al. 1990).

The spatial distribution of G_{Krest} in motoneurons is not known, but an analogous conductance in rat sympathetic ganglion cells does not appear to be uniformly distributed across the somatodendritic membrane (Redman et al. 1987). More recent evidence for a non-uniform spatial distribution of a resting leak conductance comes from simultaneous whole-cell recording from the soma and apical dendrite of rat neocortical pyramidal cells, combined with morphological reconstruction of the cells (Stuart and Spruston 1998). In these cortical cells, the leak conductance increases with distance from the soma. As described below, this distribution is opposite to that proposed for motoneurons, where the leak conductance is thought to be largest on or near the soma and smaller on the dendrites (Clements and Redman 1989; Fleshman et al. 1988; Rall et al. 1992; Rose and Vanner 1988).

The resting membrane potential of motoneurons is thought to be determined primarily by a potassium leak conductance. Although the potassium equilibrium potential has been estimated to be in the range of -80 to -100 mV in motoneurons (Barrett et al. 1980; Coombs et al. 1955; Forsythe and Redman 1988; Zhang and Krnjevic 1986), average resting potentials obtained from large samples of motoneurons recorded *in vivo* with sharp microelectrodes range from around -65 mV (Pinter et al. 1983; Zengel et al. 1985) to around -75 mV (Gustafsson and Pinter 1984a,b; Pinter et al. 1983; Zengel et al. 1985), depending upon the selection criteria used. The difference between the resting and potassium equilibrium potentials is likely to reflect some resting permeability to other ions, which may arise from a number of sources. It is possible that the leak channels are not completely selective for potassium, but also are permeable to sodium and chloride ions, so that the "true" resting potential will be more depolarized than the potassium equilibrium potential (Forsythe and Redman 1988). Also, an imperfect seal around the recording microelectrode will introduce a "shunt conductance" that is non-selective for different ions. However, the depolarizing effect of this non-specific conductance could be counteracted by additional potassium conductances activated by the electrode-induced influx of calcium and sodium ions (see below).

Four different methods have been used to estimate the magnitude of the shunt conductance around the recording micropipette. Each of these methods is based on certain untested assumptions, and consequently gives rise to very different estimates of the magnitude of the electrode-induced leak conductance. The first method is straightforward and is based on the idea that after impalement, the measured resting potential is determined by an equilibrium between current flowing across the neuron's true resting conductance (G_N) and that flowing across the leak conductance around the electrode (G_{Lelec}).

$$G_N(V_r' - V_r) = -G_{Lelec}(V_r' - V_{Lelec}), \tag{12}$$

where V_r' is the measured resting potential, V_r is the true resting potential and V_{Lelec} is the equilibrium potential for the impalement-induced leak current. If this leak conductance were relatively non-specific for different ion species, V_{Lelec} should be around 0 mV. Equation 12 dictates that the measured resting potential will approach the 'true' resting potential (V_r) when the relative magnitude of the electrode leak (G_{Lelec}/G_N) is small and will approach V_{Lelec} when G_{Lelec} is large.

Gustafsson and Pinter (Gustafsson and Pinter 1984b) estimated that if the true motoneuron resting potential were -80 mV (and $V_{Lelec} = 0$), then the range of resting potentials measured in their sample would be consistent with a maximum leak conductance that was 30% of the true input conductance, yielding estimated leak conductances on the order of 0.05–0.5 μS (resistances of 2–20 MΩ). Campbell and Rose (1997) have obtained similar estimates in cat neck motoneurons, using somewhat different techniques and assumptions. These estimated values for motoneurons are similar to previous estimates derived from experiments on frog muscle fibers (Hodgkin and Nakajima 1972; Stefani and Steinbach 1969).

A strong inverse relation between the measured values of R_N and resting potential would be expected if variations in leak conductance were the sole cause of variations in measured R_N (and if V_{Lelec} is significantly more depolarized than V_r). However, resting potential does not generally show a significant correlation with either R_N or the estimated shunt conductance (e.g., Gustafsson and Pinter 1984b; Rose and Vanner 1988; see, however, Campbell and Rose 1997) and the differences in mean R_N among different motor unit types are not associated with differences in their average resting potentials (Zengel et al. 1985). It is possible that the lack of correlation between measures of V_r and R_N is due to the fact that V_{Lelec} is close to the measured resting potential. This could occur if the electrode-induced leak led to an influx of calcium ions of sufficient magnitude to overwhelm the cell's intrinsic buffering and pumping capacity and thus activate a calcium-dependent potassium conductance. Such a mechanism has been proposed to account for the differences in membrane properties of dentate gyrus granule cells of rat hippocampus measured intracellularly with sharp electrodes versus whole-cell recordings with patch electrodes (Staley et al. 1992). In these and other hippocampal cells, the changes in resting potential and input conductance produced by lowered extracellular calcium or sodium are consistent with a resting activation of calcium- and sodium-dependent potassium conductances (for ref., see Staley et al. 1992). In contrast, neither the intracellular injection of calcium or calcium chelators markedly alters the resting potential of motoneurons (Krnjevic and Lisiewicz 1972; Krnjevic et al. 1978; Rose and Brennan 1989; Zhang and Krnjevic 1988). In rat spinal motoneurons studied *in vitro*, replacement of extracellular calcium with manganese results in a less than 1 mV hyperpolarization (Forsythe and Redman 1988). In the same study, a six-fold reduction in external sodium resulted in about a 6 mV hyperpolarization and a 25% reduction in input conductance. These effects are opposite to those expected if a sodium-activated potassium conductance made a significant contribution to the resting potential and input conductance.

The second method of estimating electrode-induced leak is based on comparisons of R_N measurements obtained with sharp and patch electrodes in the same cells under the same experimental conditions. If patch electrodes are assumed to form a perfect seal with the cell membrane, so that there is no electrode-induced leak, then the mean leak conductance induced by sharp microelectrodes would be equal to the difference between the mean values of input conductance obtained with sharp and patch electrodes. Recent comparisons of this type are available for a number of neuron types (Scroggs et al. 1994; Spruston and Johnston 1992; Staley et al. 1992), including neonatal rat lumbar motoneurons (Takahashi 1990 versus Fulton and Walton 1986), and cultured rat motoneurons (Streit and Luscher 1992 versus Ulrich, Quadroni, and Luscher 1994). Differences in the two types of input conductance measurements range from 0.006 to 0.05 μS (i.e., electrode shunt resistances of 20–167 MΩ). The sharp electrodes employed in these studies had much higher resistances (and presumably smaller tip diameters) than the intracellular electrodes typically used to record from adult cat motoneurons, so that the leak conductances introduced by electrodes in cat experiments might be even larger. However, part of the lower leak conductance observed with whole-cell recording could result from dialysis of second messengers or other substances that contribute to the resting leak conductance. In this regard, R_N values can be lower with perforated patch recordings in which dialysis of intracellular substances is minimal, as opposed to ordinary whole-cell recording (Major et al. 1992; Spruston and Johnston 1992).

A third, recently-proposed method of estimating the leak conductance induced by electrode impalement is based on the effects of applying an external DC electric field (Svirskis et al. 1997). The method uses the fact that the distribution of charge along a dendritic cable will differ depending upon whether charge is injected at one end through the recording electrode, or applied externally along the entire length of the cable. In the absence of an electrode shunt, the voltage-induced transients induced by an external electric field would be expected to decay faster than those following current injection. In contrast, in the presence of a significant shunt, the final decay rate will be the same for the two stimuli, and in addition the field-induced transient will show a significant overshoot after the offset of the pulse. Application of this method to turtle motoneurons and interneurons revealed than in many cases impalement with sharp electrodes did not introduce a detectable shunt.

The final method of estimating the leak conductance induced by electrode impalement requires anatomical reconstruction of the motoneuron in combination with measurements of R_N and τ_m, and an assumed value of specific membrane conductance (C_m). For example, Moore and Appenteng (1991) measured R_N, τ_m and cell geometry in HRP-filled rat masticatory motoneurons and then calculated R_m from τ_m and C_m (estimated at 1 μF cm^{-2}). The expected input conductance of three reconstructed cells was determined based on this calculated R_m value and the measured morphology. The calculated input conductance measurements (G_{morph}) were consistently lower than the measured values, implying electrode leak conductances of 0.39, 1.05, and 0.06 μS (resistances = 2.56, 0.95 and 16.67 MΩ).

A related method of estimating electrode-induced leak requires measured or assumed neuron morphology in combination with measurements of R_N and the two slowest membrane time constants (τ_0 and τ_1). If membrane properties are assumed to be homogeneous and there is no electrode-induced leak, then the time course of the voltage transient following the offset of an injected current pulse or step is determined by the cell's morphology together with the values of three intrinsic parameters: (1) specific resistivity (R_m), axial resistivity (R_i) and specific membrane capacitance (C_m). Figure 7 (taken from Fig. 3 of Thurbon et al. 1998) illustrates the effects of these parameters on the time course of a voltage transient evoked in the soma of a model rat spinal ventral horn neuron. The transients in the upper panels and the lower left panel were generated in models in which the membrane properties were uniform across the entire somatodendritic surface. The lower right panel shows transients from models in which the effective specific membrane resistivity was made non-uniform by introducing a shunt conductance in the soma. It was possible to fit a subset of the data derived from whole-cell recordings from neonatal rat ventral horn neurons using models with uniform membrane properties (i.e., no electrode-induced shunt conductance; Thurbon et al. 1998). In contrast, it was noted early on that the voltage transients recorded in motoneurons with sharp electrodes did not appear to be consistent with the assumption of uniform membrane properties (Iansek and Redman 1973b). In order to account for this discrepancy, the original Rall model was modified to include an additional parameter to account for the possibility of different R_m values for the soma (R_{ms}) and the dendritic cable (R_{md}; Durand 1984; Kawato 1984). Application of this model to reconstructed motoneurons suggests that R_{ms} is several-fold smaller than R_{md} (Clements and Redman 1989, Fleshman et al. 1988) and if this lower R_{ms} value is attributed entirely to an electrode-induced leak conductance, values of leak conductance of over 1 μS are often obtained. However, with the introduction of a fourth parameter into the model, there is a strong possibility of the existence of more than one parameter set that will give equally good fits to the data, unless neuronal morphology is completely and accurately specified (e.g., Holmes and Rall 1992). This problem of non-uniqueness is exacerbated by the fact that large variations in parameter values can produce relatively subtle changes in the time course of the volt-

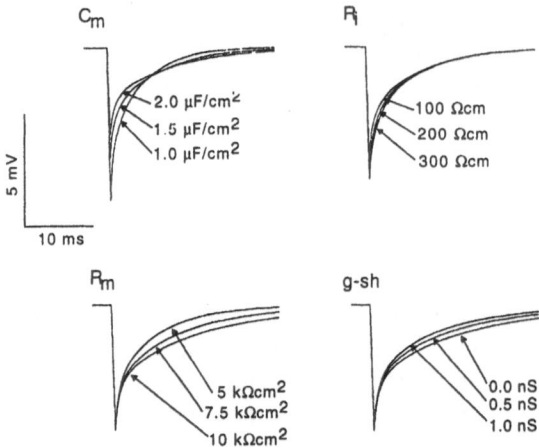

Fig. 7. Influence of specific membrane capacitance (C_m), specific membrane resistance (R_m), cytoplasmic resisitivity (R_i), the value of a somatic shunt conductance (g–sh) on the response of a model motoneuron to a current pulse (480 μs, –0.5 nA) applied to the soma. (Modified from Fig. 3 of Thurbon et al. 1998)

age transient, particularly in the presence of noise and electrode artifacts. For example, the upper right panel of Fig. 7 shows that a 3-fold change in R_i is associated with a relatively small change in the time course of the local voltage transient produced by a somatically-injected current pulse. Moreover, if dendritic resistivity is allowed to vary, then the experimental data are well fit by a model in which specific membrane resistivity exhibits a continuous increase from the soma to the distal dendrites (as opposed to the step increase implied by the somatic shunt model; Clements and Redman 1989; Fleshman et al. 1988). Recent evidence suggests that part of the lowered resistivity of the soma and perhaps of the proximal dendrites may be attributed to a potassium conductance that is active at rest and is blocked by intracellular application of cesium (Campbell and Rose 1997).

In summary, relatively small estimates of electrode-induced leak conductance are obtained from analyses based on measurements of V_r and R_N, from comparisons of R_N values obtained with sharp and patch electrodes, and based on the voltage transients induced by external polarization. In contrast, the largest estimates are based on analyses of neuron models with R_m uniform everywhere except at the impalement site. The various approaches to estimating electrode leak might be reconciled if it is assumed that even in the absence of electrode impalement R_m is non-uniform across the somatodendritic membrane, perhaps rising as a function of somatofugal distance (cf. Fleshman et al. 1988). Electrode impalement might exaggerate this non-uniform R_m by adding a non-specific leak conductance to the soma on the order of 0.05–0.1 µS. Depending on the relative magnitudes of the 'true' neuron conductance and resting potential and the electrode leak conductance and reversal potential, this leak will depolarize the neuron from resting potential and reduce the measured input resistance and membrane time constant below their true values.

How does our present understanding of the passive, linear behavior of motoneurons differ from the original Rall model (section 2.4)? The motoneuron can still be represented as a soma attached to an equivalent dendritic cylinder, but the diameter of the cylinder varies with somatofugal distance due to the observed patterns of dendritic branching, tapering and termination. R_m varies across the surface of the neuron, and although the actual values are unknown, the dendritic R_m is likely to be significantly higher than the somatic R_m. Although the tips of the longest dendritic branches in this model may be greater than 1.5 L away from the soma, the bulk of the dendritic membrane is relatively close electrically (i.e. < 1 L, cf. Rall et al. 1992).

3.3.3
Active Conductances

There is at present very little direct information regarding the membrane properties of motoneuron dendrites, due to the difficulty in obtaining dendritic patch recordings. There is only one recent study on cultured rat spinal motoneurons (Larkum et al. 1996), and a preliminary report on patch clamp recordings from the proximal dendrites of neonatal rat spinal motoneurons (Luscher et al. 1997). The lack of data arises in part from the fact that motoneuron dendrites branch profusely and the daughter branches become increasingly smaller in diameter at increasing distances from the soma (see above). This anatomy makes it extremely difficult to obtain dendritic patch recordings at significant distances from the soma. As a result, inferences regarding the properties and distribution of active dendritic conductances are based on somatic voltage- and current-clamp recordings, often in combination with additional changes in dendritic polarization induced by repetitive activation of presynaptic afferents, glutamate iontophoresis or external electric fields.

The earliest voltage-clamp recordings in motoneurons focussed on characterizing the conductances underlying action potentials (Araki and Terzuolo 1962; Nelson and Frank 1964). These demonstrated the presence of fast inward and slower outward conductances mediating the depolarizing and repolarizing phases of the action potential. Subsequent studies by Barrett and colleagues (Barrett et al. 1980; Barrett and Crill 1980) described these conductances in more detail, demonstrating a TTX-sensitive, inactivating sodium conductance and a delayed-rectifier-type potassium conductance analogous to those described in the squid giant axon (Hodgkin and Huxley 1952). The associated currents were smoothly graded, except for an all-or-none component ascribed to an uncontrolled action potential in the initial segment, since it could be occluded by an antidromic stimulus (Araki and Terzuolo 1962; Barrett and Crill 1980). The measured outward currents included not only a fast delayed-rectifier component, but also a slower component responsible for the AHP.

With the exception of the initial segment current, the currents measured under somatic voltage-clamp were considered to arise from the activation of conductances on or near the soma for three reasons: (1) they were smoothly graded with increasing depolarization, (2) the capacitative charging transient evoked by a given voltage clamp step was relatively independent of the background holding potential, indicating that the amount of current necessary to charge the dendrites did not depend upon their background level of depolarization (Barrett and Crill 1980), and (3) repetitive firing in the primary range could be approximated based on the characteristics of the measured currents and the passive impulse response of the motoneuron (Barrett et al. 1980). However, a model based on passive dendrites and the measured somatic currents could not accurately reproduce higher frequency (secondary range) repetitive firing.

The inability to recreate high frequency repetitive firing based on the voltage-clamp results of Barrett and colleagues (Barrett et al. 1980; Barrett and Crill 1980) was likely due to the omission of a persistent inward current first described by Schwindt and Crill (1977), and subsequently characterized in more detail over the next several years (Schwindt and Crill 1980a,b,c; Schwindt and Crill 1982). The rapid and relatively complete inactivation of the sodium current underlying the depolarizing phase of the action potential suggests that under steady-state or near steady-state conditions, the total membrane current will be dominated by leak and outward currents. If this were the case, then the steady-state current-voltage (I–V) curve should have a positive slope throughout the voltage range (i.e., increasing depolarization leads to increasing outward current).

An example of such an I–V relation is illustrated in Fig. 8A (taken from Schwindt and Crill 1977), obtained from the response of a cat lumbar motoneuron to a slow somatic voltage-clamp ramp. The slope of the linear relation near the resting potential (indicated by the origin on the voltage axis) reflects the leak conductance of the neuron, whereas the upward bend at more depolarized voltages results from increasing activation of additional voltage-gated potassium conductances. In the majority of motoneurons,

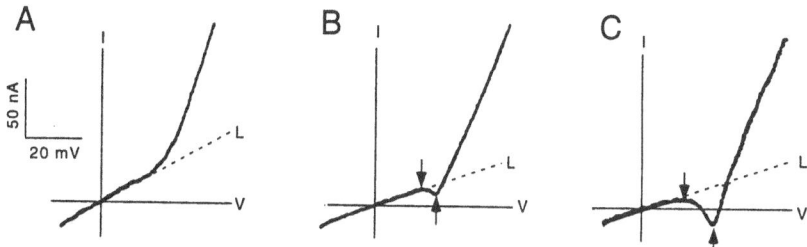

Fig. 8. Current (I) – voltage (V) relations obtained by applying slow voltage ramp commands to three different cat lumbar motoneurons. The resting potential is indicated by the intersection of the current and voltage axes. The slope of the dotted lines labeled L indicate the passive leak conductance in each cell. Downward arrows indicate the first point of zero slope, whereas upward arrows indicate the point of maximum inward current. (Modified from Fig. 1 of Schwindt and Crill 1977)

however, these two regions of positive slope were interrupted by a region of negative slope, indicating increasing activation of a persistent inward current starting at about 10 mV positive to the resting potential (Fig. 8B and C). In almost half of the motoneurons, the I–V relation exhibited a region of net inward current (Fig. 8C), so that the N-shaped relation crossed the zero-current axis at three points. Two of these points are stable equilibrium points, with the left-most point representing the resting potential and the rightmost point representing a stable, depolarized potential.

In turtle spinal motoneurons it has been shown that this persistent inward current is mediated primarily by a calcium current flowing through L-type channels (Hounsgaard and Mintz 1988), and recent work has demonstrated the presence of a similar current in mouse spinal motoneurons (Carlin et al. 2000ab). In guinea pig brainstem motoneurons the persistent inward current is either carried by sodium (Nishimura et al. 1989) or a mixture of calcium and sodium currents (Hsiao et al. 1998). Rat hypoglossal motoneurons also appear to express both sodium and calcium persistent currents (Musick 1999). In some motoneurons, activation of a calcium-dependent mixed cation conductance may lead to a persistent inward current, although multiple calcium spikes are generally required for significant activation (Perrier and Hounsgaard 1999; Rekling and Feldman 1997).

In cat motoneurons, the persistent inward current is thought to be primarily calcium-mediated since it is not eliminated by intracellular application of the lidocaine derivative QX-314 which blocks sodium currents (Schwindt and Crill 1980a,b). The persistent inward current is reduced however by QX-314 (Lee and Heckman 1999b), but this agent markedly suppresses calcium currents in addition to blocking sodium currents (Talbot and Sayer 1996). The persistent current is enhanced by extracellular barium, which is a more effective charge carrier through calcium channels than calcium itself (Schwindt and Crill 1980a). The persistent inward current appears to be mixed with outward currents over a wide range of membrane potentials, and the net inward current can be considerably enhanced by procedures which suppress outward currents, such as internal application of TEA (Schwindt and Crill 1980b; Powers and Binder 2000) or barium (Schwindt and Crill 1980a), or a depolarizing shift in the potassium equilibrium potential (Schwindt and Crill 1980c).

Schwindt and Crill (1980a,b) argued on several grounds that their experimental results were consistent with the hypothesis that the persistent inward current was primarily mediated by current flow through channels located on or near the soma. First, they did not observe voltage clamp instabilities of the type predicted for excitable cables of moderate electrotonic lengths (Jack et al. 1975). Second, if only the distal dendritic membrane displays an N-shaped I–V relation and the proximal membrane were passive, then the I–V relation obtained with a somatic voltage clamp would exhibit an attenuated negative slope region only at very large somatic depolarizations. Third, in the experiments employing extracellular iontophoresis of barium (Schwindt and Crill 1980a), the tip of the iontophoretic electrode was within 20–30 μm of the tips of the electrodes employed for the somatic voltage clamp, so that given the iontophoretic currents and barium concentrations used, barium was likely to affect calcium channels only on the soma and proximal dendrites. Finally, even if the entire dendritic membrane exhibited the same N-shaped I–V curve, the recorded clamp currents should be dominated by currents generated in the soma and proximal dendrites, particularly at more depolarized voltages where the increased membrane conductance acts to shunt out any distal dendritic component.

---→

Fig. 9. Persistent inward currents in motoneurons. A. Response of a TEA-injected cat spinal motoneuron to successive step depolarizations (upper trace) from a DC holding potential at the natural resting potential. Current response (lower trace) has been compensated for leak and capacitative currents. Note that net current remains inward after return to resting potential. (Modified from Fig. 8 of Schwindt and Crill 1980a). B. Triangular voltage clamp command and resultant membrane current measured in a cat triceps surae motoneuron. C. Membrane current in part B is plotted as a function of voltage for the ascending (thick line) and descending (thin line) portions of the voltage clamp command. The onset of the persistent inward current (I_{PIC}) is defined as the first point of zero slope on the ascending limb of the current waveform, and its offset is defined as the last point of zero slope on the descending limb. The amplitude of the initial peak is the difference between the peak inward current and the current predicted based on the leak conductance. The bistable current range is the range of membrane currents between I_{PIC} onset and offset. (Parts B and C modified from Fig. 1 of Lee and Heckman 1999a)

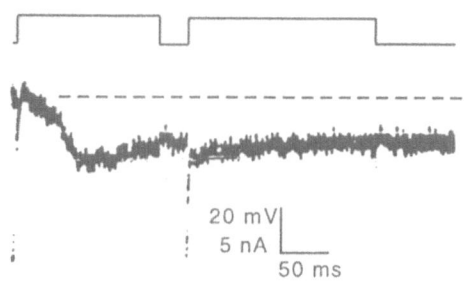

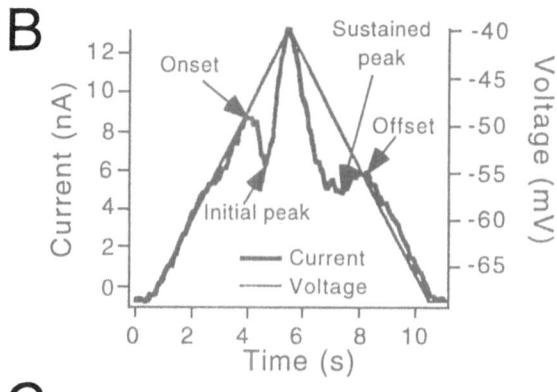

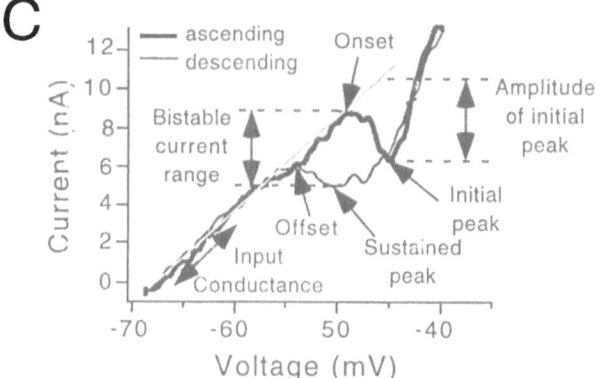

Although these considerations suggest that the measured I–V curves were dominated by current flow through proximal conductances, they do not rule out the possibility that similar conductances exist throughout the motoneuron's dendritic tree. In fact, theoretical analyses (Butrimas and Gutman 1979; Gutman 1971; Gutman 1991) have suggested that the experimental voltage-clamp results described above are best explained by postulating that the entire dendritic membrane exhibits an N-shaped I–V curve. In particular, localization of the persistent inward current in the dendrites could lead to a hysteresis in the I–V relation measured with a somatic voltage clamp. An initial somatic depolarization of sufficient magnitude would activate persistent inward currents in the dendrites, leading to a wave of depolarization that would eventually reach the distal dendritic segments. Subsequent repolarization of the soma might then be insufficient to move the distal membrane from its depolarized stable equilibrium point, leading to a steady inward current measured at the soma.

Many of the key issues addressed above are illustrated in Fig. 9. Part A of the figure (modified from Fig. 8 of Schwindt and Crill 1980b) is an example of a voltage-clamp record obtained from a cat spinal motoneuron that is consistent with the presence of persistent inward current in the distal dendrites. The initial somatic depolarization (top trace) leads to the slow onset of an inward current (bottom trace) that persists even after the somatic membrane is returned to the resting potential. More recently, Lee and Heckman (1998a, 1999a) have measured I–V relations in lumbar motoneurons in decerebrate cats using slow, triangular voltage-clamp commands. Figure 9B (from Fig.1 in Lee and Heckman 1999a) shows the time course of the voltage-clamp command and the measured clamp current in one of these motoneurons. The I–V plot shown in Fig. 9C clearly indicates that the magnitude of the persistent inward current is higher and its voltage-threshold is lower on the descending limb of the I–V curve. A similar hysteresis in the I–V relation has recently been reported in mouse motoneurons, where it has been attributed to dendritic L-type calcium currents (Carlin et al. 2000b).

Additional evidence for the presence of persistent inward currents in the dendrites of mammalian motoneurons is based on the effects of synaptic inputs on somatic membrane potential or voltage-clamp current. Following the intrathecal administration of a noradrenergic α_1 agonist, the synaptic current evoked by repetitive activation of primary spindle (Ia) afferents is larger when the somatic voltage is clamped at a depolarized level than it is at the resting potential (Lee and Heckman 1996, 2000). Since the current flowing through an excitatory synapse should normally decrease as the membrane is depolarized toward the reversal potential, and since nearly all of Ia terminals are located on the dendrites (section 3.1), this result suggests that current transfer to the soma is amplified by the activation of dendritic conductances that supply additional inward current. Under current clamp conditions, the activation of a persistent inward current produces a depolarizing "plateau potential" (Hounsgaard et al. 1988). The somatic voltage threshold for this plateau is reduced during tonic activation of Ia afferents (Bennett et al. 1998), again supporting a dendritic location for the responsible channels. Finally, persistent inward currents are enhanced by serotonergic precursors (Hounsgaard et al. 1988) and serotonergic terminals have recently been shown to be localized primarily on motoneuron dendrites (Alvarez et al. 1998).

An elegant series of studies by Hounsgaard and colleagues has provided more direct and detailed information on the properties and function of active dendritic conductances in turtle motoneurons recorded *in vitro*. In the presence of potassium channel blockers, the application of external electric fields that hyperpolarize the soma and depolarize the dendrites leads to the appearance of calcium spikes and plateau potentials at the somatic recording site (Hounsgaard and Kiehn 1993). Calcium spikes are promoted by external tetraethylammonium (TEA), whereas plateau potentials are promoted by apamin (Hounsgaard and Kiehn 1993), suggesting that the effects of dendritic calcium conductances are normally counteracted by both voltage-activated and calcium-activated potassium conductances (Rudy 1988). Calcium-mediated plateau potentials can also be facilitated by activation of presynaptic afferents running in the dorsolateral funiculus, and these facilitative effects are reduced by antagonists of $5-HT_{1A}$, group I metabotropic glutamate receptors and muscarine receptors (Delgado-Lezama et al. 1997). The results of experiments combining external electric fields with activation of either dorsolateral or mediolateral funicular afferents suggest that the intrinsic properties of different portions of the dendritic tree may be differentially controlled by different subsets of descending afferents (Delgado-Lezama et al. 1999). Finally, experiments combining focal glutamate iontophoresis with focal application of either potassium channel blockers (Skydsgaard and Hounsgaard 1994) or serotonin (Skydsgaard and Hounsgaard 1996), support the possibility that intrinsic dendritic properties can be regulated over very restricted spatial domains. The implications of this modulation for the integrative properties of dendrites will be considered later (section 4.1).

Despite the paucity of direct electrical measurements of the behavior of motoneuron dendrites, it is quite clear that they are endowed with a rich complement of voltage-gated channels. What remains to be established are the spatial distributions, selectivity, and activation-inactivation kinetics, and neuro-modulation of the various types of voltage-gated channels throughout the dendritic tree. Further, we are just beginning to study how these active dendritic conductances influence the transfer of synaptic current from the dendrites to the soma and the interactions between concurrently-active synaptic inputs (e.g., Powers and Binder 2000 and section 4.1.2).

3.4
Mechanisms Underlying Repetitive Firing

As discussed earlier (section 2.8), many of the quantitative features of repetitive discharge in motoneurons can be reproduced using a simple threshold-crossing model. In such models, action potentials are elicited when membrane potential exceeds a specified voltage-threshold and the post-spike membrane potential trajectory is determined by a leak current and a slow potassium conductance giving rise to the AHP (Kernell 1968). Although a variety of more complex models of motoneuron discharge were developed prior to 1980 (Baldissera and Gustafsson 1974b,c; Kernell and Sjoholm 1973; Traub 1977; Traub and Llinas 1977), the primary determinant of repetitive discharge in all of these models is the AHP conductance. Although more recent evidence has confirmed the importance of the AHP in the regulation of the repetitive discharge behavior of motoneurons (section 3.4.1), at least two additional factors have also been shown to influence the quantitative features of repetitive discharge: (1) the dependence of spike threshold on the prior history of membrane potential and (2) the presence of the persistent inward current described earlier. Finally, the ability to obtain patch clamp recordings and to manipulate the extracellular environment in recently developed *in vitro* preparations has led to a better understanding of the relation between the properties and spatial distributions of ion channels and repetitive discharge behavior.

3.4.1
The Afterhyperpolarization (AHP) and Its Influence on Repetitive Firing

There is now a large body of evidence that the slow potassium current underlying the medium duration afterhyperpolarization (mAHP) in motoneurons is due to an apamin-sensitive, calcium-activated potassium conductance (G_{KCa}). The mAHPs are completely or partially blocked by removing external calcium or external application of divalent cations which block calcium conductances (Chandler et al. 1994; Krnjevi'c et al. 1978; Nishimura et al. 1989; Hounsgaard and Mintz 1988; Mosfeldt-Laursen and Reckling 1989; Schwindt and Crill 1981; Viana et al. 1993; Kobayashi et al. 1997) and by apamin (Chandler et al. 1994; Hounsgaard and Mintz 1988; Viana et al. 1993; Zhang, and Krnjevic 1986, 1988; Kobayashi et al. 1997), a specific blocker of the small-conductance type calcium-activated potassium conductance (Rudy 1988). The AHP is also depressed by intracellular injection of calcium chelators (Krnjevic et al. 1978; Viana et al. 1993b; Zhang and Krnjevic 1988). The results of experiments in which the AHP conductance has been modified by affecting calcium entry or by the application of apamin have confirmed its importance in regulating repetitive discharge behavior. Comparisons of steady-state f-I relations before and after pharmacological manipulations of the AHP conductance (Chandler et al. 1994; Hounsgaard and Mintz 1988; Nishimura et al. 1989; Sawczuk et al. 1997; Viana et al. 1993b), indicate that the steady-state f-I slope is inversely proportional to the magnitude of the AHP conductance.

In neonatal rat hypoglossal motoneurons, it has been demonstrated that the calcium influx responsible for triggering the AHP enters through high threshold calcium channels that are blocked by omega-conotoxin and omega-agatoxin (Umemiya and Berger 1994; Viana et al. 1993a). The fact that the AHP is unaffected by blocking other calcium conductances suggests that these N-and P-type calcium channels must be in close proximity to the calcium-activated potassium channels, but separated from other calcium channels. The activation of G_{KCa} by calcium is therefore likely to depend upon the local concentration of calcium, which will in turn depend upon the calcium entering through site-specific calcium channels. In rat trigeminal motoneurons, only N-type channels contribute to activating the mAHP (Kobayashi et al. 1997), indicating that the spatial distributions of different types of calcium channels and their proximity to calcium-activated potassium channels may differ in different cell types.

The deactivation of G_{KCa} will depend not only upon the rate constants for calcium binding to the channel, but also on the rate at which calcium leaves the vicinity of the channel due to diffusion, binding to intracellular proteins and pumping across the membrane (cf. Sah 1996). Although approximately 98% of the calcium ions inside motoneurons appear to be taken up by endogenous buffers (Lips and Keller 1998; Palecek et al. 1999), the endogenous buffering capacity of motoneurons is actually relatively low compared to other neurons of comparable size (cf. Lips and Keller 1998; Palecek et al. 1999). Since the time course of decay of calcium transients is thought to be directly proportional to the endogenous buffering capacity of a cell (Neher 1995), it has been suggested that the low buffering capacity of motoneurons allows a rapid decay of intracellular calcium transients between bursts of rhythmic activity (Lips and Keller 1998).

The magnitude and duration of the AHP following a single spike vary according to motoneuron size (Zwaagstra and Kernell 1980a) and motor unit type: larger and longer AHPs are recorded in type S than in type F motoneurons (Kernell 1983; Zengel et al. 1985). It is not known whether these differences in AHP characteristics reflect differences in the properties of G_{KCa} channels, in their density, or in differences between the various factors controlling the time course of the calcium concentration in the vicinity of the channels. Gustafsson and Pinter (1985) proposed that the kinetics of the AHP conductance may be similar across different motoneurons, but that a more developed sag conductance (I_h) in type F motoneurons provides an inward current that decreases the measured AHP duration. However, AHP characteristics measured at the resting potential are correlated with repetitive discharge behavior, even though I_h should be largely deactivated over the range of membrane potentials encountered during steady-state repetitive discharge.

3.4.2
Voltage-and Time-Dependent Variations in Spike Threshold and Sodium Channels

The somatic voltage at which spikes are initiated is likely to depend in part upon the state of activation and inactivation of sodium channels in the soma and adjacent axon hillock and initial segment. Studies of sodium channels based on somatic voltage-clamp recordings in cat motoneurons (Barrett and Crill 1980; Schwindt and Crill 1982) revealed that the kinetics and voltage-dependency of activation and inactivation of the somatic sodium current are qualitatively similar to those described by Hodgkin and Huxley (Hodgkin and Huxley 1952) in the squid axon. The steady-state relations between voltage and both activation and inactivation are shifted to more positive values than are generally reported for axons (Barrett and Crill 1980; Schwindt and Crill 1982). Although the properties of the initial segment sodium current have not been studied directly, they can be inferred from variations in firing level (the somatic voltage at which the rapid upstroke of the action potential begins) during repetitive discharge in motoneurons and from voltage clamp studies on axons. The somatic voltage at which the initial segment spike is elicited exhibits a delayed dependence upon the value of the somatic membrane potential (Burke and Nelson 1971; Calvin 1974). This behavior has been called accommodation, and is thought to be due in part to subthreshold changes in sodium channel inactivation (Frankenhaeuser and Vallbo 1964; Schlue et al. 1974; Vallbo 1964). Experiments on cat motor axons suggest inactivation time constants in the range of 2–4 ms (Richter et al. 1974).

The voltage dependence and kinetics of this fast sodium inactivation process suggest that during repetitive discharge, significant changes in the voltage threshold for spike initiation should occur during the interspike interval as the membrane potential is initially hyperpolarized following a spike and then depolarizes at a rate of 0.1–0.2 mV/ms following the peak of the AHP (Schwindt and Calvin 1972). There have been several reports of interspike variations in the voltage threshold for spike initiation (Calvin 1974; Calvin and Stevens 1968; Powers and Binder 1996), and Fig. 10 shows examples of this phenomenon. Part A illustrates the effects of injecting brief, just-threshold current pulses at different points within the interspike inter-

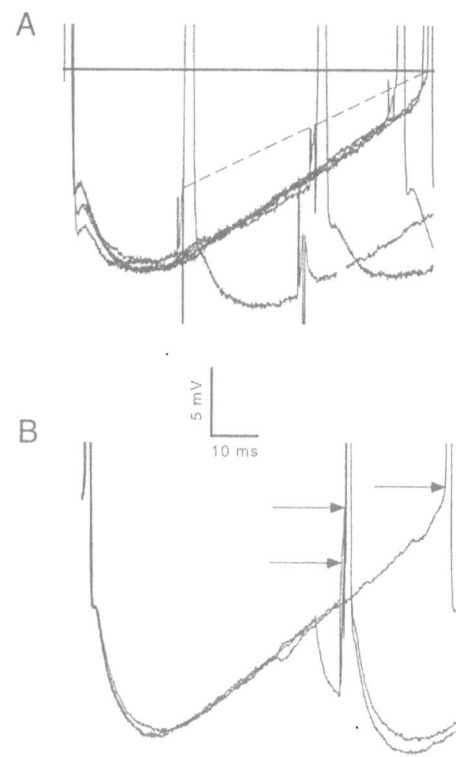

Fig. 10. Variations in spike threshold during repetitive discharge. A. Superimposed traces of interspike intervals during repetitive discharge of a cat lumbar motoneuron. In three of the traces a just suprathreshold current pulse has been superimposed at different times after the preceding spike. The dashed line shows that the voltage threshold for spike initiation varies during the interspike interval. B. Similar records from another cat lumbar motoneuron. In two of the traces a just suprathreshold pulse has been added either with (lower left arrow) or without (upper left arrow) a preceding hyperpolarizing pulse. The voltage threshold is lowered by the conditioning hyperpolarizing pulse. (Modified from Fig. 2 of Powers and Binder 1999)

val. It is clear that the voltage-threshold for spike initiation is not constant, but rather follows the time course of the interspike membrane potential trajectory, reaching a minimum early in the interval and then rising to its final value (Powers and Binder 1996). Part B shows the effects of adding a hyperpolarizing current pulse prior to a just-threshold depolarizing pulse. Spike threshold is decreased following membrane hyperpolarization, again showing the threshold is dependent on the recent time course of the somatic membrane potential.

Although the possibility of interspike variations in spike threshold has been acknowledged in a number of theoretical studies of motoneuron discharge (e.g., Ashby and Zilm 1982; Fetz et al. 1989), this variation is not generally incorporated into threshold-crossing motoneuron models. As a result, the spike-triggering efficacy of synaptic potentials may be underestimated in these models (cf. Powers and Binder 2000; Powers and Binder 1996; Turker and Powers 1999).

Slower inactivation processes have been described for sodium channels in a number of different types of cells (Brismar 1977; Colbert et al. 1997; Fleidervish et al. 1996; Howe and Ritchie 1992; Jung et al. 1997; Rudy 1981). The kinetics of the slow inactivation process are such that only a small fraction of the sodium channels enter the slow inactivation state during the depolarization associated with a single spike (Colbert et al. 1997; Jung et al. 1997). However, recovery from inactivation is also quite slow, so that there is a progressive accumulation of slow inactivation during prolonged repetitive discharge (Colbert et al. 1997; Jung et al. 1997). This accumulation of inactivation may underlie the slow increase in firing level observed during repetitive discharge in some neurons (Schwindt and Crill 1982). At low firing rates, the first spike following the onset of an injected current step is initiated at a lower somatic depolarization than subsequent spikes, which may be initiated at somatic depolarizations (from rest) of from 1.4–2.0 times that of the first spike (Schwindt and Crill 1982). At higher firing rates, the spike initiation level may continue to increase for several hundred milliseconds. As a result, the steady-state firing level may vary over a greater than two-fold range from the minimum to the maximum discharge rate (Schwindt and Crill 1982). This increase in firing level is associated with an increase in the mean level of membrane depolarization that leads to increasing activation of the persistent inward current (see below).

3.4.3
Contribution of Persistent Inward Currents to Repetitive Firing

The abrupt change in the slope of the motoneuron steady-state f-I relation (secondary range; Kernell 1965b; Schwindt 1973) was difficult to replicate in threshold-crossing models based only on the AHP conductance (e.g., Kernell 1968). Two explanations were advanced to account for secondary range firing: (1) a saturating summation of the AHP conductance (Baldissera, Gustafsson, and Parmiggiani 1976, Baldissera, Gustafsson, and Parmiggiani 1978), and (2) a plateau region in the post-spike decay of the AHP conductance (Baldissera and Gustafsson 1974a,b,c;). More recent simulations (Powers 1993) indicate that in threshold-crossing models with a fixed spike threshold, saturation of a slow post-spike potassium conductance does result in an upward curvature in the steady-state f-I relation. As previously proposed (Baldissera et al. 1976), the summation of the AHP depends on the proportion of the maximum potassium conductance that is activated by a single spike. If this proportion is relatively large, then additional spikes produce relatively little increase in potassium conductance, and the upward bend in the f-I slope occurs at low rates of repetitive discharge.

Baldissera and Gustafsson (1974 a,b,c) proposed a motoneuron model based on a measured plateau region in the post-spike decay of potassium conductance. Although this model could produce piecewise linear f-I relations similar to those observed experimentally, it was based on the assumption that the interspike conductance time course during steady-state repetitive discharge was similar to that measured after a single spike. Since the mean membrane voltage becomes increasingly depolarized at high rates of repetitive discharge as discussed earlier, any voltage-dependent currents activated positive to the resting potential will contribute to the interspike time course of membrane potential and conductance. As a result, the interspike conductance time course during repetitive discharge can differ from that measured near the resting potential following a single spike (Schwindt and Calvin 1973b; see however, Mauritz et al. 1974) and other factors must be invoked to explain secondary range discharge.

There is considerable evidence that increasing activation of persistent inward currents is responsible for secondary range discharge. The characteristics and spatial distribution of persistent inward currents (I_{PIC}; cf. Lee and Heckman 1998a) in motoneurons were described in section 3.3.3 in connection with their potential effects on dendritic processing. Since I_{PIC} is first activated below the threshold for spike initiation (cf. Lee and Heckman 1998a; Schwindt and Crill 1982), it can potentially influence the initiation of single spikes and the characteristics of repetitive discharge. As mentioned earlier, the voltage threshold for spike initiation (V_{thr}) measured in response to 50 ms current steps of threshold intensity (I_{rh}) is generally greater than the product of I_{rh} and R_N (Gustafsson and Pinter 1984a). The disparity between V_{thr} and the product of $I_{rh} * R_N$ tends to be largest in motoneurons with long AHP durations and time constants, i.e., presumably type S motoneurons (Gustafsson and Pinter 1984a). The disparity between V_{thr} and the product of $I_{rh} * R_N$ is also correlated with R_N, suggesting that the magnitude of the persistent inward current is similar in different motoneurons. Thus, the contribution of the persistent inward current to voltage threshold is most prominent in small, high R_N motoneurons (Gustafsson and Pinter 1984a), a hypothesis that is also supported by voltage-clamp data (Lee and Heckman 1998a; Schwindt and Crill 1980c).

The presence of a persistent inward current also has important effects on repetitive discharge behavior since the somatic membrane potential traverses voltage ranges over which this current is significantly activated. The increase in spike threshold at increasing rates of discharge leads to increasing activation of I_{PIC} (Powers 1993; Schwindt and Crill 1982). Figure 11 illustrates the relation between activation of I_{PIC} and steady-state repetitive discharge behavior based on current clamp and voltage clamp recordings obtained in the same motoneuron. The voltage-clamp recordings (B and D) indicate that some inward current activation takes place at the membrane voltage reached at the end of the interspike interval at the lowest firing rate (P) and that it is likely to be continuously activated throughout the interspike interval at the maximum primary range firing rate (T). The upward bend in the steady-state f-I curve (secondary range firing) may result from the predominance of the inward current.

Changes in discharge rate over time can also be related to the activation of I_{PIC}. Although firing rate typically declines following the onset of a step of injected current (spike-frequency adaptation; see above and section 4.2), if I_{PIC} predominates over outward currents, firing rate may accelerate instead (Hounsgaard et al. 1988; Hounsgaard and Kiehn 1985; Hounsgaard and Kiehn 1989; Hounsgaard and Mintz 1988; Lee and Heckman 1998b; Schwindt 1973). If the inward current component is of sufficient magnitude so that the steady-state I–V relation crosses the zero-current axis in the negative slope region, spontaneous discharge can occur in the absence of depolarizing synaptic or injected current. (Lee and Heckman 1998a; Lee and Heckman 1998b; Schwindt and Crill 1980c). Such "bistable" discharge behavior can be observed in the presence of exogenous or endogenous neuromodulators (Hounsgaard et al. 1984; Hounsgaard and Kiehn 1985; Hounsgaard and Kiehn 1989; Lee and Heckman 1996; Lee and Heckman 1998b; Lee and Heckman 1999a) or following pharmacological reduction of potassium con-

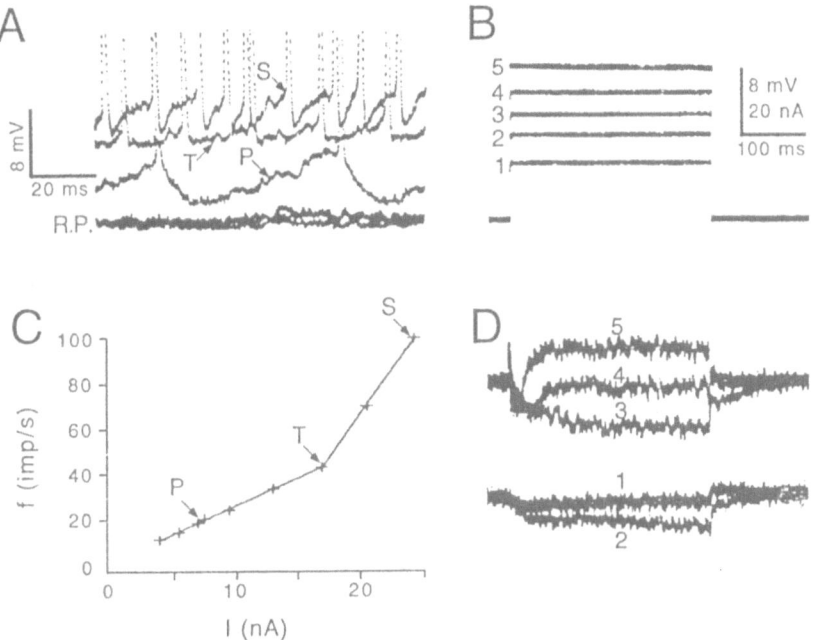

Fig. 11. Relation between the voltage-sensitivity of the persistent inward current and the voltages traversed during repetitive discharge at different rates. **A.** Sample voltage records during repetitive discharge in a cat lumbar motoneuron during discharge in the primary range (P), the transition between the primary and secondary ranges (T) and at the upper end of the secondary range (S). **B.** Somatic voltage clamp commands applied in the same motoneuron. **C.** Steady-state relation between injected current and firing rate for this motoneuron. **D.** Membrane currents (after subtraction of the leak and longest capacitative components) corresponding to the voltage commands in B. Traces 1 and 2 have been offset for clarity. (Modified from Fig. 1 of Schwindt and Crill 1982)

ductances (Hounsgaard and Mintz 1988; Schwindt and Crill 1980c). Bistable discharge behavior is sometimes spontaneously present in motoneurons in unanesthetized decerebrate cats, presumably due to tonic activity in brainstem neurons with descending serotonergic and noradrenergic fibers (cf. Hounsgaard et al. 1984, 1988; Lee and Heckman 1999a). It typically disappears following a spinal transection, but then reappears following i.v. adminstration of serotonergic precursors (5-HTP; Hounsgaard et al. 1988). It is also enhanced by intrathecal application of the noradrenergic α_1 agonist methoxamine (Lee and Heckman 1999a).

The degree to which motoneurons express bistable discharge behavior depends upon a number of factors, including the background level of depolarization (e.g., Hounsgaard et al. 1988; Hounsgaard and Kiehn 1985; Lee and Heckman 1998b), the local concentration of neuromodulators (see above), and also on motoneuron type. Many motoneurons will only discharge for a brief period of time following the offset of an excitatory input, even following the application of a noradrenergic agonist and depolarizing bias current (Lee and Heckman 1998b). These "partially" bistable motoneurons tend to have higher rheobases and conduction velocities than fully bistable cells (Lee and Heckman 1998b), suggesting that the fully bistable cells are probably type S and low threshold type FR motoneurons, whereas partially bistable cells are likely to be type FF and higher threshold type FR motoneurons. Figure 12 (derived from Figs. 1, 3 and 4 from Lee and Heckman 1998b), illustrates typical discharge characteristics for fully bistable (left) and partially bistable motoneurons (right). The middle trace in the left panel of Fig. 12A illustrates the effects of applying a steady excitatory synaptic input from Ia afferents onto a fully bistable motoneuron. The motoneuron continues to discharge after the excitatory input is terminated, and this discharge continues indefinitely if a depolarizing bias current is superimposed (upper trace). The prolonged discharge is not due to a continued synaptic input, since the depolarization produced by the Ia input has a crisp offset when the cell is hyperpolarized. In contrast, in the partially bistable motoneuron illustrated in the right panel of Fig. 12A, only a few spikes occur after the offset of the Ia input, even when a depolarizing bias current brings the membrane just below firing threshold (upper trace).

The remaining panels in Fig. 12 illustrate other distinguishing characteristics of fully and partially bistable motoneurons, based on their responses to injected current steps (B) and triangular ramps (C and D). In a fully bistable motoneuron (Fig. 12B, left), an acceleration in firing rate occurs in

---

Fig. 12. Differences in the discharge behavior of fully and partially bistable motoneurons. A. Responses to repetitive activation of Ia spindle afferents produced by 160 Hz 80 μm peak-to-peak longitudinal vibration of the triceps surae muscles. Different levels of background depolarization were produced by levels of DC current injection (current amplitudes indicated in the figure). B. Examples of firing rate acceleration in fully and partially bistable motoneurons. Upper traces show instantaneous firing rate (dotted) and a moving average of firing rate (solid). Lower traces show a 'stairstep of injected current', in which each step increase is 3 nA. In the fully bistable cell firing rate acceleration occurs close to threshold, whereas in the partially bistable cell, the acceleration occurs only at higher levels of injected current and is followed by a decrease in firing rate. C and D. Responses of fully and partially bistable cells to triangular waveforms of injected current. Part C shows both firing rate and injected current as a function of time, whereas part D shows firing rate as a function of current. (Modified from Figs. 1, 3 and 4 of Lee and Heckman 1998a)

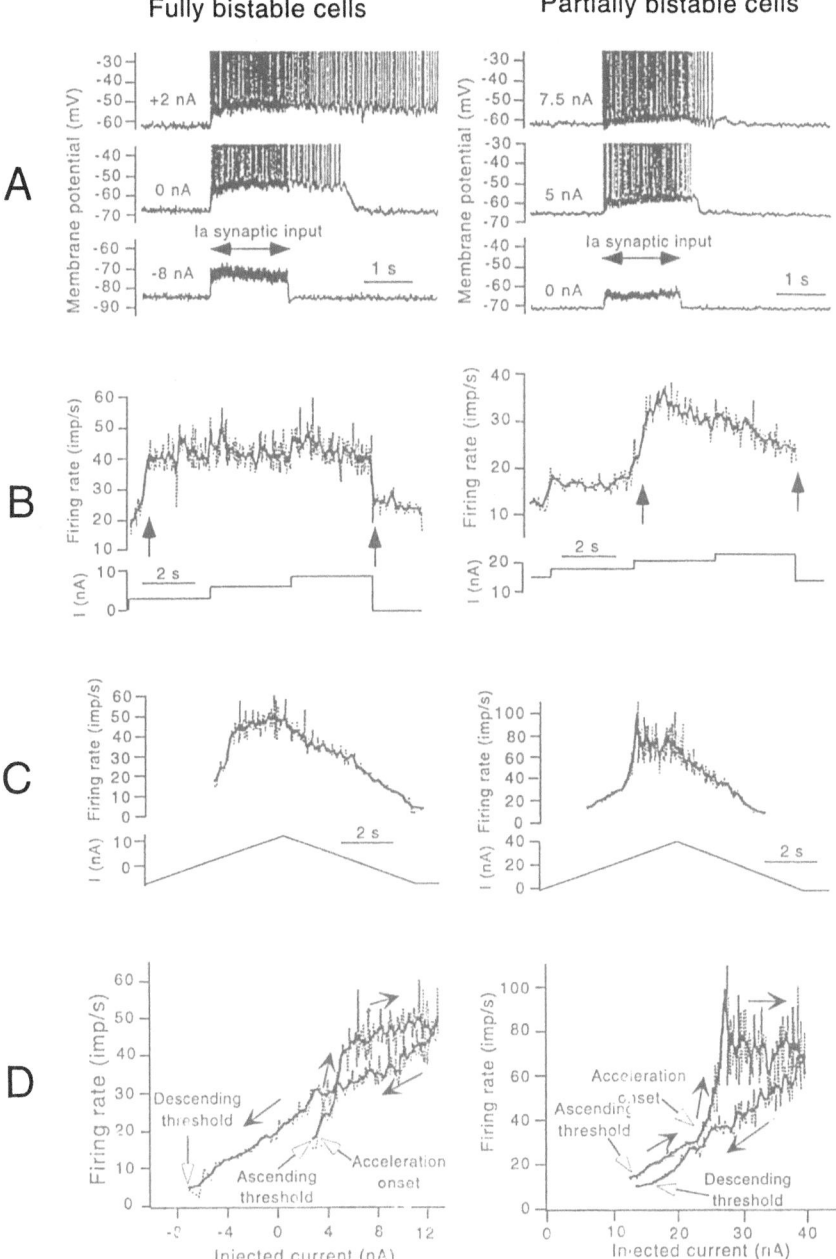

response to the first step of injected current (left arrow), and further in-creases in current elicit relatively small increases in firing rate. In this moto-neuron, discharge continues after the offset of injected current (right arrow). In contrast, in a partially bistable motoneuron (Fig. 12B, right), a higher level of injected current was required to elicit firing rate acceleration (left arrow). The firing rate exhibited a noticeable decline from its peak value, and discharge ceased after the injected current was reduced below the initial level required for repetitive discharge (right arrow). In response to a trian-gular waveform of injected current, fully bistable motoneurons exhibit firing rate acceleration shortly after current onset and continue to discharge on the descending limb of the injected current waveform, even when the current level drops below that initially required to elicit discharge (Fig. 12C and D, left panels). Similar patterns of firing rate acceleration and continued dis-charge on the descending limb of the f-I relation have been reported by other investigators, although in many cases the firing rate on the descending limb may be consistently higher than that on the ascending limb for matched current levels (e.g., Bennett et al. 1998b; Hounsgaard et al. 1984, 1988). In partially bistable cells, the increase in the slope of the f-I relation occurs at a much higher level of injected current relative to that required to initiate discharge. The firing rate may decline in these cells at the higher levels of injected current, and discharge on the descending limb of the f-I relation stops at about the same level of current as that required to initiate discharge (Fig. 12 C and D, right panels).

The differences in rhythmic discharge behavior between low and high threshold motoneurons are associated with differences in the properties of I_{PIC} (Lee and Heckman 1998a). Figure 13 (adapted from Fig. 8A and B of Lee and Heckman 1998a) presents a schematic representation of the major dif-ferences in the voltage-clamp data obtained in fully (solid lines) and par-tially bistable cells (dotted lines). Figure 13A illustrates the average relations between total membrane current (I) and somatic voltage (V) during an as-cending voltage-ramp command. The average I–V functions for both fully and partially bistable cells exhibit a similar shape, but the I–V function for the partially bistable cell is shifted upward and to the right. As a result, the point at which the average I–V curve for partially bistable cells first exhibits a negative slope (left open inverted triangle) occurs at a higher level of membrane current and a more depolarized voltage than in the fully bistable cell (left filled inverted triangle). In fully bistable motoneurons, the mem-brane potential at which the negative slope region begins is below the threshold for spike initiation, so that under current clamp conditions firing rate begins to accelerate at about the same point that discharge begins on

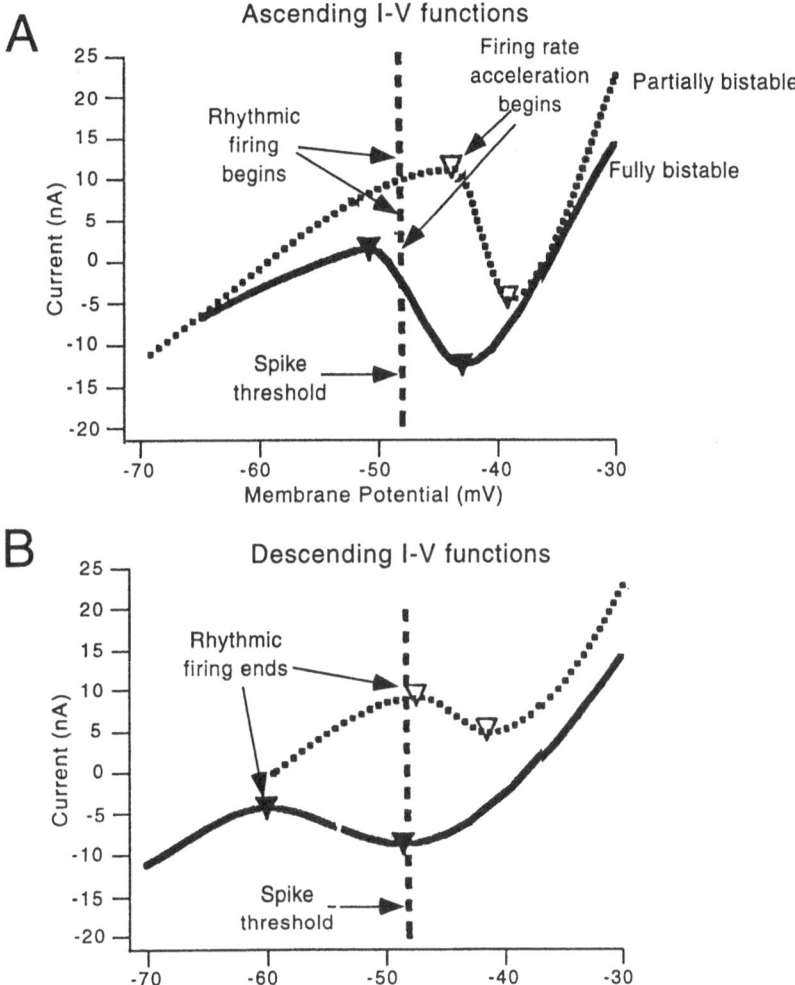

Fig. 13. Relation between discharge behavior and I–V functions. Average I–V functions measured during the application of a triangular voltage-clamp command to fully (solid) and partially (dotted) bistable motoneurons. **A.** I–V relations obtained during the ascending portion of the voltage clamp command. Filled and open inverted triangles indicate the average onset and peak inward current for fully and partially bistable cells. **B.** I–V relation obtained during the descending portion of the voltage clamp command. Filled and open inverted indicate average offset of the inward current and peak value during the descending voltage ramp. (Modified from Fig. 8 of Lee and Heckman 1998b)

the ascending phase of the injected current waveform (Fig. 12D). On the descending phase of the voltage-ramp command, the negative slope region also continues below spike threshold in fully bistable cells, and at this point the I–V slope is negative (Fig.13B, left filled triangle), leading to continued discharge under current clamp conditions (Fig. 12D). In contrast, in partially bistable cells the negative slope region of the ascending I–V curve begins at voltages positive to spike threshold (left open inverted triangle in Fig. 13A), and this region also ends above spike threshold on the descending I–V curve (left open inverted triangle in Fig. 13B). These features are correlated with two characteristic differences in the f-I relations of partially bistable cells: (1) the later onset of firing rate acceleration on the ascending limb of the f-I relation, and (2) similar current thresholds for discharge for ascending and descending current ramps (Fig. 12D).

The response of motoneurons to a particular waveform of injected current can also depend upon the presence of synaptic input and the prior history of the motoneuron's membrane potential (Bennett et al. 1998a, 1998b). As mentioned earlier, the threshold for activation of I_{PIC} can be altered in the presence of tonic synaptic input, supporting a dendritic location for the responsible channels. As a consequence of this shift in activation threshold, the point at which discharge rate accelerates during an ascending injected current ramp changes in the presence of synaptic input. The middle panel in Fig. 14A (from Bennett et al. 1998b) illustrates the response of a cat triceps surae motoneuron to a triangular waveform of injected current in the absence of applied synaptic input. The thin diagonal lines represent the increase in discharge rate above that expected based on the initial relation between injected current and firing rate on the ascending limb of the injected current waveform. The discharge rate was higher than expected starting at about two-thirds of the way along the ascending phase of the

---

Fig. 14. Discharge evoked in cat lumbar motoneurons by triangular waveforms of injected current, with or without additional tonic synaptic input. A. Instantaneous discharge rate in a cat triceps surae motoneuron and injected current waveform at rest (middle traces), during a steady 10 mm stretch of the triceps surae muscle (right traces), and during repetitive stimulation of the common peroneal (CP) nerve (100 Hz at 2 times threshold for the lowest threshold fibers). B. Composite results from A, showing discharge rate as a function of current during the ascending phase of the current waveform. (A and B modified from Fig. 2 of Bennett et al. 1998a) C. Instantaneous discharge rate as a function of current during injection of successive triangular current waveforms into a cat tibial motoneuron. The lower right plot shows the ascending phases of the first three responses. (Part C modified from Fig. 1 of Bennett et al. 1998b)

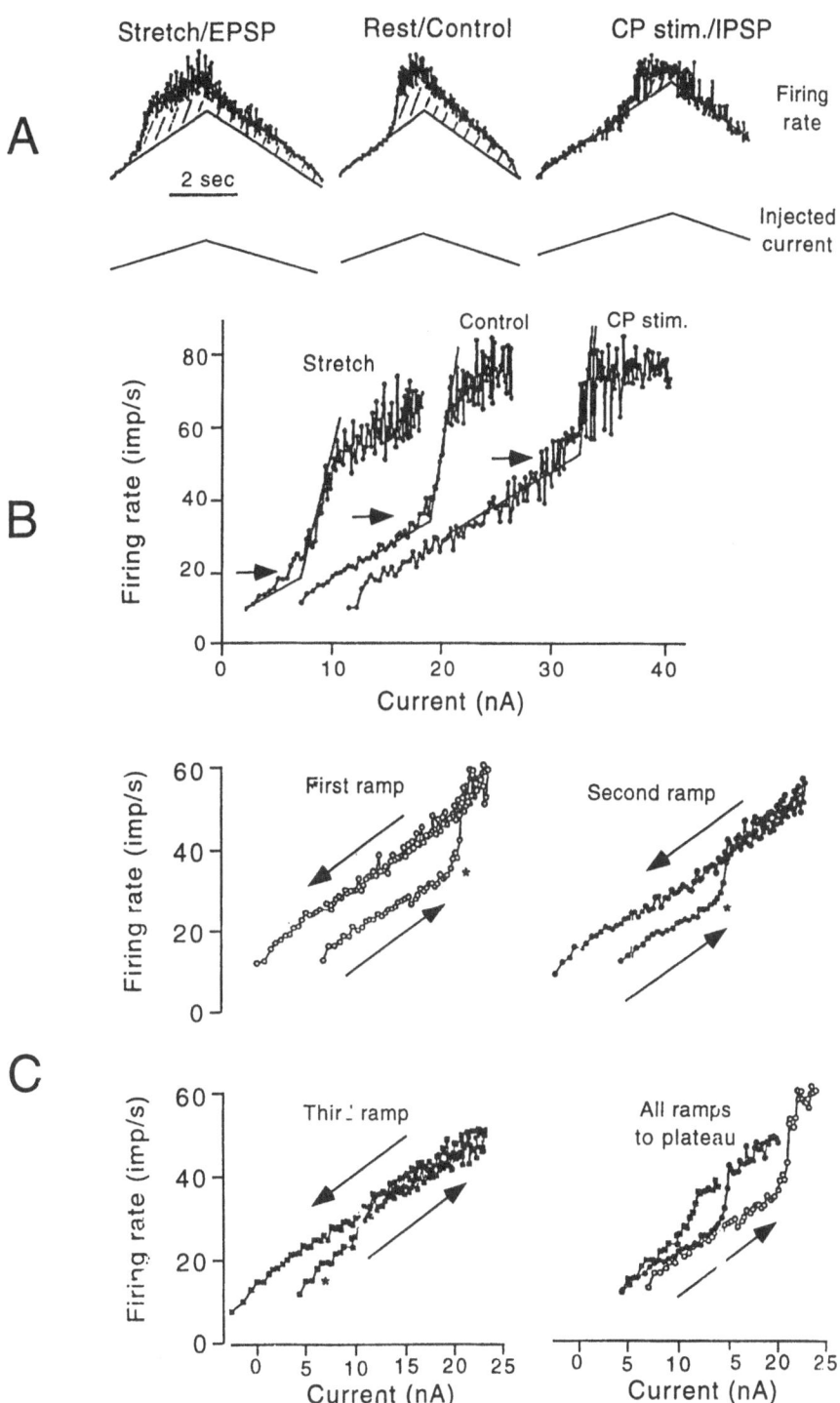

Stretch/EPSP   Rest/Control   CP stim./IPSP

Firing rate

A   2 sec

Injected current

B

Stretch   Control   CP stim.

Firing rate (imp/s)

Current (nA)

First ramp   Second ramp

Firing rate (imp/s)

C

Third ramp   All ramps to plateau

Firing rate (imp/s)

Current (nA)   Current (nA)

injected current waveform and then remains elevated throughout the descending phase, reflecting the effects of I_{PIC}. When tonic excitatory Ia input was produced by stretch of the triceps surae muscle, the increase in discharge rate started much closer to the recruitment threshold (Fig. 14A left panel), whereas the application of a tonic inhibitory input (by 100 Hz stimulation of the common peroneal nerve) delayed the onset of the firing rate acceleration and decreased its magnitude (Fig. 14A right panel).

The effects of tonic synaptic input on the motoneuron discharge records shown in Fig. 14A cannot be fully explained by the addition of a constant amount of excitatory or inhibitory effective synaptic current. If this were the only effect of the synaptic input, the f-I relations in the presence of synaptic input would be identical to those in its absence, except for a parallel shift along the current axis (cf. Section 2.8). Figure 14B illustrates the relation between discharge rate and current intensity for the ascending portion of the current waveforms shown in part A. Excitatory input does indeed shift the f-I relation as expected, but the transition to a steeper f-I slope (left arrow) occurs at a lower frequency than in the control case (middle arrow). Inhibitory input has the opposite effect. These results are consistent with the idea that synaptic input can shift the somatic voltage threshold for activation of I_{PIC} by changing the relative depolarization of the dendrites (Bennett et al. 1998b).

The threshold for I_{PIC} activation can also be altered by the prior history of membrane depolarization (Bennett et al. 1998a). In turtle motoneurons, this phenomenon is caused by a depolarization-induced facilitation of L-type calcium channels (Svirskis and Hounsgaard 1997). A similar mechanism may underlie history-dependent effects on the f-I function measured with current ramps. Figure 14C shows f-I relations obtained in a cat spinal motoneuron in response to a series of triangular injected current ramps separated by 3–4 seconds. In the first response (upper left panel) firing rate accelerates near the end of the ascending portion of the current waveform (asterisk). The acceleration occurs progressively earlier on subsequent trials, suggesting a progressive facilitation of I_{PIC}.

3.5
Effects of 'Non-Classical' Transmitters on Motoneuron Behavior

Transmitter release from many of the presynaptic afferent terminals influences motoneuron discharge by causing current flow through ionotropic receptor-gated channels whose behavior is governed by equation 4. The voltage-dependence of the synaptic current is derived solely from the changes in the driving force for synaptic current flow induced by changing the background membrane potential. The effects of the excitatory neuro-transmitter glutamate are mediated by two classes of ionotropic receptors: N-methyl-D-aspartate (NMDA) receptors and non-NMDA receptors (Collingridge and Lester 1989). The latter class behaves in the manner described above, i.e., increasing membrane depolarization reduces synaptic current flow due to a decrease in the driving force. Under steady-state conditions, synaptic current flow through these receptors is a positive, linear function of membrane voltage (i.e., less inward current flows as the membrane is depolarized). In contrast, NMDA receptor activation can induce a region of negative-slope conductance in the neuron's I–V relation (Flatman et al. 1986). This is due to the fact that at relatively hyperpolarized voltages, NMDA-activated receptors are blocked by magnesium, and this block is relieved by depolarization (Ascher and Nowak 1988; Mayer and Westbrook 1987).

In addition to the direct gating of ion channels by ionotropic receptors, the activation of metabotropic receptors can alter the properties of ion channels indirectly through membrane-bound guanosine triphosphate binding proteins (G-proteins) or second-messenger systems activated by G-proteins (rev. in Hille 1992). A variety of transmitters have been shown to exert such neuromodulatory effects on motoneuron properties. The next two sections will review the alterations in motoneuron properties produced by activation of ionotropic NMDA receptors, followed by a review of the effects produced by metabotropic receptors activated by common neuro-modulatory agents including achetylcholine, glutamate, noradrenaline, serotonin, substance P and thyrotropin releasing factor (TRH).

3.5.1
NMDA-Receptor Mediated Effects on Motoneurons

The effects of NMDA on motoneuron behavior have been studied extensively with *in vitro* preparations, in which rhythmic motoneuron discharge or rhythmic oscillations in membrane potential can be induced by bath application of NMDA (e.g., (Guertin and Hounsgaard 1998a; Hochman et al. 1994; Wallen and Grillner 1987). These behaviors arise in part from intrinsic oscillatory membrane properties in motoneurons, since NMDA-induced membrane oscillations occur in motoneurons in the presence of TTX (Guertin and Hounsgaard 1998b; Hochman et al. 1994; Wallen and Grillner 1987). As mentioned above, the voltage-dependence of the excitatory current through NMDA-activated channels acts like certain intrinsic voltage-dependent inward conductances by inducing a region of negative-slope conductance in a neuron's I–V relation (Flatman et al. 1986; Kim and Chandler 1995; Wallen and Grillner 1987). Further, NMDA-activated channels are permeable to calcium as well as sodium (Mayer and Westbrook 1987), and calcium entry can activate a calcium-dependent potassium conductance (Wallen and Grillner 1987). The alternation between depolarization and calcium entry via NMDA channels and hyperpolarization by calcium-activated potassium channels is thought to underlie NMDA-induced membrane potential oscillations in lamprey motoneurons, since replacement of external calcium with barium eliminates them (Wallen and Grillner 1987). In turtle motoneurons, depolarization and calcium entry through L-type calcium channels also contributes to NMDA-induced oscillations, since these are blocked by application of L-type calcium channel antagonists (Guertin and Hounsgaard 1998b).

The effects of NMDA receptor activation on the behavior of adult mammalian motoneurons is less well understood, in part because most of the *in vitro* studies of NMDA-induced rhythmic behavior have employed neonatal or immature animals (e.g. Cazalets et al. 1992; Hochman and Schmidt 1998; Smith and Feldman 1987; see however, Kim and Chandler 1995), and the expression of NMDA receptors in motoneurons may change during development (Hori and Kanda 1996; Kalb et al. 1992; Palecek et al. 1999; Turman et al. 1999). For example, the monosynaptic Ia input is probably not mediated by NMDA receptors in the adult cat spinal cord (Miller et al. 1997), although there is evidence for NMDA-receptor involvement in neonates (see below). Nonetheless, NMDA receptor activation is likely to be important in the generation of rhythmic motoneuron activity in the intact adult nervous system, since extracellular iontophoresis of NMDA can induce oscillations in adult motoneurons (Durand 1991, 1993), and rhythmic motoneuron activity recorded in adult mammals *in vivo* can be disrupted by NMDA antagonists (Bohmer et al. 1991; Chitravanshi and Sapru 1996; Katakura and Chandler 1990).

Activation of NMDA receptors in motoneurons is not likely to be limited to situations in which rhythmic discharge is elicited, but may be a more common feature of excitatory synaptic control of motoneuron activity. The monosynaptic excitatory connection from Ia afferents to spinal motoneurons is partially mediated by NMDA receptors in neonatal animals (Pinco and Lev-Tov 1993; Ziskind-Conhaim 1990), and NMDA and non-NMDA receptors are co-localized in neonatal rat hypoglossal motoneurons (O'Brien et al. 1997). Monosynaptic EPSPs evoked in adult guinea pig trigeminal motoneurons have also been shown to be composed of both an NMDA and a non-NMDA component (Trueblood et al. 1996). As in other neurons (McBain and Mayer 1994), the non-NMDA component has a slower time course, which together with its voltage-dependence could act to enhance temporal summation of excitatory inputs (Trueblood et al. 1996). Activation of both NMDA and non-NMDA receptors could reduce the voltage-dependence of the total synaptic current, since the depolarization-induced decrease in synaptic current through non-NMDA receptor channels would be counteracted by increased flow through NMDA-activated channels (Cook and Johnston 1999).

3.5.2
Effects of Metabotropic Receptor Activation on Motoneurons

Both the subthreshold and suprathreshold behavior of motoneurons can be influenced by a variety of neuromodulators. As mentioned earlier (section 3.1), motoneurons in both the spinal cord and brainstem are densely inner-vated by both noradrenergic and serotonergic terminals. The serotonergic terminals often contain the peptides thyrotropin-releasing hormone (TRH) and substance P (Arvidsson et al. 1990) as well. Although the effects of these various substances on motoneuron behavior have been assessed *in vivo* by using extracellular iontophoresis (rev. in White et al. 1996), their mecha-nisms of action have been more fully explored in a variety of *in vitro* prepa-rations. However, as mentioned earlier, the *in vitro* preparations often util-ize immature animals, and since both the intrinsic properties of motoneu-rons and the projections of presynaptic neuromodulatory fibers undergo significant changes during the early postnatal period (e.g. Bayliss et al. 1994; Nunez-Abades and Cameron 1995; Nunez-Abades et al. 1993; Viana et al. 1995; Ziskind-Conhaim et al. 1993), the results of many of these studies may not apply to adult motoneurons. Moreover, both the magnitude and type of effects produced by a particular neuromodulator can vary between different species and different motoneuron populations in the same species (see be-low).

The most extensively documented action of metabotropic receptor acti-vation on the subthreshold behavior of vertebrate motoneurons is a reduc-tion of the resting, barium-sensitive, potassium leak conductance (G_{Krest}). Reductions in the resting leak conductance of motoneurons have been re-ported following the application of the following agents: the metabotropic glutamate receptor agonist (1S,3R)-1-amino-1,3-cyclopentane-dicarboxylic acid ((1S, 3R)-ACPD; Del Negro and Chandler 1998; Dong and Feldman 1999; Svirskis and Hounsgaard 1998), substance P (Fisher and Nistri 1993), muscarine (Svirskis and Hounsgaard 1998), thyrotropin-releasing hormone (TRH; Bayliss et al. 1992; Fisher and Nistri 1993; Rekling 1990), norepineph-rine (NE; Elliott and Wallis 1992; Larkman and Kelly 1992; Parkis et al. 1995), and serotonin (Elliott and Wallis 1992; Hsiao et al. 1997; Inoue et al.

1999; Larkman and Kelly 1992; Lindsay and Feldman 1993). The reductions in G_{Krest} are associated with membrane depolarization under current clamp conditions, and together these effects reduce the amount of synaptic or injected current required to produce motoneuron discharge (e.g., Bayliss et al. 1992; Parkis et al. 1995; Hsiao et al. 1997). The reduction of the resting potassium conductance by serotonin may also depend on blockade of an inwardly rectifying potassium current (Larkman and Kelly 1998). In addition, in some cases metabotropic agonists can also depolarize motoneurons via a barium-insensitive inward current carried predominantly by sodium ions (Bayliss et al. 1992; Hsiao et al. 1997; Parkis et al. 1995). Finally, serotonin can enhance the hyperpolarization-activated inward conductance (G_h) in both adult and neonatal mammalian motoneurons (Takahashi 1990; Larkman and Kelly 1992; Hsiao et al. 1997). The enhancement of the current mediated by G_h (I_h) may be associated either with depolarizing shift in its half-activation voltage (Larkman and Kelly 1992), or an increase in maximal conductance (Hsiao et al. 1997), leading to an increase in the resting level of I_h with a consequent membrane depolarization (Larkman and Kelly 1992).

Serotonin can also produce a decrease in the amplitude of the medium duration afterhyperpolarization (mAHP) in motoneurons of a number of different species (Bayliss et al. 1995; Berger et al. 1992; Hsiao et al. 1997; Inoue et al. 1999; Kiehn and Harris-Warrick 1992; Wallen et al. 1989; White and Fung 1989; Wu et al. 1991). In neonatal rat hypoglossal motoneurons, this effect of serotonin appears to result from inhibition of the calcium channels responsible for activation of the calcium-activated potassium conductance underlying the mAHP (Bayliss et al. 1995). The effects of serotonin on the mAHP vary with developmental stage, species and the particular motoneuron population. For example, the serotonin-induced reduction of the mAHP in rat hypoglossal motoneurons disappears after postnatal day 20, and this phenomenon appears to be due to reduced expression of the 5-HT1A receptors responsible for the effect (Talley et al. 1997). In juvenile guinea pig trigeminal motoneurons, concentrations of serotonin as low as 10 µM produce a significant reduction in the mAHP (Hsiao et al. 1997), whereas much higher concentrations of serotonin or an increase in external calcium are required to produce similar effects in rat trigeminal motoneurons (Inoue et al. 1999). It is possible that these species differences also reflect a difference in receptor expression.

The net effect of a given metabotropic agent on motoneuron input-output behavior reflects the combined effects of the increase in input resistance, membrane depolarization, and the reduction of the mAHP. In addition, acetylcholine (Bellingham and Berger 1996; Jiang and Dun 1986), glutamate (Del Negro and Chandler 1998; Dong and Feldman 1999) and serotonin (Elliott and Wallis 1992; Lindsay and Feldman 1993; Ziskind-Conhaim et al. 1993) have all been shown to reduce excitatory synaptic transmission via a presynaptic mechanism. The increase in input resistance and membrane depolarization reduce the amount of injected or synaptic current needed to produce repetitive discharge leading to a leftward shift in the frequency-current relation (e.g., Inoue et al. 1999; Lindsay and Feldman 1993; Parkis et al. 1995). When the mAHP is significantly reduced, there is an increase in the slope of the f-I relation (Bayliss et al. 1995; Berger et al. 1992; Hounsgaard and Kiehn 1989; Hsiao et al. 1997; Inoue et al. 1999; Parkis et al. 1995).

The reduction of resting conductances and calcium-activated potassium conductances described above may act in concert with an enhancement of persistent inward currents (Hsiao et al. 1998; Svirskis and Hounsgaard 1998) to produce (Hsiao et al. 1998) or accentuate (Lee and Heckman 1999a) a region of negative slope resistance in the steady-state I–V relation. As previously discussed, this type of I–V relation can produce two stable membrane potentials: one near the resting potential and one at a more depolarized level. Under current clamp conditions, if the motoneuron is depolarized sufficiently to bring the membrane potential into the region over which the slope of the I–V relation is negative, then the membrane potential will gravitate toward the more depolarized stable state. If action potentials are blocked, this regenerative depolarization leads to a plateau potential (cf. Hultborn and Kiehn 1992; Kiehn 1991; Kiehn and Eken 1998). When action potentials are not blocked, the initiation of a plateau potential can either produce a sustained jump in discharge rate if the motoneuron is already firing (bistable firing) or lead to sustained firing if the motoneuron is initially quiescent (self-sustained firing; Kiehn and Eken 1998). If the voltage threshold for plateau initiation is higher than the threshold for action potential initiation, then bistable firing would be expected, whereas if the plateau is initiated at less depolarized voltages then self-sustained firing will occur (Kiehn and Eken 1998).

The relative prevalence of bistable and self-sustained firing is likely to depend upon a number of factors, including the level of activity in brainstem neurons with monoaminergic projections to motoneurons, the source of depolarizing current, and motoneuron type. The activity of brainstem neurons with descending serotonergic projections suggests that these neurons facilitate motor output in both postural and rhythmic tasks (Jacobs and Fornal 1993). In contrast, noradrenergic cells in the locus coeruleus of the cat fire at the highest frequencies under stressful conditions (Levine et al. 1990). The original descriptions of bistable discharge behavior in mammalian motoneurons were obtained in the decerebrate cat (Hounsgaard et al.

1984), and this behavior is thought to depend upon tonic activity in brainstem neurons with monaminergic projections to the spinal cord, since it is abolished by spinalization but reinstated by the adminstration of serotonergic or noradrenergic precursors to spinalized animals (Conway et al. 1988; Hounsgaard et al. 1988).

Recent evidence suggests that excitation of motoneurons by synaptic input (as opposed to somatically-injected current) lowers the somatic voltage threshold for plateau potentials (Bennett et al. 1998b), consistent with a dendritic location for the channels responsible for the plateau (section 3.3.3). This drop in "plateau threshold" favors self-sustained discharge as opposed to bistable discharge, i.e., the production of sustained discharge in an initially quiescent motoneuron by a transient excitatory input rather than a sustained change in discharge rate in a tonically discharging cell. This finding may explain the fact that although bistable discharge behavior can be produced by current injection (e.g., Conway et al. 1988; Hounsgaard et al. 1984; Hounsgaard et al. 1988; Hounsgaard and Kiehn 1989)), it is much rarer during natural synaptic activation of motoneurons (Eken 1998; Eken and Kiehn 1989; Kiehn and Eken 1997). In contrast, there have been several recent reports of self-sustained firing during quiet standing and slow locomotion in rats (Eken 1998; Gorassini et al. 1999), and during weak isometric contractions in humans (Gorassini et al. 1998; Kiehn and Eken 1997).

The clearest examples of self-sustained firing come from paired motor unit recordings showing that a transient synaptic input can recruit a previously silent motor unit into tonic activity without producing a sustained change in firing rate in a previously active motor unit (Gorassini et al. 1998; Kiehn and Eken 1997). Figure 15A (taken from Fig. 1A of Gorassini et al 1998) illustrates an example of this phenomenon. The subject produced a weak isometric contraction of the tibialis anterior (TA) muscle, producing about 3% of the force of a maximal voluntary contraction (top trace), and a steady-discharge rate in the 'control' motor unit of about 10 imp/s (bottom trace). A transient Ia excitatory input was then produced by vibration of the TA tendon. This input recruited a second motor unit (unit #2, third trace) and produced a transient increase in the discharge of the 'control' unit. Following the cessation of vibration, the firing rate of the 'control' unit returned to its pre-vibration level, and yet the newly recruited unit continued to discharge.

Figure 15B and C show two other types of motor unit behavior attributed to plateau activation. Figure 15B (taken from Fig. 4 of Kiehn and Eken 1997) shows the behavior of three simultaneously recorded motor units during a voluntary isometric ramp contraction of the human TA muscle. In all three motor units, the discharge rates increase abruptly shortly after recruitment but then vary relatively little during the subsequent increase in muscle force. This nonlinear behavior, which has also been reported during ramp contractions in other muscles (e.g., De Luca et al. 1982), is consistent with the activation of a plateau near the recruitment threshold of a motor unit. Finally, Fig. 15C (taken from Fig 2. of Eken 1998) shows rectified gross EMG activity of the rat soleus muscle (top trace), together with the activity of several different motor units (bottom two traces) obtained during a prolonged period of quiet standing. The expanded records of motor units activity (bottom traces) shows that each segment of tonic activity featured two or three motor units firing at similar and stable frequencies, and that different sets of units were active in different segments. The classical view of motor unit recruitment via the 'size principle' (section 2.1) would predict that units that are active at the lowest levels of motoneuron pool activity (segment 4) should continue to discharge at higher activation levels. Instead, the activity appears to 'cycle' between different sets of motor units over time, and this phenomenon has been attributed to the onset and offset of plateaus in motor units with similar recruitment thresholds (Eken 1998).

The functional role of plateau potentials in normal motor unit behavior is not yet clear. It has been suggested that plateau activation and its control by activity in descending monoaminergic systems is important during prolonged, low-level tonic activity (Kiehn and Eken 1998). The recent demonstration that selective depletion of spinal monoamines (Kiehn et al. 1996) results in a dramatic decrease in the tonic activity of soleus supports this view. It has also been suggested that the increase in membrane conductance during plateau activation could uncouple spike generation from the variability in synaptic input (Kiehn and Eken 1998), thus reducing firing rate variability. However, this increase in membrane conductance could also limit the amount of synaptic current that can be delivered to the spike generating conductances, making it difficult to achieve maximal rates of discharge (see section 4.1).

Metabotropic receptor control of plateau activation may also act to support rhythmic patterns of motoneuron activation. During fictive locomotion in the decerebrate cat, AHPs are reduced and alterations in discharge behavior occur that are consistent with the predominance of an active, inward current (Brownstone et al. 1992). It is possible that the combined activation of serotonin and NMDA receptors facilitate rhythmic oscillations in motoneuron discharge rate. Modeling work (Brodin et al. 1991; Ekeberg et al. 1991; Wallen et al. 1992) suggests that rhythmic oscillations in lamprey neurons during fictive locomotion depend upon calcium entry through NMDA channels and the consequent activation of G_{KCa}, and that serotonin affects

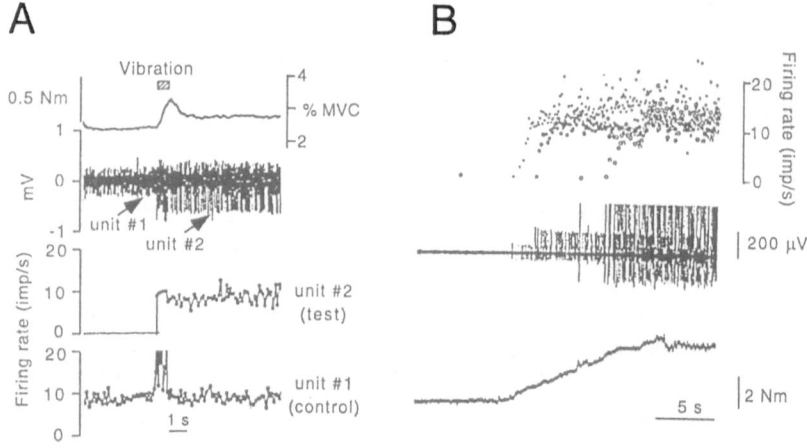

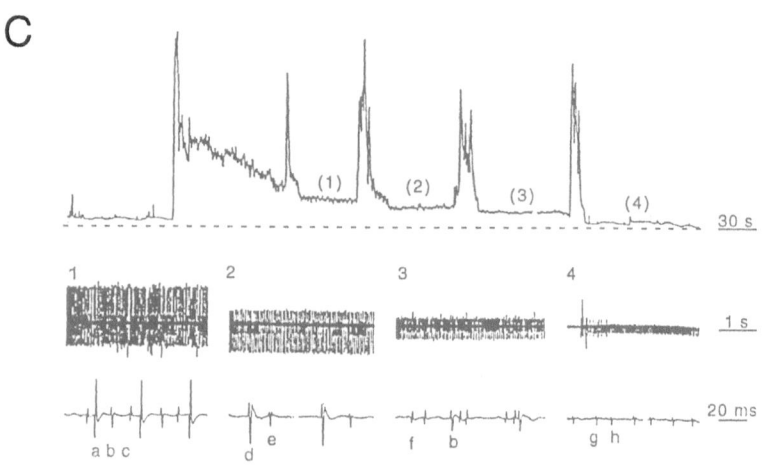

Fig. 15. Examples of motor unit discharge which may reflect the actions of plateau potentials. A. Self-sustained firing of a human tibialis anterior (TA) motor unit recruited by a brief vibration of the TA tendon. The top trace shows dorsiflexion torque, the next trace the raw extracellular record of the discharge of two motor units, and the bottom two traces the instantaneous frequency of each unit. Unit #2 is recruited by the vibration and continues to discharge after vibration offset even though the discharge rate of unit #1 returns to its pre-vibration level. (Modified from Fig. 1 of Gorassini et al. 1998) B. Firing behavior of three human TA units during isometric ramp contractions. Instantaneous discharge rate of the three units (upper), raw record of motor unit activity (middle) and dorsiflexion torque (lower

the frequency of oscillation through its effects on G_{KCa}. It is not known whether similar mechanisms contribute to locomotion in higher vertebrates. Bath applied 5-HT facilitates NMDA-induced rhythmic oscillations in neonatal rat spinal motoneurons (Schmidt et al. 1998), and fictive locomotion can be induced in spinal cats by the administration of noradrenergic precursors (Rossignol et al. 1998). In guinea pig trigeminal motoneurons, slow bursting can be induced in the presence of serotonin, and in these cells G_{KCa} also plays a role in burst termination (Del Negro et al. 1999).

4
Unresolved Issues Regarding Motoneuron Input-Output Functions

Although the past twenty years of work have produced considerable advances in our understanding of the biophysical mechanisms governing motoneuron function, there are a number of unresolved issues. The following sections focus on three different aspects of motoneuron input-output properties that are still not well understood: (1) the role of dendritic conductances in governing the delivery of synaptic current to the soma, (2) the biophysical mechanisms responsible for spike-frequency adaptation, and (3) the relation between the magnitude and time course of synaptic current transients and their effects on discharge probability.

4.1
Integration of Concurrent Synaptic Inputs:
Potential Contributions of Active Dendritic Conductances

The amount of synaptic current transferred to the soma from a given set of activated synapses will be affected by concurrent transmitter release from other synapses due to changes in the driving force for synaptic current flow and to an increase in membrane conductance. Both of these influences are likely to lead to sub-linear summation of the currents produced by two sepa-

---

trace). All three units showed an initial steep increase in discharge rate upon recruitment, followed by much slower increases in rate over the remainder of the ramp. (Modified from Fig. 4 of Kiehn and Eken 1997) . C. Cycling of activity between different motor units in a rat soleus muscle during quiet standing. The top trace shows integrated rectified gross-EMG activity with four different tonic segments labeled. The middle traces show the unprocessed EMG signal underlying each of the four segments. The bottom trace shows parts of these records expanded in time to reveal the activity of distinct sets of motor units (labeled a–h). (Modified from Fig. 2 of Eken and Kiehn 1989) rate sets of synapses (Jack et al. 1975; Rall 1964, 1967). The magnitude of the difference between the summed effects of two inputs and their expected linear sum will depend both upon the relative proximity of the two sets of synapses (e.g., Rall 1964, 1967; Rall et al. 1967), and upon the extent to which the transfer of synaptic current to the soma is modified by voltage-sensitive dendritic conductances (Bernander et al. 1994; Clements et al. 1986).

The vast majority of experimental and theoretical work on summation of synaptic inputs has concentrated on the effects of transient activation of a relatively small subset of the synaptic terminals on a particular neuron (e.g., Barrett and Crill 1974a; Burke 1967, Kuno and Miyahara 1969). Due to the high input resistance of the distal dendritic segments, even unitary synaptic potentials may produce a large decrease in driving force, leading to sub-linear summation of the voltage changes produced by two inputs (Barrett and Crill 1974a; Kuno and Miyahara 1969). However, when tens to hundreds of presynaptic fibers are activated, the widespread distribution of synaptic terminals on the dendritic tree may help isolate synapses from one another, thus minimizing nonlinear interactions. Indeed, measurements of summation of monosynaptic Ia excitatory postsynaptic potentials (PSPs) in mammalian motoneurons indicate that linear summation is often observed and departures from linearity are relatively small (i.e., generally less than 10%; Burke 1967). In contrast, physiological activation of neurons is achieved by repetitive discharge of thousands of presynaptic fibers. In passive dendrites, less-than-linear summation might be expected to be more common in the case of repetitive synaptic activation than for transient synaptic inputs for two reasons. First, the steady-state change in voltage produced at one synaptic site will decay less with distance than would a transient change in voltage, leading to a larger change in driving force at other synaptic sites (Jack et al. 1975). Second, the steady-state increase in membrane conductance during synaptic activation will reduce the amount of current reaching the soma from more distal sites (Barrett 1975; Barrett and Crill 1974a; Jack et al. 1975).

The presence of active dendritic conductances has been shown in other neurons to either exaggerate the sub-linear summation of synaptic inputs expected in passive dendrites (Urban and Barrionuevo 1998), to allow linear summation (Cash and Yuste 1998) or to mediate supra-linear summation (Wessel et al. 1999). The net effect of active conductances is likely to depend upon their distribution, voltage-sensitivity and kinetics as well as upon on the local voltage changes produced by the synaptic inputs (cf. Bernander et al. 1994; Cook and Johnston 1999). Unfortunately, as discussed above (see section 3.3.3), there is relatively little information about the nature and distribution of active conductances on motoneuron dendrites. The next three sections will present some recent experimental work on summation of synaptic inputs under conditions in which the effects of persistent inward dendritic currents are either suppressed or enhanced, followed by some modeling work on the relation between dendritic properties and the transfer of synaptic current to the soma.

4.1.1
Summation of Synaptic Inputs on 'Passive' Dendritic Tree

A recent set of experimental findings concerning the summation of synaptic inputs was obtained in spinal motoneurons studied in intact, pentobarbital-anesthetized cats (Powers and Binder 2000). Under these conditions, there is likely to be little or no activity in descending monoaminergic fibers (Hultborn and Kiehn 1992), and as a result, the extracellular concentration of serotonin and noradrenaline will be very low. As discussed above, these substances normally act to favor the expression of persistent inward currents, both by direct facilitating action and by suppressing leak and other potassium currents (see section 3.5.2). In addition, pentobarbital has recently been shown to directly suppress L-type calcium channels (Guertin and Hounsgaard 1999), which have been shown to contribute to the persistent inward current in turtle spinal motoneurons (Hounsgaard and Mintz 1988) and guinea pig trigeminal motoneurons (Hsiao et al. 1998), and may also contribute in cat spinal motoneurons (Schwindt and Crill 1980a). The behavior of motoneuron dendrites in the anesthetized cat is thus likely to be dominated by their passive properties and by voltage-activated potassium conductances.

Summation of synaptic inputs has been studied by comparing the steady-state synaptic current reaching the soma (effective synaptic current; I_N, see section 3.2), during repetitive activation of different sets of presynaptic fibers (Powers and Binder 2000). As discussed earlier (section 3.2) and elsewhere (e.g., Binder et al. 1993; 1996; Powers and Binder 1995), measurements of effective synaptic currents include an estimate of the effective synaptic current flowing at the resting potential (I_{Nrest}), the change in input conductance produced by the synaptic input, and the effective synaptic current flowing at the threshold for repetitive discharge ($I_{Nthresh}$). This last quantity, together with the slope of the motoneuron's steady-state frequency-current relation (f/I) can be used to estimate the change in firing rate produced by a steady-state synaptic input (Powers and Binder 1995 and section 3.2).

In general, when two different sets of afferent fibers are activated concurrently, the amplitude of both the total effective synaptic current and the change in motoneuron firing rate are generally equal to or slightly less than the linear sum of the effects produced by activating each input alone. Figure 16A (modified from Powers and Binder 2000) shows that the total effective synaptic current estimated at threshold (observed $I_{Nthresh}$) during concurrent activation of two different inputs was quite close to the linear sum of their individual effects (predicted linear sum). This general finding applied during activation of many different combinations of inputs (Ia afferents: Ia, descending rubrospinal fibers: RN, descending fibers from Dieter's nucleus: DN, descending pyramidal tract fibers: PT, cutaneous afferents from the sural nerve: SN, mixed afferents from the common peroneal nerve: CP). Even when synaptic activation produced a large (>20%) decrease in the input resistance of the cell (open symbols), summation of effective synaptic currents was close to linear. As expected from these results, Fig. 16B shows summation of synaptically-evoked changes in firing rate was also close to linear, although sublinear summation was more common (14 of 18 points fall below the line of identity).

The near-linear summation of synaptic currents and synaptically-evoked changes in firing rate described above may result from the fact that inputs used may still represent a relatively small fraction of the total input to motoneurons. The individual inputs generally produced effective synaptic currents of ± 10 nA, whereas synaptic inputs required to produce maximal output of the motoneuron pool may be up to an order of magnitude larger (Binder et al. 1996). Thus, a relatively sparse, widespread distribution of active synapses may help minimize nonlinear interactions even during repetitive activation. Simulation results (Rose and Cushing 1999 and section

---

Fig. 16. Summation of effective synaptic currents and firing rate modulation in cat triceps surae motoneurons. A. Effective synaptic current at threshold during concurrent activation of two different inputs versus the linear sum of the currents produced by each input alone. B. Changes in discharge rate produced by concurrent activation of two inputs versus the linear sum of the changes produced by each input alone. Open symbols refer to cases in which synaptic activation produced a greater than 20% decrease in the input resistance of the cell, whereas filled symbols refer to cases in which the synaptic input produced less than a 20% change in input resistance. Different symbol types refer to different sources of afferent input. Ia: primary muscle spindle afferents, CP: low and high threshold afferents in the mixed common peroneal nerve, SN: low and high threshold cutaneous afferents from the sural nerve, RN: descending fibers from the contralateral red nucleus, DN: descending fibers from the ipsilateral Dieter's nucleus, PT: descending fibers from the ipsilateral pyramidal tract. The half-filled circles in B indicate two motoneurons in which the summation of Ia and RN effects on firing rate were measured, but their effects on membrane conductance were not. (Modified from Figs. 2 and 7 of Powers and Binder 2000)

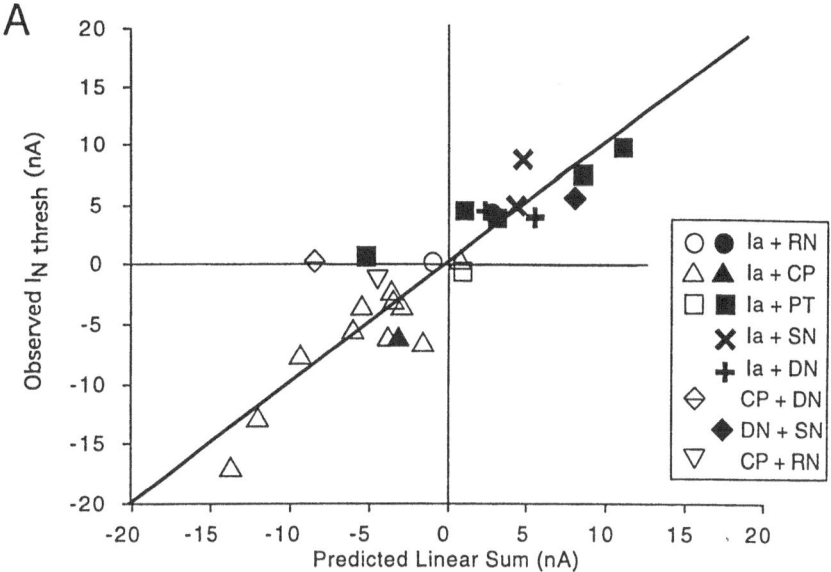

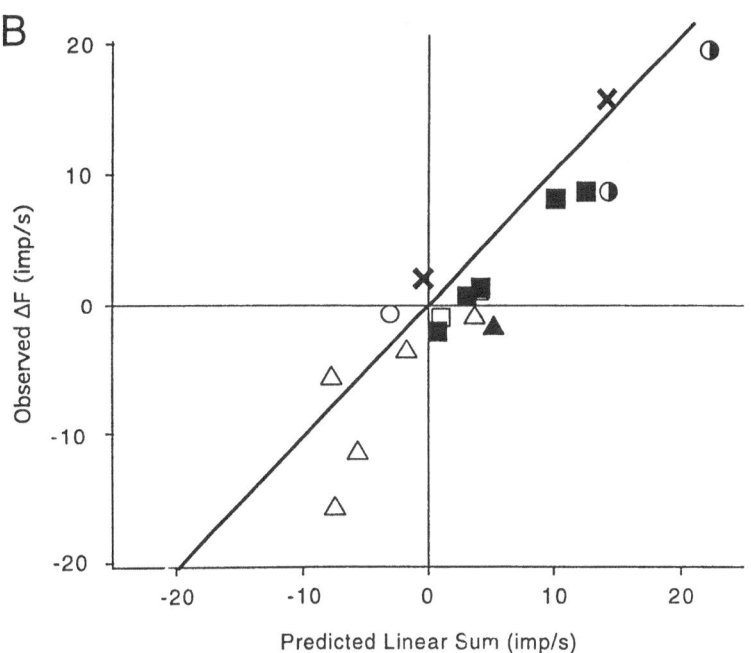

4.1.3) suggest that nonlinear summation should be more profound at higher levels of presynaptic activity. An alternative explanation for the near-linear summation observed above is that activation of dendritic conductances helps compensate for the changes in driving force and increased conductance produced by concurrent activation of two sets of synapses (section 4.1.2).

4.1.2
Summation of Synaptic Inputs with "Active" Dendrites

A number of lines of evidence suggest that even in the absence of significant metabotropic receptor activation, motoneuron dendrites do not behave as passive structures. The steady-state I–V relations obtained in motoneurons recorded in anesthetized animals can be N-shaped, reflecting the activation of a persistent inward current (Schwindt and Crill 1977). As argued earlier (section 3.3.3), the I–V relation measured during somatic voltage clamp is likely to reflect activation of inward currents on the dendrites. The net inward current measured at the soma is enhanced by procedures that suppress outward currents (Schwindt and Crill 1980a–c), and this effect may also arise from modification of dendritic channels. This view is supported by the finding that Ia EPSPs in cat spinal motoneurons increase in amplitude and duration when K^+ channels are blocked with internal TEA (Clements et al. 1986). Finally, depolarization of the dendrites of turtle motoneurons leads to Ca^{2+} -mediated spikes and plateau potentials when external K^+ blockers are applied (Hounsgaard and Kiehn 1993). These results can be replicated in a model with both calcium and calcium-activated potassium conductances on the dendritic compartment (Booth et al. 1997).

It has recently been demonstrated that plateau potentials can be produced in spinal motoneurons recorded in intact, pentobarbital-anesthetized cats following internal blockade of potassium channels (Powers and Binder 2000). These potentials generally decayed within a few hundred ms following the end of current injection, except in a few cases in cells impaled with electrodes containing 3M cesium. Figure 17A (taken from Fig. 9B of Powers and Binder 2000) shows examples of plateau potentials evoked by somatically-injected current after impalement with a microelectrode containing cesium and QX-314 to block action potentials. The initial voltage response to the current step (1) declined back to the resting potential within about 0.5 s following the offset of the current. Six minutes later the voltage response to the same injected current step (2) remained at a depolarized level for over 15 s following the offset of the current step, and the subsequent response (3) exhibited a voltage plateau lasting more than 30 s following current offset.

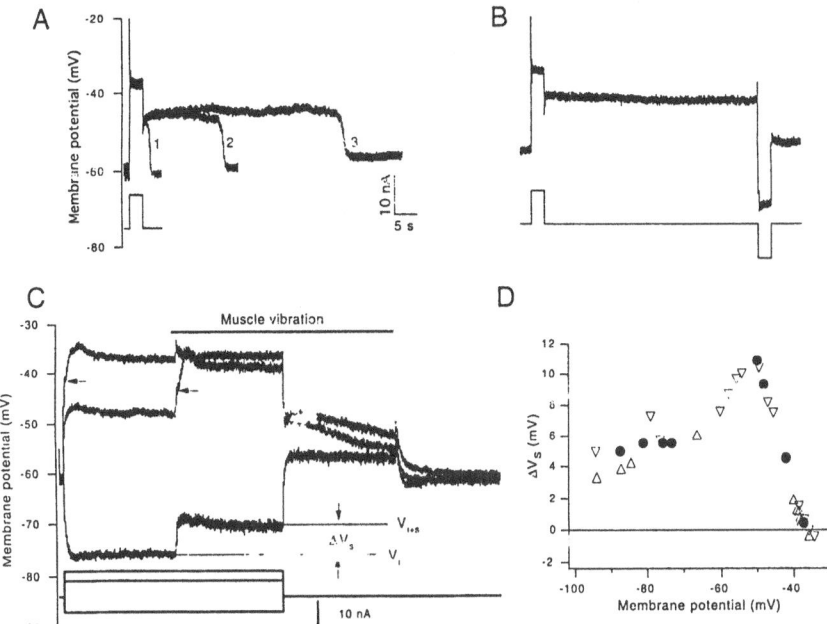

Fig. 17. Plateau potentials in cat triceps surae motoneurons impaled with electrodes containing potassium channel blockers. **A.** Long-lasting plateau potentials evoked by successive injections of a 2.5 s 8.9 nA step of current obtained from 33–39 minutes following impalement with an electrode containing 3M CsCl. **B.** In the same cell, the same stimulus applied 2.5 minutes later evokes a stable plateau that is only terminated following a hyperpolarizing current pulse. (A and B modified from Fig. 9B of Powers and Binder 2000) **C.** Voltage-dependent amplification of excitatory synaptic input in a motoneuron impaled with an electrode containing 1M TEA-Cl. The top traces show the voltage response to a series of 1 s injected current steps of different amplitude (bottom traces) in combination with muscle vibration. The amplitude of the steady-state depolarization produced by vibration (ΔV$_s$) clearly depends upon the background level of membrane potential. At depolarized levels plateau potentials are initiated (arrows). **D.** Relation between ΔV$_s$ and the background membrane potential. (C and D modified from Fig. 8A and C of Powers and Binder 2000)

Figure 17B shows the response to the same stimulus obtained 2.5 min later, and in this case the voltage plateau could be terminated only by applying a hyperpolarizing current step. This stable plateau state indicates that potassium currents have been sufficiently blocked to lead to a region of net inward current in the cell's steady-state current-voltage (I–V) relation (Schwindt and Crill 1980c).

Even when potassium channels were not sufficiently blocked to produce stable plateau potentials, the change in motoneuron properties led to a voltage-dependent amplification of synaptic input. Figure 17C (taken from Fig. 8A of Powers and Binder 2000) illustrates the steady-state synaptic potentials produced by repetitive activation of Ia afferents (via vibration of the triceps surae muscle) at three different levels of somatic membrane potential (produced by different amounts of injected current – lower traces). When the membrane is hyperpolarized relative to the resting potential (lower voltage trace), the amplitude of the synaptic potential (ΔV_s) is about 5 mV. The amplitude is clearly increased when Ia activation is superimposed upon the depolarization produced by a +6 nA injected current step (middle voltage trace). The initial response to the Ia input (right horizontal arrow) is similar in amplitude to that seen at the hyperpolarized membrane potential. This is followed by a further depolarization that leads to a total change in voltage that is nearly twice that seen at the hyperpolarized membrane potential. A further increase in the magnitude of depolarizing injected current leads to amplification prior to the onset of muscle vibration (left horizontal arrow). The additional depolarization produced by Ia input is quite small in this case. The membrane potential thus appears to be clamped at a more depolarized level, as previously reported for plateau potentials obtained in the decerebrate cat preparation (Hounsgaard et al. 1984; Hounsgaard et al. 1988).

The steady-state amplitude of the Ia synaptic potential exhibited a steep dependence upon membrane potential, first increasing as the membrane was depolarized from the resting potential and then decreasing sharply once a critical level of depolarization was reached. The resting potential often became more depolarized over time, and this may have reflected increasing block of a resting potassium conductance (cf. Campbell and Rose 1997). Figure 17D illustrates this relation for the same motoneuron. The filled circles represent synaptic potentials recorded from 11–16 minutes after impalement, when the resting membrane potential remained fairly stable (-61.2 to -60.7 mV), and the open circles represent data taken over a wider range of resting potentials. The peak amplitude of the steady-state synaptic depolarization was 10.6 mV, which was 2.8 times greater than the mean amplitude recorded at hyperpolarized membrane potentials. This amount of amplification was typical for the sample of motoneurons studied with internal potassium channel blockers (range 1.6–7.3 times control value, mean = 2.8 ± 1.8, n=10). The synaptic current flowing during maximum amplification could be calculated by dividing the steady-state synaptic potential amplitude by estimated input resistance of the motoneuron during synaptic activation. This estimated value ranged from 0.6–5.1 (mean 2.3 ± 1.5) times greater than the synaptic current estimated at the resting potential.

The pattern of voltage-dependent amplification illustrated in Fig. 17D suggests that summation of synaptic potentials can range from sub-linear to supra-linear, depending upon the background membrane voltage. Over the range in which the relation between background membrane potential and ΔV_s is flat, the depolarization produced by one excitatory input will not affect that produced by another, and summation will be linear. Over the range in which the slope of the relation is increasing, supra-linear summation might be expected, whereas a decreased slope will be associated with sub-linear summation. Thus for the relation illustrated in Fig. 17D, the combined depolarization produced by two excitatory synaptic potentials might be near the linear sum of their individual effects at hyperpolarized membrane potentials, greater-than-linear at slightly more depolarized potentials and less-than-linear at potentials more depolarized than about -50 mV.

The modification of channel properties by internal potassium channel blockers and QX-314 is likely to be very different from that produced by physiological activation of metabotropic receptors. Internal TEA and Cs are relatively non-selective for different potassium channel types (Hille 1992), and QX-314, in addition to blocking sodium channels, also blocks calcium channels (Talbot and Sayer 1996) and the channels mediating I_h (Perkins and Wong 1995). However, activation of metabotropic receptors by monoamines has been shown to reduce calcium-activated potassium conductances, a resting potassium leak conductance, and a hyperpolarization-activated mixed-cation conductance in a variety of types of motoneurons (see section 3.33). Further, serotonergic terminals have also recently been shown to be localized primarily on motoneuron dendrites (Alvarez et al. 1998). Thus, the effects of neuromodulators may be largely mediated by their action on dendritic potassium and mixed-cation conductances. For this reason, it is not surprising that the effects of internally-applied potassium channel blockers on motoneuron behavior seen here are qualitatively similar to those of neuromodulators. Recent somatic voltage-clamp data obtained in spinal motoneurons of decerebrate cats show a voltage-dependent amplification of Ia synaptic currents similar to that shown for Ia synaptic potentials in Fig. 17D (cf. Fig. 3 in Lee and Heckman 2000). In both cases, there is voltage-dependent amplification of synaptic input starting at voltages slightly depolarized to the resting potential. Quantitative differences may arise due to the different spatial distributions of the modulation of channel behavior: potassium channel blockers diffusing from the somatic site of electrode impalement should have the most marked effect on proximally located channels, whereas the widespread distribution of serotonergic terminals (Alvarez et al. 1998) suggests that exogenous neuromodulators may exert more spatially-uniform effects on dendritic channels. Further, neuromodulators have been shown in some cases to directly enhance the persistent inward current (Hsiao et al. 1998; Svirskis and Hounsgaard 1998), whereas QX-314 reduces the magnitude of this current (Lee and Heckman 1999b).

In spite of the unphysiological nature of internal channel blockade, it is likely that even under more physiological conditions the summation of synaptic currents and their effects on motoneuron firing rate will depend upon the state of active dendritic conductances. However, the presence of these conductances does not necessarily preclude a simple prediction of the effects of synaptic inputs on firing rate. Our prediction of the effects of synaptic current on firing rate (Powers and Binder 1995) is based on an estimate of the effective synaptic current flowing during repetitive discharge ($I_{Nthresh}$). In dendrites with significant voltage-activated inward currents, $I_{Nthresh}$ will be greater than the current measured at the resting potential (I_{Nrest}). Nonetheless, the predicted change in firing rate (ΔF) produced by this amplified current is still the product of $I_{Nthresh}$ and f/I, and this prediction has been recently shown to hold for neocortical pyramidal cells, even though persistent inward currents on the dendrites amplify the effective synaptic current in a voltage-dependent manner (Schwindt and Crill 1996).

The effects of amplification of synaptic currents may depend both on the strength of the synaptic input and on the voltage range over which the persistent, inward dendritic currents are activated (Lipowsky et al. 1996). If the dendritic voltages reached during combined synaptic activation and repetitive discharge already lead to near maximal activation of these inward currents, then further changes in depolarization associated with increasing firing rates may not lead to any further amplification. In contrast, if individual excitatory synaptic inputs produce a level of dendritic depolarization that is subthreshold for activation of a persistent inward current, but in combination produce a suprathreshold depolarization, then their combined effective synaptic current and effects on firing rate will be greater than the linear sum of their individual effects.

Recent work on the effects of synaptic input on motoneuron firing rate (Prather et al., 2000) suggests that in the decerebrate cat, synaptic currents are amplified by an intrinsic mechanism, and yet amplified currents from two different sets of afferents exhibit linear summation. This surprising result is difficult to reconcile with the presence of dendritic plateau potentials produced by activation of a persistent inward current. If the combination of synaptic and voltage-activated inward current is of sufficient magnitude to produce a dendritic plateau potential, this high conductance state could shunt other synaptic inputs, leading to less-than-linear summation

(Kiehn and Eken 1998; Oakley et al. 1999). The linearity of current summation observed in the presence of amplification suggests that synaptic integration processes are undisturbed by the activation of intrinsic inward currents (cf. Binder et al. 1993). The simulation results described in the next section suggest that the activation of dendritic inward currents should be able to amplify synaptic currents only when the level of excitatory input to a motoneuron is low, and that saturation and less-than-linear summation is expected to occur at higher input levels.

4.1.3
Computer Simulations of the Transfer of Synaptic Current from Dendrites to Soma

The high input resistance of distal dendritic branches (Rall and Rinzel 1973) suggests that even modest levels of excitatory input should produce large local depolarizations of the dendritic membrane. Previous simulation work suggested that the depolarization produced by even a single synaptic input onto a distal branch results in a significant decrease in the driving force for synaptic current flow (Barrett and Crill 1974a; Ulrich et al. 1994), leading to less-than-linear summation of the synaptic potentials produced by two synapses activated on the same branch (Barrett and Crill 1974a). However, there has been relatively little quantitative analysis of the effects of these types of interactions on the ability of branched dendritic trees to deliver synaptic current to the soma during activation of large numbers of synapses (Korogod and Kulagina 1998; Rose and Cushing 1999). In a recent simulation study, compartmental models of three anatomically-reconstructed cat cervical motoneurons were used to estimate the synaptic current delivered to the soma during different levels of activation of excitatory synapses (Rose and Cushing 1999). The synaptic conductance change (g_{syn}) produced in a given compartment by repetitive activity in a group of presynaptic afferents was calculated according to equation 13:

$$g_{syn} = g_{peak} * \exp(1) * t_{peak} * n * f * P \tag{13}$$

The product ($g_{peak} * \exp(1) * t_{peak}$) represents the area under a quantal synaptic conductance change, where g_{peak} and t_{peak} are the peak conductance and time to peak conductance, respectively (Finkel and Redman 1983). The number of synaptic boutons on a given compartment (n) was estimated from published anatomical data on synaptic density (Rose and Neuber 1991), the firing frequency (f) was assumed to be the same for all active in-

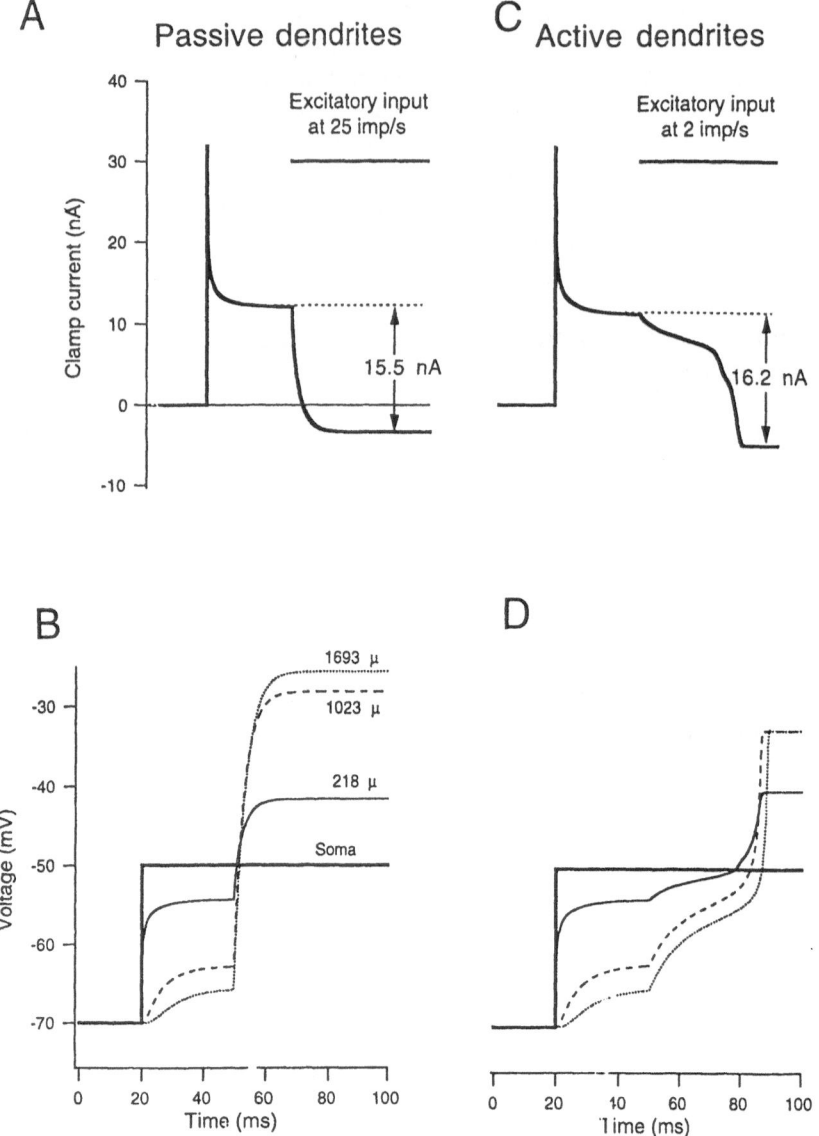

Fig. 18. Effects of active dendritic conductances on the behavior of a multi-compartmental motoneuron model. Model consisted of a cylindrical soma (length = diameter = 50 μm) connected to six identical dendritic trees. The dendritic tree was based on the published reconstruction of the dendritic tree of a cat triceps surae type FR motoneuron (motoneuron 42/4, dendrite E in Fleshman et al. 1988). The total surface area of the model was 644,766 μm², which is near the top end of the range

puts, and each terminal was assumed to release transmitter with an average probability P of 0.5 (Redman and Walmsley 1983a). When different proportions of the total number of excitatory synapses were activated at 100 imp/s, the total current delivered to the soma increased in a less-than-linear fashion with increasing excitatory input, and even maximal activation resulted in much less current than is typically needed to produce maximal firing rates. A qualitatively similar result was obtained in all three motoneuron models and was not appreciably changed by changing specific membrane resistivity between 6,000 and 60,000 Ωcm^2.

These simulation results suggest that passive motoneuron dendrites may be incapable of generating sufficient current to produce the firing rates required to fully activate the innervated muscle. It is possible that the 'missing' current may be generated by activation of persistent inward current on the dendrites. To examine this possibility, we constructed a compartmental motoneuron model consisting of a soma with six identical dendritic branching trees attached to it. The structure of each tree was derived from the published description of reconstructed dendritic tree from a type FR cat triceps surae motoneuron (Cell 42/4, Dendrite E from Fig. 6 of Fleshman et al. 1988). In order to estimate the average effective synaptic current flowing during repetitive discharge the soma membrane potential was clamped at -50 mV, which represents a typical average value during repetitive discharge in cat spinal motoneurons (cf. Schwindt and Crill 1982). Repetitive activation of excitatory afferents was simulated according to equation 13. Using a peak quantal conductance change of 5 nS and a time to peak of 0.2 ms (Finkel and Redman 1983), a release probability of 0.5 and a excitatory synaptic density of 5 boutons per 100 μm^2 (e.g., Brannstrom 1993; Bras et al. 1987), repetitive discharge in all of the excitatory inputs at a rate of 10 imp/s produces a synaptic conductance change of $6.75 * 10^{-8}$ S/cm^2.

Figure 18A shows the somatic voltage clamp current obtained in a model motoneuron with passive dendrites in response to a voltage command step from -70 to -50 mV, followed by the onset of a steady synaptic conductance of $1.69 * 10^{-7}$ S/cm^2 applied to the entire dendritic surface (corresponding to

---

reported for cat triceps surae motoneurons in Cullheim et al. 1987a. The cytoplasmic resistivity was 70 Ωcm, the specific membrane capacitance was 1 $\mu F/cm^2$ and the specific membrane resistivity was 225 Ωcm^2 in the soma and 11000 Ωcm^2 elsewhere. A. Voltage clamp current recorded during a somatic voltage clamp command from -70 to -50 mV. Excitatory synaptic input was applied to the dendrites during the time indicated by the bar. B. Somatic voltage and voltages on different parts of the dendritic tree at the specified distances from the soma. C and D. Same as A and B, except that a smaller excitatory input was applied to a model with active dendrites. See text for further details

a discharge rate of 25 imp/s in all of the excitatory terminals). The effective synaptic current (i.e., clamp current at the soma) was 15.5 nA, which would be slightly suprathreshold for repetitive discharge in a typical cat triceps surae type FR motoneuron (Kernell and Monster 1981). However, further increases in excitation produce relatively small increments in effective synaptic current, because as shown in Fig. 18B, the dendritic membrane potential is relatively close to the excitatory equilibrium potential, particularly in the distal dendrites.

Figure 18C and D illustrate the behavior of a model with active dendrites. Voltage-sensitive conductances mediating inward and outward currents were inserted uniformly on the dendritic membrane, and the voltage-activation curves and maximal conductances were adjusted to reproduce typical steady-state I-V curves as measured during a somatic voltage clamp (see below). Much lower levels of excitatory input are required to deliver a suprathreshold current to the soma because the depolarization produced by the synaptic input is sufficient to initiate dendritic plateau potentials (Fig. 18D). However, the large dendritic depolarizations and the high conductance during the plateau potentials makes further increments in excitatory input relatively ineffective.

The limited current delivering capacity of active dendrites was relatively independent of the spatial distribution of the conductances mediating inward and outward currents, provided that the somatic voltage-clamp behavior was similar. Figure 19A shows voltage-clamp currents obtained during a slow voltage ramp in the different models. The overall shapes of the functions are similar to those reported for spinal motoneurons recorded in the decerebrate cat (Lee and Heckman 1999a). In different models, the spatial distribution of the conductances mediating the voltage-dependent inward (g_i) and outward currents (g_o) could either be uniformly distributed (U), decrease in a sigmoidal fashion with distance from the soma (proximally weighted, P) or increase with distance (distally weighted, D). In each case the amplitudes and half-activation voltages of the conductances were adjusted so that overall I-V curves were similar (Fig. 19A). Figure 19B shows that at

Fig. 19. I-V curves and effective synaptic current in a models with active dendrites and different spatial distributions of conductances mediating inward (g_i) and outward currents (g_o). U: uniform spatial distribution, P: proximally weighted distribution, D: distally weighted distribution. A. Clamp currents measured during slow somatic voltage ramp commands. B. Effective synaptic current measured at a somatic voltage of -50 mV as a function of the strength of excitatory input. See text for further details

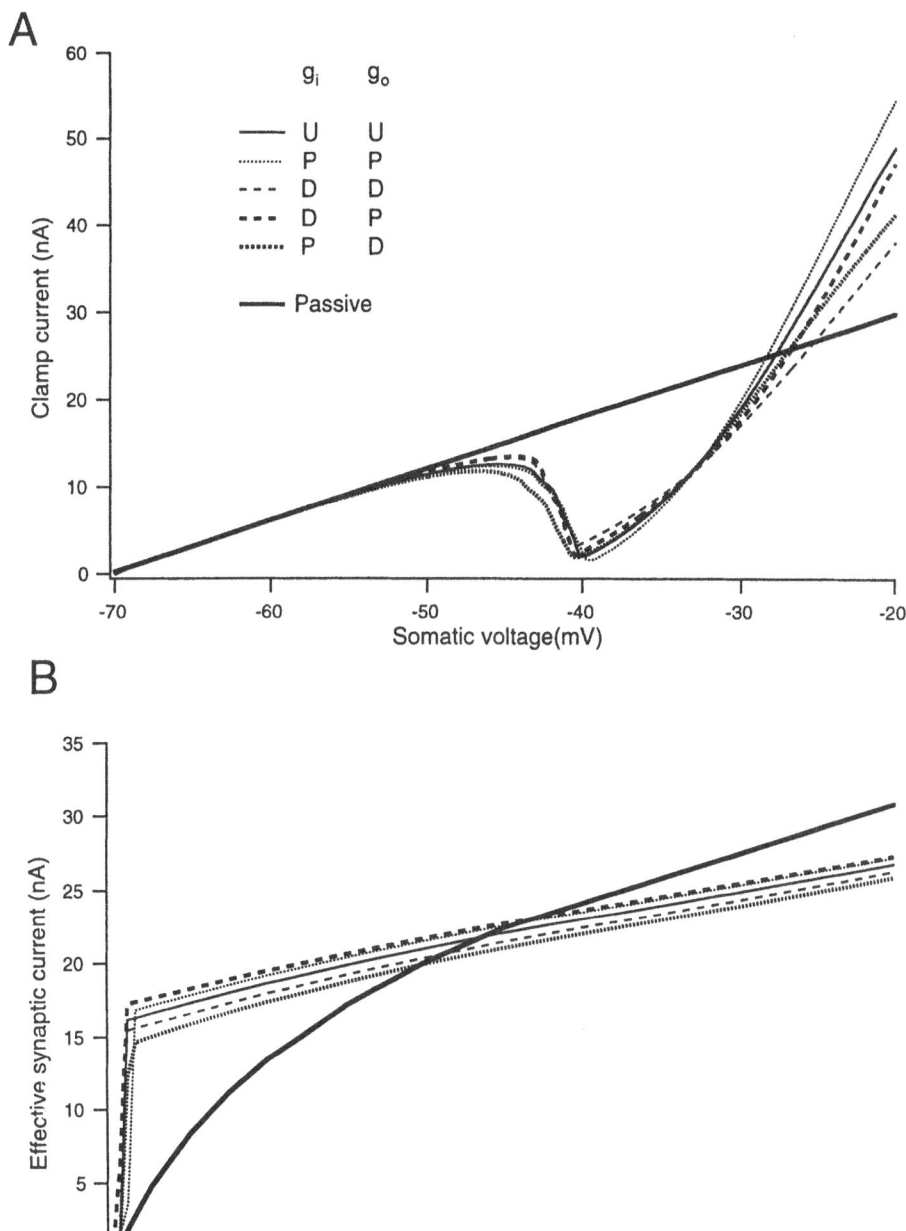

low levels of excitatory input, all of the active models produced more effective synaptic current than the passive model. However, due to the shunting effect of dendritic plateaus, further increases in synaptic input produced relatively little increase in the current delivered to the soma. In fact, above input rates of about 50 imp/s, the active dendrites delivered less current than the passive dendrites.

The presence of dendritic potassium conductances (g_o) will thus limit the ability of active dendrites to deliver synaptic current to the soma (see also, Korogod and Kulagina 1998). It is possible to produce whole-motoneuron I–V curves similar to those shown in Fig. 19 in a model with potassium conductances confined to the soma. In this case, active dendrites can transmit higher levels of synaptic current to the soma than passive dendrites (simulations not shown). However, the available experimental evidence suggests that at least part of the outward current measured during somatic voltage-clamp is derived from dendritic potassium channels. First, the amplitude of somatically-recorded current produced by glutamate iontophoresis onto the dendrites of turtle motoneurons is altered by the extracellular application of channel blockers and neuromodulators (Skydsgaard and Hounsgaard 1994; Skydsgaard and Hounsgaard 1996). Second, activation of metabotropic receptors by serotonin can modify resting leak and calcium-activated potassium currents (see section 3.5.2), and serotonergic boutons are distributed throughout the dendrites of cat motoneurons (Alvarez et al. 1998). Finally, outward currents recorded during somatic voltage-clamp of rat hypoglossal motoneurons are only partially blocked by internal cesium, and are eliminated only by the further addition of external potassium channel blockers (Musick 1999). This result suggests that in absence of external K^+ channel blockers, the internal concentration of cesium in the dendrites is apparently insufficient to completely block dendritic K^+ channels.

In summary, it is difficult to reconcile the 'all-or-none' nature of dendritic plateau potentials with the smooth gradation of motoneuron firing rate by synaptic inputs. Theoretical work has suggested that an appropriate combination of an inward current and an inactivating outward current could lead to both amplification of synaptic input and linear summation of synaptic inputs (Bernander et al. 1994). In addition, it has been proposed that the voltage-dependence of NMDA-receptor-activated current might enable synapses that utilize a combination of NMDA and non-NMDA receptors to provide a net synaptic current that is relatively independent of membrane voltage over a wide range of membrane potentials (Cook and Johnston 1999). Finally, even though the subthreshold responses of reconstructed motoneurons can be replicated with models in which the leak conductance is uniformly low throughout the dendrites or decreases monotonically with distance from the soma (Clements and Redman 1989; Fleshman et al. 1988), it is possible that the leak conductance is actually high in the distal dendrites. This condition has been shown to improve signal transfer by reducing the change in driving force during distal synaptic inputs (London et al. 1999). Assessing the potential contribution of these different mechanisms to synaptic integration in motoneurons will require more direct information on the properties of motoneuron dendrites.

4.2
Mechanisms Underlying Spike-Frequency Adaptation

The discharge rate of motoneurons is a function not only of the amplitude of the synaptic or injected current delivered to the soma, but also of its time course. In response to a step of current, the discharge rate typically declines as a function of time since current onset, a process known as spike-frequency adaptation. Adaptation has a rapid initial phase (Kernell 1965a; Sawczuk et al. 1995b, 1997), followed by a slow, gradual decline that continues throughout the duration of firing (late adaptation; Granit et al. 1963; Kernell 1965a, Kernell and Monster 1982b; Spielmann et al. 1993). The short interspike intervals that typically occur at the onset of an excitatory input to a motoneuron increases the speed of force development in its innervated muscle fibers (Baldissera et al. 1987). The later slow decline in firing rate can allow motor units to maintain a steady force by matching increases in twitch contraction times with a decrease in activation rate (Bigland-Ritchie et al. 1983; Sawczuk et al. 1995a), although it may eventually lead to a significant decline in force output (Kernell and Monster 1982a).

In spite of the functional importance of spike-frequency adaptation, its underlying biophysical mechanisms are not well understood. A recent quantitative analysis of the time course of spike-frequency adaptation in rat hypoglossal motoneurons recorded *in vitro* with sharp electrodes (Sawczuk et al. 1995b) revealed that most cells responded to a long step of injected current with three, temporally distinct phases of adaptation: initial, early and late. The initial phase consists of a rapid drop in frequency that is a linear function of time. This phase is followed by a more gradual decline that is generally fit by the sum of two exponential functions. The early process has a time constant on the order of 250 ms, and is followed by a slower process that has a time constant on the order of 10–20 s. It is not clear whether or not there are different biophysical mechanisms underlying the different phases of adaptation (rev. in Powers et al. 1999).

Much of the previous research on the initial and early phases of adaptation has examined the role of the medium-duration afterhyperpolarization (mAHP) which follows each action potential. As discussed earlier (section 3.4.1), the mAHP results from a calcium-sensitive potassium conductance activated by calcium entry during the action potential. If two action potentials are evoked in rapid succession, the second spike may occur before the calcium concentration and the calcium-activated potassium conductance have decayed to resting levels and the result is a summation of this potassium conductance. The increase in potassium conductance will be associated with an increase in membrane hyperpolarization and a decrease in firing rate. As discussed earlier (section 3.4.1), the mAHP is tied to calcium entry through N- and P-type calcium channels, but not T-type calcium channels (Kobayashi et al. 1997; Umemiya and Berger 1994; Viana et al. 1993b). This selectivity, together with the relatively rapid onset of the mAHP (Sah 1996), suggests that the calcium-activated potassium channels responsible for the mAHP are in close proximity to the N- (and P-) type calcium channels. The calcium concentration near the G_{KCa} channels will represent a balance between influx through the adjacent calcium channels, diffusion, extrusion via membrane-bound pumps and binding to intracellular calcium buffers (cf. Sah 1992, 1996). As discussed above (section 3.4.1), the endogenous buffering of motoneurons is relatively low. During long-lasting repetitive discharge it is therefore possible that saturation of calcium buffers would lead to a larger and longer local calcium transient following each action potential.

Earlier work indicated that AHP summation plays a prominent role in initial adaptation (Baldissera and Gustafsson 1974b; Baldissera et al. 1978; Kernell 1968). However, more recent experimental and simulation results suggest that AHP summation is probably not responsible for the later phases of adaptation (Powers et al. 1999). Figure 20 illustrates AHP summation in two different threshold-crossing motoneuron models (A), along with AHP summation measured in a rat hypoglossal motoneuron during two different stimulation protocols (B and C). The threshold-crossing models were single compartment models with a specified resistance and capacitance. The AHPs were produced following a threshold-crossing either by activating a potassium conductance with a slow exponential decay (G_K panel A1) or two-different potassium conductances (panel A2): one rapidly decaying conductance that repolarized the spike (G_{Kf}) and a calcium-activated potassium conductance (G_{KCa}). The model parameters could be tuned to replicate the pattern of AHP summation observed in real motoneurons (see Powers et al. 1999 for details) .

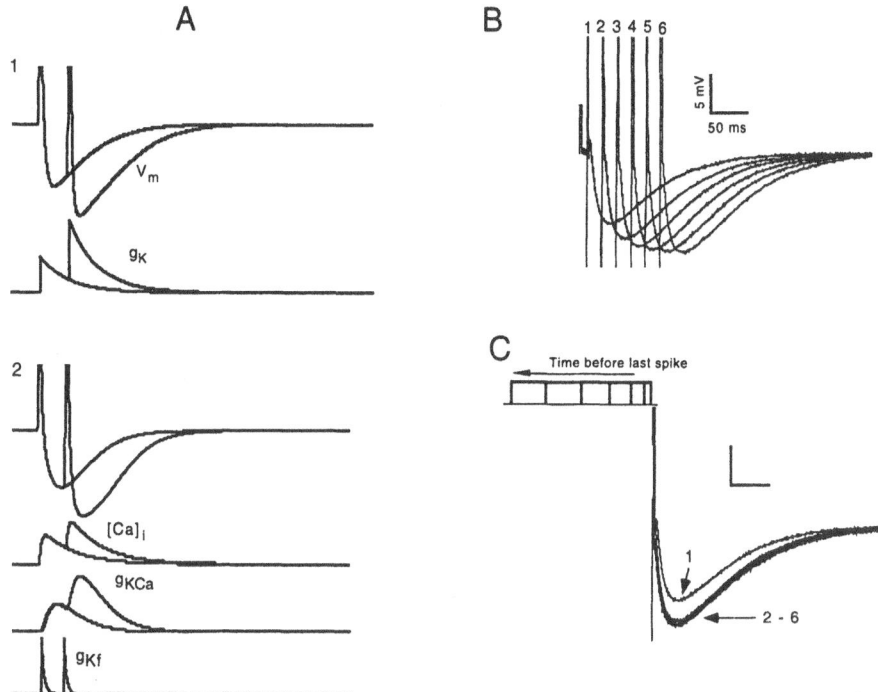

Fig. 20. AHP summation in real and model motoneurons. **A.** AHPs following two spikes in two different threshold crossing motoneuron models. Model 1 has a single, time-dependent potassium conductance (G_K), whereas model 2 has a fast potassium conductance (G_{Kf}) and a calcium-activated potassium conductance (G_{KCa}). **B.** AHPs following 1–6 spikes evoked in a rat hypoglossal motoneuron by 1 ms suprathreshold current pulses applied at 20 ms intervals. **C.** AHPs in the same motoneuron following 1–6 spikes evoked by suprathreshold current steps of different duration. Note that the AHP following two spikes was larger than the single spike AHP, but that the AHP remained nearly unchanged as more than two spikes were evoked. (Modified from Figs. 2 and 3 of Powers et al. 1999)

Similar patterns of AHP summation were observed in rat hypoglossal motoneurons studied *in vitro* and cat lumbar motoneurons recorded *in vivo* (Powers et al. 1999). Regardless of whether action potentials were evoked by a series of short, suprathreshold pulses at a fixed interpulse interval (Fig. 20B) or a constant amplitude current step of variable duration (Fig. 20C), the largest degree of AHP summation occurred between the first and second action potentials, and relatively little further increase in AHP amplitude occurred thereafter. As spike-frequency adaptation continues well beyond the first interspike interval, these results indicate that AHP summation contributes to the initial but not to the later phases of adaptation (i.e., early and late adaptation). Further support for this view comes from the finding that although the threshold-crossing models described above can mimic the AHP summation seen in real motoneurons (particularly model 2), they reproduce only the initial phase of spike-frequency adaptation, not the later phases (Powers et al. 1999). Finally, early and late adaptation are still present when the mAHP is eliminated by replacing external calcium with manganese (Musick 1999; Powers et al. 1999; Sawczuk et al. 1997).

A slowly-activating G_{KC_a} conductance that is pharmacologically distinct from that underlying the mAHP has been described in mammalian neocortical and hippocampal neurons (Madison and Nicoll 1984; Schwindt et al. 1988; Storm 1990). This G_{KC_a} conductance contributes both to a long-lasting AHP (sAHP) following repetitive discharge and to slow spike-frequency adaptation in these cells. A sodium-activated calcium-conductance may also contribute to the sAHP in neocortical neurons (Schwindt et al. 1989). The time constant of decay of the sAHP in neocortical neurons is roughly an order of magnitude longer than that of the mAHP and its amplitude following 20–100 spikes is 2–6 mV; cf. (Schwindt et al. 1988). A similar sAHP has been described in vagal motoneurons (Sah 1996), but there is less evidence for a sAHP in motoneurons innervating skeletal muscle. Sodium-activated potassium channels are present in neonatal rat lumbar motoneurones (Safronov and Vogel 1996), and sAHPs can be recorded in these cells following multiple spikes. However, there is at present no convincing evidence for functionally important sAHPs in mature motoneurones. The presence of a sAHP has been detected in both rat hypoglossal (Viana et al. 1993b) and guinea pig facial motoneurones (Nishimura et al. 1989), but both of these studies reported that sAHPs were observed only occasionally. We have found no evidence of AHPs with these characteristics in either cat spinal motoneurones or rat hypoglossal motoneurones. In these cells, the decay of the AHP following multiple spikes is characterized by a single exponential time course (Powers et al. 1999), unlike neocortical neurons in which the AHP decays with two distinct components reflecting the mAHP and sAHP.

There is at present little evidence that other slowly-developing outward currents contribute to spike-frequency adaptation in motoneurons. The later phases of adaptation cannot be attributed to the activation of an electrogenic sodium-potassium pump, since partial blockade of this pump with ouabain has no detectable effect on spike-frequency adaptation (Sawczuk et al. 1997). The voltage-dependence and kinetics of voltage-sensitive potassium conductances in motoneurons make it unlikely that they contribute directly to adaptation. Delayed rectifier potassium currents have been characterized in cultured and neonatal motoneurons (Alessandri-Haber et al. 1999; Gao and Ziskind-Conhaim 1998; Lape and Nistri 1999; McLarnon 1995;

Safronov and Vogel 1995). The characteristics of these non- or slowly-inactivating currents differ between different cell types, but in general the rapid kinetics and high activation threshold of delayed rectifier currents make it likely that they primarily contribute to the repolarizing phase of action potentials (see however, Lape and Nistri 1999). Transient (A-type) currents have also been described in these same studies. These currents may be significantly activated by the membrane voltages traversed during the interspike interval, but they are also generally completely inactivated by sustained voltages in this range, suggesting that may make little contribution during prolonged repetitive discharge. However, it has been suggested that a fast, transient potassium current might contribute to the initial and part of the early phase of adaptation (Lape and Nistri 1999).

Spike-frequency adaptation in both cat and rat motoneurons is associated with changes in the shape of action potentials. During long periods of repetitive discharge, there is a progressive decrease in spike height, an increase in spike duration, and a decrease in both the maximum rate of depolarization and rate of repolarization of the action potential (Musick 1999; Sawczuk et al. 1995b; and Powers and Binder, unpublished observations). The decrease in spike height reflects both an increase in the voltage threshold for spike initiation and a decrease in the peak spike voltage. Since spikes are thought to be initiated in the initial segment (see section 2.7), the increase in voltage threshold is likely to reflect inactivation of initial segment sodium channels. Figure 21A shows two 200 ms segments from a 2 s epoch of repetitive discharge recorded in a cat lumbar motoneuron. The lower thick line indicates the voltage threshold for spike initiation determined as the point at which the rate of change of membrane potential first exceeds a specified value. The threshold rises continuously, with the most marked increase occurring over the first few spikes. Although the absolute value of voltage threshold is difficult to measure accurately from a single electrode using a bridge circuit, recordings obtained with separate current passing and voltage-recording electrodes indicate a similar change in spike threshold during repetitive discharge (Schwindt and Crill 1982). A relatively small increase in the threshold for spike initiation could a produce significant decrease in firing rate. For example, given a typical rate of rise of membrane potential of 0.1–0.2 mV/ms over the latter half of the interspike interval (cf. Schwindt and Calvin 1972), a 1 mV rise in voltage threshold would increase the interspike interval by 5–10 ms.

In addition to the effects of changes in spike threshold, other changes in spike shape can influence firing rate by altering the amount of calcium that enters during the spike and is available to activate the calcium-activated

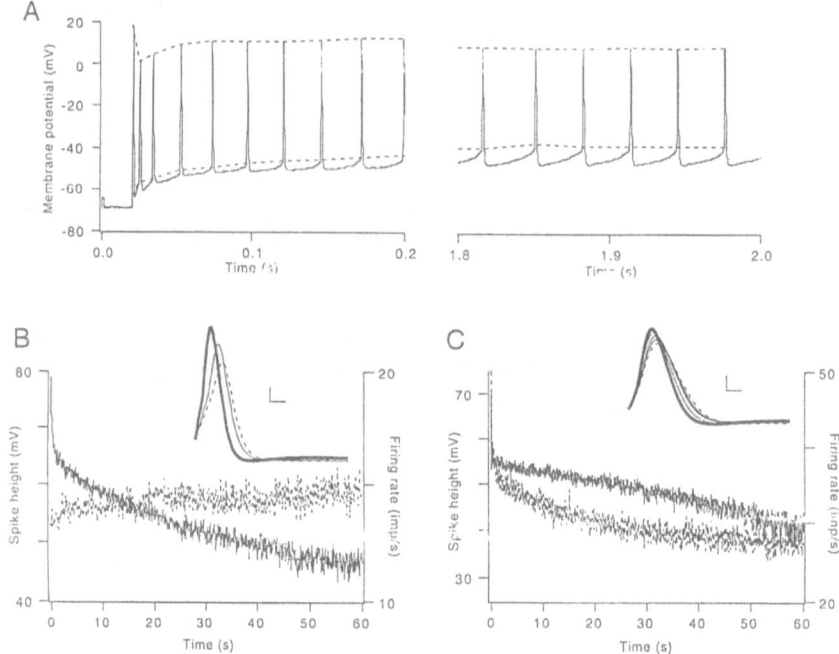

Fig. 21. Changes in spike shape during repetitive discharge in motoneurons. **A.** 200 ms portions of a 2 s record of repetitive discharge in a cat lumbar motoneuron. Lower dotted line marks spike threshold, upper dotted line indicates peak spike voltage (Powers et al. 1999). **B.** and **C.** Spike shape and spike frequency changes during 60 s epochs of discharge in two different rat hypoglossal motoneurons, one recorded in whole-cell mode (B) and one with a sharp, intracellular electrode (C). In both cells, spike height declines continuously (solid traces). Instantaneous frequency (dotted traces) increases with time in the cell shown in B, but decreases with time in the cell shown in part C. Insets show spikes recorded at different times during the discharge (Musick, Powers and Binder, unpublished; Calibration bars: 10 mV, 1 ms.)

potassium channels responsible for the mAHP. Figure 21B and C illustrate progressive changes in spike shape during 60 s of repetitive discharge in two different rat hypoglossal motoneurons recorded with either a whole-cell patch (B) or sharp microelectrode (C). The bold trace in the inset of B is the initial action potential of the record, whereas the thin solid and dotted traces show the action potentials taken a 10 and 60 s respectively after the onset of discharge. Both the rate of depolarization afd the rate of repolarization decrease progressively, and as a result the spike duration increases. The inset in C shows the changes in spike shape during late adaptation in more detail by illustrating spikes taken at 10 s intervals starting at 10 s after discharge onset. In contrast to previous reports (Kernell and Monster 1982b), there are progressive changes in spike shape during late adaptation.

The progressive changes in spike shape described above are a common feature during prolonged repetitive discharge in motoneurons. However, these changes in spike shape are not always well correlated with changes in discharge rate. In the motoneuron illustrated in Fig. 21B, spike height declines continuously during a 60 s epoch of repetitive discharge (solid trace), even though the discharge rate increases slightly over the same period (dotted trace). In the motoneuron in Fig. 21C, both spike height and firing rate decline continuously, although the time courses of decline are somewhat different. The variable relation between spike shape and firing rate changes is not surprising, given the multitude of factors that could affect firing rate. Further, there is likely to be a complex relationship between spike shape and the amount of calcium available to bind to the G_{KCa} channels responsible for the mAHP.

An increase in spike duration may not necessarily lead to increased G_{KCa} activation, if it is accompanied by significant decreases in spike height. For example, at the jellyfish neuromuscular junction, large brief action potentials lead to sharper, larger calcium transients and larger junctional potentials than do broader action potentials (Spencer et al. 1989). Some insight into the functional significance of changes in spike shape could be gained by 'replaying' measured action potentials as voltage-clamp commands and then measuring whole-cell calcium currents during 'action potentials' of different shape (e.g., McCobb and Beam 1991). However, due to space-clamp problems, this experiment would work only on electrotonically compact cells (i.e., either neonatal motoneurons or dissociated cells).

The final potential contributor to spike-frequency adaptation is a slowly-inactivating persistent inward current. As discussed earlier (section 3.3.3), there are both persistent sodium and persistent calcium currents activated in the voltage range between resting potential and spike threshold. Progressive inactivation of these currents would lead to a decrease in firing rate. The low-threshold calcium current thought to contribute to the development of plateau potentials generally shows facilitation rather than inactivation (Bennett et al. 1998a; Svirskis and Hounsgaard 1997). The net inward current observed during triangular somatic voltage-clamp commands in cat spinal motoneurons is generally greater during the ascending voltage ramp than during the subsequent descending phase, particularly in high threshold motoneurons (Lee and Heckman 1998a, 1999b), which could be explained by progressive inactivation of an inward current. However, recent evidence suggests that this phenomenon may instead reflect progressive activation of outward currents (Lee and Heckman 1999b). A persistent sodium current has been found in a variety of cells (Crill 1996), including some motoneurons (Hsiao et al. 1998; Mosfeldt-Laursen and Rekling 1989; Nishimura et al. 1989), and this current has been shown to undergo slow inactivation in neocortical neurons (Fleidervish and Gutnick 1996). We have recently characterized a persistent sodium current in rat hypoglossal motoneurons, but have found no convincing evidence for slow inactivation (Musick 1999).

In summary, there are probably a number of cellular mechanisms involved in spike-frequency adaptation, and their relative contribution may differ depending upon the phase of adaptation (i.e., initial, early or late), the extracellular environment and developmental stage. A quantitative analysis of the effects of changes in spike shape on calcium entry and subsequent activation of calcium-activated potassium conductances has not yet been achieved due to inadequate space clamp of the underlying membrane. Thus, untangling the relative contribution of different mechanisms will probably require improved voltage-clamp data from dissociated cells in combination with computer simulation of current clamp behavior.

4.3
Dynamic Input-Output Functions

In section 2.8, we discussed the general finding that under steady-state conditions, motoneurons encode the synaptic inputs they receive into a repetitive train of action potentials, the frequency of which is linearly related to the magnitude of the input (rev. in Binder et al. 1996). However, a general, dynamic input-output function for motoneurons that accurately describes their responses to transient synaptic inputs has eluded us for more than 30 years. The classical approach to this problem has been to record postsynaptic potentials (PSPs) in the soma of a motoneuron and then assess their effects on firing probability by compiling peristimulus time histograms (PSTH; rev. in Kirkwood 1979). The change in firing probability associated with the arrival of a synaptic input at the soma has been called the primary correlation kernel (Knox 1974), and it has been proposed that the primary correlation kernel of a neuron can be derived from a linear combination of the PSP and its derivative (Kirkwood and Sears 1978). However, for excitatory inputs (EPSPs), the relative contributions that the EPSP and its derivative make to the primary correlation kernel depend on the background synaptic noise (Cope et al. 1987; Fetz and Gustafsson 1983; Gustafsson and McCrea 1984; Poliakov et al. 1996). Moreover, this derivation does not apply to large inhibitory inputs (Poliakov et al. 1997; Fetz and Gustafsson 1983).

An alternative method of deriving a dynamic input-output function is to relate the changes in the discharge of a cell to the input current rather than to the PSP. Synaptic inputs that reach the soma can be simulated by intracellular current injection under standard current-clamp conditions (Poliakov et al. 1996, 1997; Powers and Binder 1996; Reyes and Fetz 1993). The injected current transients can be constructed to reproduce the amplitude and time-courses of real synaptic inputs, including the filtering and attenuation processes that are normally carried out by the dendrites as they transfer synaptic currents to the soma (Poliakov et al. 1997). A mathematical expression relating the effects of synaptic current on discharge probability can then be derived by examining a wide array of injected current waveforms.

An efficient means of deriving a general input-output transform is through the application of the white noise method of system identification (rev. in Marmarelis and Marmarelis 1978; Sakai 1992), a method particularly useful for the analysis of nonlinear systems. We have recently used this method to characterize the input-output transforms of mammalian motoneurons in a study in which the white noise stimulus contained trains of specific current transients that mimicked those underlying individual PSPs recorded in the soma (Poliakov et al. 1997). The spike discharge of the motoneurons in response to this input was used to compute the zero-, first- and second-order Wiener kernels (Bryant and Segundo 1976; Lee and Schetzen 1965). A series of orthogonal functionals derived from these kernels provided a good approximation of the motoneuron's input-output function that predicts the motoneuron's response to any arbitrary input (Hunter and Korenberg 1986; Marmarelis and Marmarelis 1978; Sakai 1992).

The first-order Wiener kernel, $h_1(\tau)$, obtained in a cat lumbar motoneuron, is shown in Fig. 22A. The first-order kernel represents the changes in firing rate elicited by a brief pulse of current whose area is 1 nA-ms. This kernel has two phases: an initial sharp increase in firing rate, followed by a shallower but more prolonged decrease in firing rate. The convolution of the first-order kernel with the input gives the best-fit, linear model of the neuron's response to this range of input signals (Marmarelis and Marmarelis 1978). If the input signal is small and the neuron exhibits near-linear behavior, this model accurately predicts the output. In particular, if the input is a brief pulse of area A at time 0, the system response ($r(t)$) is well approximated by the following function:

$$r(t) = h_0 + A\,h_1(\tau). \tag{14}$$

The second-order Wiener kernel, $h_2(\tau_1, \tau_2)$ is calculated by a second-order crosscorrelation between the motoneuron spikes and the input signal

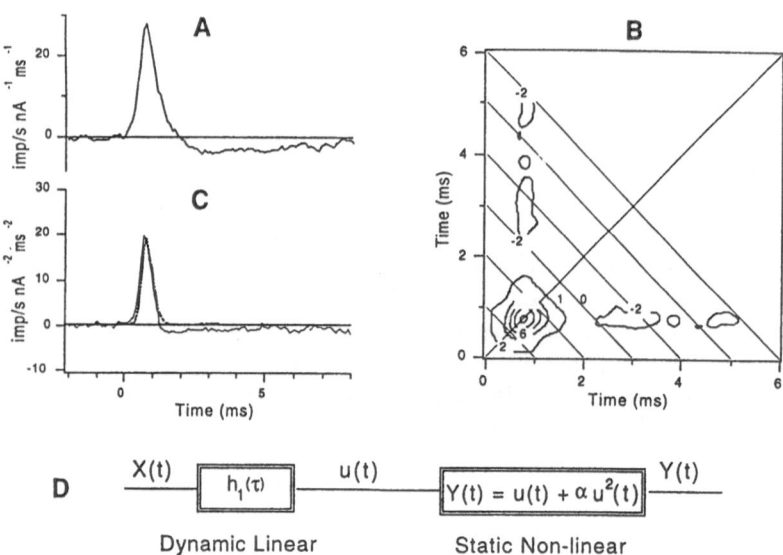

Fig. 22. The input-output transform of a cat lumbar motoneuron identified by the white-noise method A. The first order Wiener kernel, $h_1(\tau)$. B. Contour plot of the second order Wiener kernel $h_2(\tau_1,\tau_2)$ at levels -2, 2, 4, 6, 10, 14 and 18 imp/s nA^{-2} ms^{-2}. C. The diagonal of the second-order kernel is shown with the solid line. This is compared with an approximation of the second order kernel, $\alpha\, h_1(\tau)\, h_1(\tau)$, shown with the dotted line ($\alpha = 0.025$ s). D. The input-output function of the motoneuron can be represented as a second-order Wiener model: a cascade of a dynamic, linear transform, described by the first-order kernel $h_1(\tau)$, followed by a static non-linearity. (Modified from Figs. 5 and 6 in Poliakov et al. 1997)

at two different time lags (τ_1 and τ_2). The result is a function of two variables (τ_1 and τ_2) and is symmetric with respect to the main diagonal $\tau_1 = \tau_2$. In Fig. 22B a contour plot is shown of the second-order kernel for the same motoneuron shown in Fig. 22A. The prominent features of this kernel are a peak with a maximal value at the point $\tau_1 = \tau_2 = 1.0$ ms and two symmetric depressions at lags of about 3 to 6 ms along the lines $\tau_1 = 1.0$ ms and $\tau_2 = 1.0$ ms. The second-order kernel describes the deviation of the output from that predicted by the first-order model (equation 14). The second-order prediction of the neuron's response r(t) to a brief pulse of area A at time zero is:

$$r(t) = h_0 + A\, h_1(\tau) + 2\, A^2\, h_2(\tau,\tau) \qquad (15)$$

where $h_2(\tau,\tau)$ represents the values of the second-order kernel along the main diagonal (this dissection is shown in Fig. 22C), and the term 2 A^2

$h_2(\tau,\tau)$ represents the deviation from linearity. Unlike the first-order model, this expression gives asymmetric responses to positive and negative pulses, because the coefficient $2\,A^2$ is positive in both cases. For positive pulses, the response amplitude and area would increase in a greater-than-linear fashion with respect to A. For negative pulses, the response amplitude and area will decline less than linearly, reach a minimum, and start increasing. Thus, the short-latency, positive peak in the second-order kernel represents rectification of the input signal. The second-order Wiener model eliminates the erroneous negative response values for large hyperpolarizing inputs predicted by the first-order model (cf. Fig. 23).

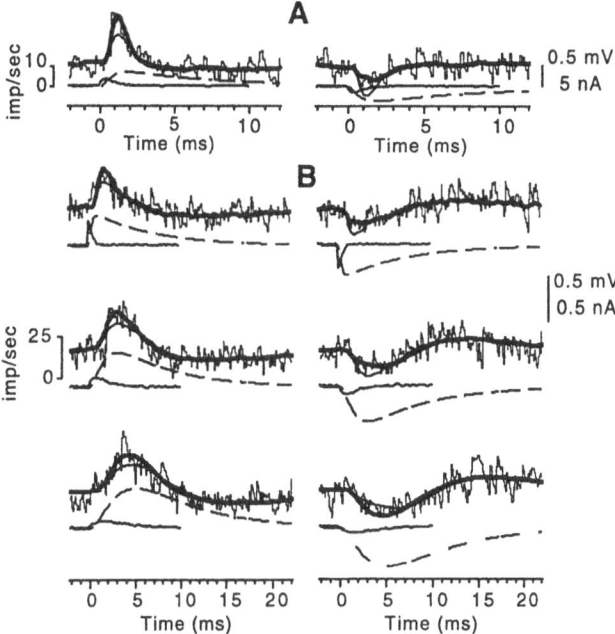

Fig. 23. Post-stimulus time histograms (PSTHs) predicted by first- and second order Wiener models. A. The PSTHs show the responses of a cat lumbar motoneuron to a depolarizing (left) and symmetric hyperpolarizing (right) current transient. The superimposed thin lines show the predicted PSTH based on the first-order Wiener model (i.e. the sum of the zero- and first-order Wiener functionals). The predicted PSTHs are symmetrical and feature a "negative firing rate" in response to the inhibitory transient. The predictions of a second-order model (i.e., sum of the zero-, first- and second order Wiener functionals) are shown by the thick lines and provide a better match to the actual PSTHs. B. Analogous results for a rat hypoglossal motoneuron. The underlying excitatory (left column) and inhibitory (right column) current transients had times to peak of 0.2, 0.8 and 1.6 ms, top to bottom. The calculated PSPs are superimposed on the current transients as broken lines. (Modified from Fig. 8 in Poliakov et al. 1997)

The depressions in the contour plot of the second-order kernel represent nonlinear interactions between inputs occurring at different times. For example, a motoneuron's response to a pair of identical pulses separated by 4 ms differs from the linear sum of the effects of each pulse acting in isolation. The response r(t) can be expressed as follows:

$$r(t) = h_0 + Ah_1(\tau) + Ah_1(\tau - 4) + 2 A^2 h_2(\tau,\tau - 4) \qquad (16)$$

where $h_0 + h_1(\tau) + h_1(\tau - 4)$ is the response of a linear system to a pair of identical pulses at times 0 and 4 ms and the term $2 A^2 h_2(\tau,\tau - 4)$ represents deviations from this sum due to nonlinear interaction between the pulses. The interaction effects can be predicted from a kernel "dissection" along a line starting at a point 4 ms along either the τ_1 or τ_2 axis, and running parallel to the main diagonal. Since this line crosses the "depression zone" of the contour plot, the second-order Wiener model predicts that the response to the second pulse will be smaller than that to the first, as has been observed experimentally (Fetz and Gustafsson 1983).

The peak along the main diagonal of the second-order kernel occurs at about the same latency as the peak in the first-order kernel, whereas the symmetric depressions off the main diagonal occur at lags corresponding to the duration of the trough in the first-order kernel. This suggests that the second-order kernel might be approximated by a scaled version of the product of the values of the first-order kernel at delays τ_1, τ_2:

$$h_2(\tau_1, \tau_2) \approx \alpha\, h_1(\tau_1)h_1(\tau_2) \qquad (17)$$

where α is the scaling coefficient. The value of the coefficient can be estimated by substituting the maximal values of the first- and second-order kernels into equation 17.

The accuracy of this approximation is illustrated in Fig. 22C that shows the "dissection" of the second-order kernel contour plot along its main diagonal. The thin line represents the actual values of the kernel, whereas the thick line represents the approximation of the second-order kernel based on equation 17. The approximated and calculated kernel values are generally quite close, although the approximation does not account for the small depression along the main diagonal at lags of 2 to 5 ms. This form of the second-order kernel is characteristic of a system consisting of a cascade of a dynamic linear component followed by a static nonlinear component (Hunter and Korenberg 1986). This so called "Wiener cascade model" (Sakai 1992) is schematized in Fig. 22D.

The response of such a system, Y (t), to any input, X (t – τ) can be predicted as follows:

$$u(t) = \int_0^\infty h_1(\tau) X(t - \tau)\, d\tau \qquad (18a)$$

$$Y(t) = h_0 + u(t) + \alpha\, u^2(t) \qquad (18b)$$

According to this Wiener cascade model, the input X (t) is first convolved with the kernel $h_1(\tau)$ to produce an intermediate variable, u(t), an operation known as dynamic linear filtering (equation 18a). The output value, Y (t), is then computed from u (t) using a parabolic transform (equation 18b). This type of function is known as a static nonlinearity because the output value is computed from the intermediate value at only a single moment in time. The static nonlinearity can be described by this parabolic function, or some other, as long as its value and its first and second derivative match those of the parabolic function over the range of interest.

The Wiener models described above predict a motoneuron's response to any input within a white noise waveform. The accuracy of these models for the current transients (CTs) embedded within the white noise to mimic PSPs is easily tested. Figure 23 illustrates the PSTHs obtained in response to one set of symmetric depolarizing and hyperpolarizing CTs in a cat spinal motoneuron (part A), and in response to three sets of CTs in a rat hypoglossal motoneuron (part B). The CTs are presented as the solid traces below the PSTHs and the resultant PSP as dashed lines. The thin lines superimposed upon the PSTHs represent the best first-order approximations (i.e., the sum of zero- and first-order Wiener functionals) of the response to the input CTs. The best second-order approximations (i.e., the sum of the zero-, first- and second-order Wiener functionals) are represented by the superimposed thick lines. In these and other cases, the first-order model tends to underestimate the amplitude of the response to excitatory CTs and overestimate the minimal values of PSTH troughs produced by an inhibitory CT, often predicting negative values, which, of course, the real PSTHs could not attain. The predictions that include the second-order Wiener functional are asymmetric and provide a better match to the PSTHs for both excitatory and inhibitory CTs.

The results summarized above indicate that even the first-order Wiener model provides a good prediction of the PSTH features produced by a fairly wide range of current transients. Considering that a single set of model parameters is used to predict a range of responses, this represents a considerable improvement over previously-proposed linear models of spike encoding. One common characteristic of previously-proposed linear models is that the predicted value of motoneuron firing probability at a given time following a PSP depends only on the value of the PSP (and/or its derivative) at a single point in time; i.e., they are all static models. In contrast, the first-order model based on the Wiener kernels is dynamic; i.e., the probability of spike generation at a given lag depends on the PSP values at a number of points in time.

In conclusion, compiling peristimulus time histograms (PSTHs) between one or even a few synaptic inputs and the discharge of a neuron is not sufficient to describe spike encoding quantitatively. However, the white noise method of systems identification looks quite promising in that it yields both the best linear approximation of a neuron's input-output function, as well as a more complete, higher-order description of its spike-encoding behavior. It has been found that truncating the Wiener series at the second-order functional was sufficient to capture both the linear and principal nonlinear components of spike encoding. Further, the contribution of the second-order Wiener functional could be approximated simply by a static nonlinearity (Fig. 22D and equation 18b). This Wiener cascade model accurately predicted the responses of motoneurons to a wide range of synaptic inputs and provided a substantial improvement over the best-fit linear model. Thus, by calculating the mean firing rate of a neuron (i.e., the zero-order kernel) and crosscorrelating its spike train with the white noise input, one can derive a general expression for spike encoding.

5
Summary and Conclusions

Our intent in this review was to consider the relationship between the biophysical properties of motoneurons and the mechanisms by which they transduce the synaptic inputs they receive into changes in their firing rates. Our emphasis has been on experimental results obtained over the past twenty years, which have shown that motoneurons are just as complex and interesting as other central neurons. This work has shown that motoneurons are endowed with a rich complement of active dendritic conductances, and flexible control of both somatic and dendritic channels by endogenous neuromodulators. Although this new information requires some revision of the simple view of motoneuron input-output properties that was prevalent in the early 1980's (see sections 2.3 and 2.10), the basic aspects of synaptic transduction by motoneurons can still be captured by a relatively simple input-output model (see section 2.3, equations 1–3).

It remains valid to describe motoneuron recruitment as a product of the total synaptic current delivered to the soma, the effective input resistance of the motoneuron and the somatic voltage threshold for spike initiation (equations 1 and 2). However, because of the presence of active channels activated in the subthreshold range, both the delivery of synaptic current and the effective input resistance depend upon membrane potential. In addition, activation of metabotropic receptors by achetylcholine, glutamate, noradrenaline, serotonin, substance P and thyrotropin releasing factor (TRH) can alter the properties of various voltage- and calcium-sensitive channels and thereby affect synaptic current delivery and input resistance. Once motoneurons are activated, their steady-state rate of repetitive discharge is linearly related to the amount of injected or synaptic current reaching the soma (equation 3). However, the slope of this relation, the minimum discharge rate and the threshold current for repetitive discharge are all subject to neuromodulatory control.

There are still a number of unresolved issues concerning the control of motoneuron discharge by synaptic inputs. Under dynamic conditions, when synaptic input is rapidly changing, time- and activity-dependent changes in the state of ionic channels will alter both synaptic current delivery to the spike-generating conductances and the relation between synaptic current and discharge rate. There is at present no general quantitative expression for motoneuron input-output properties under dynamic conditions. Even under steady-state conditions, the biophysical mechanisms underlying the transfer of synaptic current from the dendrites to the soma are not well understood, due to the paucity of direct recordings from motoneuron dendrites. It seems likely that resolving these important issues will keep motoneuron afficiandoes well occupied during the next twenty years.

Acknowledgments. We thank Drs. C.J. Heckman, Robert H. Lee, Lorne M. Mendell and John B. Munson for reviewing this manuscript. Our work is supported by grants NS-26840 and NS-31925 from the National Institute of Neurological Diseases and Stroke and grant IBN-9986167 from the National Science Foundation

References

Alessandri-Haber, N., C. Paillart, C. Arsac, M. Gola, F. Couraud, and M. Crest (1999) Specific distribution of sodium channels in axons of rat embryo spinal motoneurones. J Physiol (Lond) 518:203-14

Alvarez, F. J., J. C. Pearson, D. Harrington, D. Dewey, L. Torbeck, and R. E. W. Fyffe (1998) Distribution of 5-hydroxytryptamine-immunoreactive boutons on alpha-motoneurons in the lumbar spinal cord of adult cats. J Comp Neurol 393:69-83

Araki, T., and C. A. Terzuolo (1962) Membrane currents in spinal motoneurons associated with the action potential and synaptic activity. J Neurophysiol 25:772-789

Arvidsson, U., S. Cullheim, B. Ulfhake, G. W. Bennett, K. C. Fone, A. C. Cuello, A. A. Verhofstad, T. J. Visser, and T. Hokfelt (1990) 5-Hydroxytryptamine, substance P, and thyrotropin-releasing hormone in the adult cat spinal cord segment L7: immunohistochemical and chemical studies. Synapse 6:237-70

Ascher, P., and L. Nowak (1988) The role of divalent cations in the N-methyl-D-aspartate responses of mouse central neurons in culture. J Physiol (Lond) 399:247-266

Ashby, P., and D. Zilm (1982) Relationship between EPSP shape and cross-correlation profile explored by computer simulation for studies on human motoneurons. Exp Brain Res 47:33-40

Baldissera, F., P. Campadelli, and L. Piccinelli (1987) The dynamic response of cat gastrocnemius motor units investigated by ramp-current injection into their motoneurones. J Physiol (Lond) 387:317-30

Baldissera, F., and B. Gustafsson (1974a) Afterhyperpolarization time course in lumbar motoneurones of the cat. Acta Physiol Scand 91:512-527

Baldissera, F., and B. Gustafsson (1974b) Firing behaviour of a neuron model based on the afterhyperpolarization conductance time-course and algebraical summation. Adaptation and steady state firing. Acta Physiol Scand 92:27-47

Baldissera, F., and B. Gustafsson (1974c) Firing behaviour of a neuron model based on the afterhyperpolarization conductance time-course. First interval firing. Acta Physiol Scand 91:528-544

Baldissera, F., B. Gustafsson, and F. Parmiggiani (1976) A model for refractoriness accumulation and secondary range firing in spinal motoneurones. Biol Cybern 24:61-65

Baldissera, F., B. Gustafsson, and F. Parmiggiani (1978) Saturating summation of the afterhyperpolarization conductance in spinal motoneurones: a mechanism for 'secondary range' repetitive firing. Brain Res 146:69-82

Barrett, E. F., J. N. Barrett, and W. E. Crill (1980) Voltage-sensitive outward currents in cat motoneurones. J Physiol (Lond) 304:251-76

Barrett, J. N. (1975) Motoneuron dendrites: role in synaptic integration. Fed Proc 34:1398-1407

Barrett, J. N., and W. E. Crill (1974a) Influence of dendritic location and membrane properties on the effectiveness of synapses on cat motoneurones. J Physiol (Lond) 239:325-45

Barrett, J. N., and W. E. Crill (1974b) Specific membrane properties of cat motoneurones. J Physiol (Lond) 239:301-324

Barrett, J. N., and W. E. Crill (1980) Voltage clamp of cat motoneurone somata: properties of the fast inward current. J Physiol (Lond) 304:231-49

Bayliss, D. A., M. Umemiya, and A. J. Berger (1995) Inhibition of N- and P-type calcium currents and the after- hyperpolarization in rat motoneurones by serotonin. J Physiol (Lond) 485:635-47

Bayliss, D. A., F. Viana, and A. J. Berger (1992) Mechanisms underlying excitatory effects of thyrotropin-releasing hormone on rat hypoglossal motoneurons *in vitro*. J Neurophysiol 68:1733-45

Bellingham, M. C., and A. J. Berger (1996) Presynaptic depression of excitatory synaptic inputs to rat hypoglossal motoneurons by muscarinic M2 receptors. J Neurophysiol 76:3758-70

Bennett, D. J., H. Hultborn, B. Fedirchuk, and M. Gorassini (1998a) Short-term plasticity in hindlimb motoneurons of decerebrate cats. J Neurophysiol 80:2038-2045

Bennett, D. J., H. Hultborn, B. Fedirchuk, and M. Gorassini (1998b) Synaptic activation of plateaus in hindlimb motoneurons of decerebrate cats. J Neurophysiol 80:2023-2037

Berger, A. J., D. A. Bayliss, and F. Viana (1992) Modulation of neonatal rat hypoglossal motoneuron excitability by serotonin. Neurosci Lett 143:164-8

Bernander, O., C. Koch, and R. J. Douglas (1994) Amplification and linearization of distal synaptic input to cortical pyramidal cells. J Neurophysiol 72:2743-2753

Bigland-Ritchie, B., R. Johansson, O. C. J. Lippold, S. Smith, and J. J. Woods (1983) Changes in motoneurone firing rates during sustained maximal voluntary contractions. J Physiol (Lond) 340:335-346

Binder, M.D. (2000) Comparison of effective synaptic currents generated in spinal motoneurons by activating different input systems. In: Biomechanics and Neural Control of Posture and Movement. J.M. Winters and P.E. Crago (eds) Springer-Verlag, New York, 74-81

Binder, M. D., C. J. Heckman, and R. K. Powers (1993) How different afferent inputs control motoneuron discharge and the output of the motoneuron pool. Curr Op Neurobiol 3:1028-1034

Binder, M. D., C. J. Heckman, and R. K. Powers (1996) The physiological control of motoneuron activity. In: L. B. Rowell and J. T. Shepherd (eds.). Handbook of Physiology. Section 12. Exercise: Regulation and Integration of Multiple Systems. New York: Oxford University Press, pp 3-53

Binder, M. D., and L. M. Mendell (1990) The Segmental Motor System. New York: Oxford University Press, pp Pages

Binder, M. D., F. R. Robinson, and R. K. Powers (1998) Distribution of effective synaptic currents in triceps surae motoneurons. VI. Contralateral pyramidal tract. J Neurophysiol 80:241-298

Bohmer, G., K. Schmid, and W. Schauer (1991) Evidence for an involvement of NMDA and non-NMDA receptors in synaptic excitation of phrenic motoneurons in the rabbit. Neurosci Lett 130:271-4

Booth, V., J. Rinzel, and O. Kiehn (1997) Compartmental model of vertebrate motoneurons for Ca^{2+}-dependent spiking and plateau potentials under pharmacological treatment. J Neurophysiol 78:3371-3385

Botterman, B. R., G. A. Iwamoto, and W. J. Gonyea (1986) Gradation of isometric tension by different activation rates in motor units of cat flexor carpi radialis muscle. J Neurophysiol 56:494-506

Brannstrom, T. (1993) Quantitative synaptology of functionally different types of cat medial gastrocnemius alpha-motoneurons. J Comp Neurol 330:439-54

Bras, H., J. Destombes, P. Gogan, and D. S. Tyc (1987) The dendrites of single brain-stem motoneurons intracellularly labelled with horseradish peroxidase in the cat. An ultrastructural analysis of the synaptic covering and the microenvironment. Neuroscience 22:971-81

Bras, H., P. Gogan, and D. S. Tyc (1987) The dendrites of single brain-stem motoneurons intracellularly labelled with horseradish peroxidase in the cat. Morphological and electrical differences. Neuroscience 22:947-70

Bras, H., S. Korogod, Y. Driencourt, P. Gogan, and S. Tycdumont (1993) Stochastic Geometry and Electronic Architecture of Dendritic Arborization of Brain Stem Motoneuron. Eur Jour Neurosci 5:1485-1493

Brismar, T. (1977) Slow mechanism for sodium permeability inactivation in myelinated nerve fibre of Xenopus laevis. J Physiol (Lond) 270:283-297

Brock, L. G., J. S. Coombs, and J. C. Eccles (1951) Action potentials of motoneurons with intracellular electrode. Proc Otago Med Sch 29:14-15

Brock, L. G., J. S. Coombs, and J. C. Eccles (1952) The recording of potentials from motoneurones with an intracellular electrode. J Physiol (Lond) 117:431-460

Brock, L. G., J. S. Coombs, and J. C. Eccles (1953) Intracellular recording from antidromically activated motoneurons. J Physiol (Lond) 122:429-461

Brodin, L., H. G. Trav'en, A. Lansner, P. Wall'en, O. Ekeberg, and S. Grillner (1991) Computer simulations of N-methyl-D-aspartate receptor-induced membrane properties in a neuron model. J Neurophysiol 66:473-84

Brown, A. G. (1981) Organization in the Spinal Cord. Berlin: Springer, pp Pages

Brown, A. G., and R. E. Fyffe (1981) Direct observations on the contacts made between Ia afferent fibres and alpha-motoneurones in the cat's lumbosacral spinal cord. J Physiol (Lond) 313:121-40

Brownstone, R. M., L. M. Jordan, D. J. Kriellaars, B. R. Noga, and S. J. Shefchyk (1992) On the regulation of repetitive firing in lumbar motoneurones during fictive locomotion in the cat. Exp Brain Res 90:441-55

Bryant, H. L., and J. P. Segundo (1976) Spike initiation by transmembrane current: a white-noise analysis,. J Physiol (Lond) 260:279-314

Burke, R. E. (1967) Composite nature of the monosynaptic excitatory postsynaptic potential. J Neurophysiol 30:1114-37

Burke, R. E. (1981) Motor units: anatomy, physiology, and functional organization. In: V. B. Brooks (eds.). Handbook of Physiology, The Nervous System, Motor Control. Bethesda, MD: American Physiological Society, pp 345-422

Burke, R. E., R. P. Dum, J. W. Fleshman, L. L. Glenn, T. A. Lev, M. J. O'Donovan, and M. J. Pinter (1982) A HRP study of the relation between cell size and motor unit type in cat ankle extensor motoneurons. J Comp Neurol 209:17-28

Burke, R. E., L. Fedina, and A. Lundberg (1971) Spatial synaptic distribution of recurrent and group Ia inhibitory systems in cat spinal motoneurones. J Physiol (Lond) 214:305-26

Burke, R. E., J. W. Fleshman, and I. Segev (1988) Factors that control the efficacy of group Ia synapses in alpha-motoneurons. J Physiol Paris 83:133-40

Burke, R. E., and L. L. Glenn (1996) Horseradish peroxidase study of the spatial and electrotonic distribution of group Ia synapses on type-identified ankle extensor motoneurons in the cat. J Comp Neurol 372:465-485

Burke, R. E., E. Jankowska, and G. t. Bruggencate (1970) A comparison of peripheral and rubrospinal synaptic input to slow and fast twitch motor units of triceps surae. J Physiol (Lond) 207:709-32

Burke, R. E., D. N. Levine, P. Tsairis, and F. E. Zajac (1974) Physiological types and histochemical profiles in motor units of the cat gastrocnemius. J Physiol (Lond) 234:723-748

Burke, R. E., and P. G. Nelson (1971) Accommodation to current ramps in motoneurons of fast and slow twitch motor units. Int J Neurosci 1:347-356

Burke, R. E., W. Z. Rymer, and J. V. Walsh (1976) Relative strength of synaptic input from short-latency pathways to motor units of defined type in cat medial gastrocnemius. J Neurophysiol 39:447-58

Burke, R. E., and G. ten Bruggencate (1971) Electrotonic characteristics of alpha motoneurones of varying size. J Physiol (Lond) 212:1-20

Burke, R. E., B. Walmsley, and J. A. Hodgson (1979) HRP anatomy of group Ia afferent contacts on alpha motoneurones. Brain Res 160:347-52

Butrimas, P., and A. Gutman (1979) Theoretical analysis of an experiment with voltage clamping in the motoneurone. Proof of the N-shape pattern of the steady voltage-current characteristic of the dendrite membrane. Biophys 23:897-904

Calvin, W. H. (1974) Three modes of repetitive firing and the role of threshold time course between spikes. Brain Res 59:341-346

Calvin, W. H., and C. F. Stevens (1968) Synaptic noise and other sources of randomness in motoneuron interspike intervals. J Neurophysiol 31:574-87

Cameron, W. E., D. B. Averill, and A. J. Berger (1983) Morphology of cat phrenic motoneurons as revealed by intracellular injection of horseradish peroxidase. J Comp Neurol 219:70-80

Cameron, W. E., D. B. Averill, and A. J. Berger (1985) Quantitative analysis of the dendrites of cat phrenic motoneurons stained intracellularly with horseradish peroxidase. J Comp Neurol 231:91-101

Campbell, D. M., and P. K. Rose (1997) Contribution of voltage-dependent potassium channels to the somatic shunt in neck motoneurons of the cat. J Neurophysiol 77:1470-1486

Carlin, K. P., Jiang, Z. and Brownstone, R. M. (2000a) Characterization of calcium currents in functionally mature mouse spinal motoneurons. Eur J Neurosci 12:1624-1634.

Carlin, K. P., Jones, K. E., Jiang, Z., Jordan, L. M. and Brownstone, R. M. (2000b) Dendritic L-type calcium currents in mouse spinal motoneurons: implications for bistability. Eur J Neurosci 12: 1635-1646.

Cash, S., and R. Yuste (1998) Input summation by cultured pyramidal neurons is linear and position-independent. J Neurosci 18:10-15

Chandler, S. H., C.-F. Hsaio, T. Inoue, and L. J. Goldberg (1994) Electrophysiological properties of guinea pig trigeminal motoneurons recorded *in vitro*. J Neurophysiol 71:129-145

Chitravanshi, V. C., and H. N. Sapru (1996) NMDA as well as non-NMDA receptors mediate the neurotransmission of inspiratory drive to phrenic motoneurons in the adult rat. Brain Res 715:104-12

Clements, J. D., P. G. Nelson, and S. J. Redman (1986) Intracellular tetraethylammonium ions enhance group Ia excitatory post-synaptic potentials evoked in cat motoneurones. J Physiol (Lond) 377:267-82

Clements, J. D., and S. J. Redman (1989) Cable properties of cat spinal motoneurones measured by combining voltage clamp, current clamp and intracellular staining. J Physiol (Lond) 409:63-87

Colbert, C. M., J. C. Magee, D. A. Hoffman, and D. Johnston (1997) Slow recovery from inactivation of Na+ channels underlies the activity-dependent attenuation of dendritic action potentials in hippocampal CA1 pyramidal neurons. J Neurosci 17:6512-6521

Collingridge, G. L., and R. A. J. Lester (1989) Excitatory amino acid receptors in the vertebrate central nervous system. Pharmacol Rev 40:143-210

Conradi, S. (1969) Ultrastructure and distribution of neuronal and glial elements on the motoneuron surface in the lumbosacral spinal cord of the adult cat. Acta Physiol Scand [Suppl.] 332:5-48

Conradi, S., S. Cullheim, L. Gollvik, and J. O. Kellerth (1983) Electron microscopic observations on the synaptic contacts of group Ia muscle spindle afferents in the cat lumbosacral spinal cord. Brain Res 265:31-9

Conradi, S., J.-O. Kellerth, C.-H. Berthold, and C. Hammarberg (1979) Electron microscopic studies of serially sectioned cat spinal α-motoneurons: IV. Motoneurons innervating slow-twitch (Type S) units of the soleus muscle. J Comp Neurol 184:

Conway, B. A., H. Hultborn, O. Kiehn, and I. Mintz (1988) Plateau potentials in alpha-motoneurones induced by intravenous injection of L-dopa and clonidine in the spinal cat. J Physiol (Lond) 405:369-84

Cook, E. P., and D. Johnston (1999) Voltage-dependent properties of dendrites that eliminate location- dependent variability of synaptic input [In Process Citation]. J Neurophysiol 81:535-43

Coombs, J. S., D. R. Curtis, and J. C. Eccles (1957a) The generation of impulses in motoneurones. J Physiol (Lond) 139:232-249

Coombs, J. S., D. R. Curtis, and J. C. Eccles (1957b) The interpretation of spike potentials of motoneurones. J Physiol (Lond) 139:198-231

Coombs, J. S., J. C. Eccles, and P. Fatt (1955) The electrical properties of the motoneurone membrane. J Physiol (Lond) 130:291-325

Cope, T. C., and B.D. Clark (1995) Are there important exceptions to the size principle of α-motoneurone recruitment? In: A. Taylor, M.H. Gladden, and R. Dur baba (Eds.), Alpha and Gamma Motor Systems. Plenum Press, New York, pp 71-78

Cope, T. C., E. E. Fetz, and M. Matsumura (1987) Cross-correlation assessment of synaptic strength of single Ia fibre connections with triceps surae motoneurones in cats. J Physiol (Lond) 390:161-188

Cope, T. C., and A. J. Sokoloff (1999) Orderly recruitment tested across muscle boundaries. In: M. D. Binder (eds.). Peripheral and Spinal Mechanisms in the Neural Control of Movement. Amsterdam: Elsevier, pp 177-190

Cordo, P. J., and W. Z. Rymer (1982) Motor-unit activation patterns in lengthening and isometric contractions of hindlimb extensor muscles in the decerebrate cat. J Neurophysiol 47:782-796

Crill, W. E. (1996) Persistent sodium current in mammalian central neurons. Annu Rev Physiol 58:349-362

Cullheim, S. (1978) Relations between cell body size, axon diameter and axon conduction velocity of cat sciatic α-motoneurons stained with horseradish peroxidase. Neurosci Lett 8:17-20

Cullheim, S., J. W. Fleshman, L. L. Glenn, and R. E. Burke (1987a) Membrane area and dendritic structure in type-identified triceps surae alpha motoneurons. J Comp Neurol 255:68-81

Cullheim, S., J. W. Fleshman, L. L. Glenn, and R. E. Burke (1987b) Three-dimensional architecture of dendritic trees in type-identified alpha-motoneurons. J Comp Neurol 255:82-96

Cullheim, S., and J.-O. Kellerth (1978) A morphological study of the axons and recurrent axon collaterals of cat α-motoneurones supplying different hindlimb
muscles. J Physiol (Lond) 281:285-299

De Luca, C. J., R. S. LeFever, M. P. McCue, and A. P. Xenakis (1982) Behavior of
human motor units in different muscles during linearly varying contractions. J
Physiol (Lond) 329:113-128

Del Negro, C. A., and S. H. Chandler (1998) Regulation of intrinsic and synaptic
properties of neonatal rat trigeminal motoneurons by metabotropic glutamate
receptors. J Neurosci 18:9216-26

Del Negro, C. A., C. F. Hsiao, and S. H. Chandler (1999) Outward currents influencing bursting dynamics in guinea pig trigeminal motoneurons. J Neurophysiol
81:1478-85

Delgado-Lezama, R., J. F. Perrier, and J. Hounsgaard (1999) Local facilitation of
plateau potentials in dendrites of turtle motoneurones by synaptic activation of
metabotropic receptors. J Physiol (Lond) 515:203-7

Delgado-Lezama, R., J. F. Perrier, S. Nedergaard, G. Svirskis, and J. Hounsgaard
(1997) Metabotropic synaptic regulation of intrinsic response properties of turtle
spinal motoneurones. J Physiol (Lond) 504:97-102

Dememes, D., and J. Raymond (1982) Radioautographic identification of
[3H]glutamic acid labeled nerve endings in the cat oculomotor nucleus. Brain
Res 231:433-7

Destombes, J., B. G. Horcholle, and D. Thiesson (1992) Distribution of glycinergic
terminals on lumbar motoneurons of the adult cat: an ultrastructural study.
Brain Res 599:353-60

Dong, X. W., and J. L. Feldman (1999) Distinct subtypes of metabotropic glutamate
receptors mediate differential actions on excitability of spinal respiratory motoneurons. J Neurosci 19:5173-84

Dum, R. P., and T. T. Kennedy (1980) Synaptic organization of defined motor unit
types in cat tibialis anterior. J Neurophysiol 43:1631-1644

Durand, D. (1984) The somatic shunt cable model for neurons. Biophys J 46:645-53

Durand, J. (1991) NMDA actions on rat abducens motoneurones. Eur J Neurosci
3:621-633

Durand, J. (1993) Synaptic excitation triggers oscillations during NMDA receptor
activation in rat abducens motoneurons. Eur J Neurosci 5:1389-1397

Durand, J., I. Engberg, and D. S. Tyc (1987) L-glutamate and N-methyl-D-asparatate
actions on membrane potential and conductance of cat abducens motoneurones.
Neurosci Lett 79:295-300

Eccles, J. C. (1957) The Physiology of Nerve Cells. Baltimore: Johns Hopkins Press
pp Pages

Eccles, J. C. (1964) The Physiology of Synapses. Berlin: Springer-Verlag, pp Pages

Eccles, J. C., R. M. Eccles, A. Iggo, and M. Ito (1961) Distribution of recurrent inhibition among motoneurones. J Physiol (Lond) 159:479-499

Eccles, J. C., R. M. Eccles, and A. Lundberg (1957) The convergence of monosynaptic
excitatory afferents on to many different species of alpha motoneurones. J
Physiol (Lond) 137:22-50

Egger, M. D., N. C. Freeman, and E. Proshansky (1980) Morphology of spinal motoneurones mediating a cutaneous spinal reflex in the cat. J Physiol (Lond)
306:349-63

Ekeberg, O., P. Wall'en, A. Lansner, H. Trav'en, L. Brodin, and S. Grillner (1991) A computer based model for realistic simulations of neural networks. I. The single neuron and synaptic interaction. Biol Cybern 65:81-90

Eken, T. (1998) Spontaneous electromyographic activity in adult rat soleus muscle. J Neurophysiol 80:365-376

Eken, T., and O. Kiehn (1989) Bistable firing properties of soleus motor units in unrestrained rats. Acta Physiol Scand 136:383-94

Elliott, P., and D. I. Wallis (1992) Serotonin and L-norepinephrine as mediators of altered excitability in neonatal rat motoneurons studied *in vitro*. Neuroscience 47:533-44

Endo, K., T. Araki, and Y. Kawai (1975) Contra- and ipsilateral cortical and rubral effects on fast and slow spinal motoneurons of the cat. Brain Res 88:91-98

Fetz, E. E., P. D. Cheney, K. Mewes, and S. Palmer (1989) Control of forelimb muscle activity by populations of corticomotoneuronal and rubromotoneuronal cells. Progr Brain Res 80:437-449

Fetz, E. E., and B. Gustafsson (1983) Relation between shapes of post-synaptic potentials and changes in firing probability of cat motoneurones. J Physiol (Lond) 341:387-410

Finkel, A. S., and S. J. Redman (1983) The synaptic current evoked in cat spinal motoneurones by impulses in single group 1a axons. J Physiol (Lond) 342:615-32

Fisher, N. D., and A. Nistri (1993) Substance P and TRH share a common effector pathway in rat spinal motoneurones: an *in vitro* electrophysiological investigation. Neurosci Lett 153:115-9

Flatman, J. A., P. C. Schwindt, and W. E. Crill (1986) The induction and modification of voltage sensitive responses in cat neocortical neurons by N-methyl-D-aspartate. Brain Res 363:62-77

Fleidervish, I. A., A. Friedman, and M. J. Gutnick (1996) Slow inactivation of Na+ current and slow cumulative spike adaptation in mouse and guinea-pig neocortical neurones in slices. J Physiol (Lond) 493:83-97

Fleidervish, I. A., and M. J. Gutnick (1996) Kinetics of slow inactivation of persistent sodium current in layer V neurons of mouse neocortical slices. J Neurophysiol 76:2125-2130

Fleshman, J. W., J. B. Munson, G. W. Sypert, and W. A. Friedman (1981) Rheobase, input resistance, and motor-unit type in medial gastrocnemius motoneurons in the cat. J Neurophysiol 46:1326-38

Fleshman, J. W., I. Segev, and R. B. Burke (1988) Electrotonic architecture of type-identified alpha-motoneurons in the cat spinal cord. J Neurophysiol 60:60-85

Forsythe, I. D., and S. J. Redman (1988) The dependence of motoneurone membrane potential on extracellular ion concentrations studied in isolated rat spinal cord. J Physiol (Lond) 404:83-99

Frankenhaeuser, B., and A. B. Vallbo (1964) Accomodation in myelinated nerve fibres of Xenopus laevis as computed on the basis of voltage clamp data. Acta Physiol Scand 63:1-20

Friedman, W. A., G. W. Sypert, J. B. Munson, and J. W. Fleshman (1981) Recurrent inhibition in type-identified motoneurons. J Neurophysiol 46:1349-1359

Fukushima, K., B. W. Peterson, and V. J. Wilson (1979) Vestibulospinal, reticulospinal and interstitiospinal pathways in the cat. Prog Brain Res 50:121-136

Fulton, B. P., and K. Walton (1986) Electrophysiological properties of neonatal rat motoneurones studied *in vitro*. J Physiol (Lond) 370:651-78

Fuortes, M. G. F., K. Frank, and M. C. Becker (1957) Steps in the production of motoneuron spikes. J Gen Physiol 40:735-752

Fyffe, R. E. (1991) Spatial distribution of recurrent inhibitory synapses on spinal motoneurons in the cat. J Neurophysiol 65:1134-49

Fyffe, R. E., and A. R. Light (1984) The ultrastructure of group Ia afferent fiber synapses in the lumbosacral spinal cord of the cat. Brain Res 300:201-9

Fyffe, R. E. W., F. J. Alvarez, J. C. Pearson, D. Harrington, and D. E. Dewey (1993) Modulation of motoneuron activity: Distribution of glycine receptors and serotonergic inputs on motoneuron dendrites. The Physiologist 36:A-11

Gao, B. X., and L. Ziskind-Conhaim (1998) Development of ionic currents underlying changes in action potential waveforms in rat spinal motoneurons. J Neurophysiol 80:3047-61

Glenn, L. L. (1988) Overestimation of the electrical length of neuron dendrites and synaptic electrotonic attenuation. Neurosci Lett 91:112-9

Gorassini, M., D. J. Bennett, O. Kiehn, T. Eken, and H. Hultborn (1999) Activation patterns of hindlimb motor units in the awake rat and their relation to motoneuron intrinsic properties. J Neurophysiol 82:709-17

Gorassini, M. A., D. J. Bennett, and J. F. Yang (1998) Self-sustained firing of human motor units. Neurosci Lett 247:13-16

Granit, R., D. Kernell, and Y. Lamarre (1966) Algebraical summation in synaptic activation of motoneurones firing within the 'primary range' to injected currents. J Physiol (Lond) 187:379-99

Granit, R., D. Kernell, and G. K. Shortess (1963) Quantitative aspects of repetitive firing of mammalian motoneurones, caused by injected currents. J Physiol (Lond) 168:911-931

Granit, R., and B. Renkin (1961) Net depolarization and discharge rate of motoneurones, as measured by recurrent inhibition. J Physiol (Lond) 158:461-475

Grillner, S., T. Hongo, and S. Lund (1970) The vestibulospinal tract. Effects on alpha-motoneurones in the lumbosacral spinal cord in the cat. Exp Brain Res 10:94-120

Guertin, P. A., and J. Hounsgaard (1998a) Chemical and electrical stimulatin induce rhythmic motor activity in an *in vitro* preparation of the spinal cord from adult turtles. Neurosci Lett 245:5-8

Guertin, P. A., and J. Hounsgaard (1998b) NMDA-Induced intrinsic voltage oscillations depend on L-type calcium channels in spinal motoneurons of adult turtles. J Neurophysiol 80:3380-2

Guertin, P. A., and J. Hounsgaard (1999) Non-volatile general anaesthetics reduce spinal activity by suppressing plateau potentials. Neuroscience 88:353-8

Gustafsson, B., and D. McCrea (1984) Influence of stretch-evoked synaptic potentials on firing probability of cat spinal motoneurones. J Physiol (Lond) 347:431-51

Gustafsson, B., and M. J. Pinter (1984a) An investigation of threshold properties among cat spinal alpha-motoneurones. J Physiol (Lond) 357:453-83

Gustafsson, B., and M. J. Pinter (1984b) Relations among passive electrical properties of lumbar alpha-motoneurones of the cat. J Physiol (Lond) 356:401-31

Gustafsson, B., and M. J. Pinter (1985) Factors determining the variation of the afterhyperpolarization duration in cat lumbar alpha-motoneurones. Brain Res 326:392-5

Gutman, A. M. (1971) Further remarks on the effectiveness of dendrite synapses. Biophysics 16:131-138

Gutman, A. M. (1991) Bistability of dendrites. Int J Neural Sys 1:291-304

Gydikov, A., and D. Kosarov (1973) Physiological characteristics of the tonic and phasic motor units in human muscles. In: A. Gydikov, N. Tankov and D. Kosarov (eds.). Motor Control. New York: Plenum Press, pp 75-94

Harada, Y., and T. Takahashi (1983) The calcium component of the action potential in spinal motoneurones of the rat. J Physiol (Lond) 335:89-100

Harrison, P. J., and A. Taylor (1981) Individual excitatory post-synaptic potentials due to muscle spindle Ia afferents in cat triceps surae motoneurones. J Physiol (Lond) 312:455-470

Heckman, C. J. (1994) Computer simulations of the effects of different synaptic input systems on the steady-state input-output structure of the motoneuron pool. J Neurophysiol 71:1727-1739

Heckman, C. J., and M. D. Binder (1988) Analysis of effective synaptic currents generated by homonymous Ia afferent fibers in motoneurons of the cat. J Neurophysiol 60:1946-66

Heckman, C. J., and M. D. Binder (1990) Neural mechanisms underlying the orderly recruitment of motoneurons. In: M. D. Binder and L. M. Mendell (eds). The Segmental Motor System. New York: Oxford University Press, pp 182-204

Heckman, C. J., and M. D. Binder (1991a) Analysis of Ia-inhibitory synaptic input to cat spinal motoneurons evoked by vibration of antagonist muscles. J Neurophysiol 66:1888-1893

Heckman, C. J., and M. D. Binder (1991b) Computer simulation of the steady-state input-output function of the cat medial gastrocnemius motoneuron pool. J Neurophysiol 65:952-67

Heckman, C. J., and M. D. Binder (1993a) Computer simulations of motoneuron firing rate modulation. J Neurophysiol 69:1005-8

Heckman, C. J., and M. D. Binder (1993b) Computer simulations of the effects of different synaptic input systems on motor unit recruitment. J Neurophysiol 70:1827-1840

Henneman, E. (1957) Relation between size of neurons and their susceptibility to discharge. Science 126:1345–1347

Henneman, E., and L. M. Mendell (1981) Functional organization of motoneuron pool and its inputs. In: V. B. Brooks (eds). Handbook of Physiology, The Nervous System, Motor Control. Bethesda, MD: American Physiological Society, pp 423-507

Henneman, E., G. Somjen, and D. O. Carpenter (1965) Excitability and inhibitability of motoneurons of different sizes. J Neurophysiol 28:599-620

Hille, B. (1992) Ionic Channels of Excitable Membranes. 2nd ed. Sunderland, MA: Sinauer Assoc. Inc.

Hochman, S., L. M. Jordan, and B. J. Schmidt (1994) TTX-resistant NMDA receptor-mediated voltage oscillations in mammalian lumbar motoneurons. J Neurophysiol 72:2559-62

Hodgkin, A. L., and A. F. Huxley (1952) A quantitative description of membrane current and its application to conduction and excitation in nerve. J Physiol (Lond) 116:500-544

Hodgkin, A. L., and S. Nakajima (1972) The effect of diameter on the electrical constants of frog skeletal muscle fibres. J Physiol (Lond) 221:105-120

Holmes, W. R., I. Segev, and W. Rall (1992) Interpretation of time constant and electrotonic length estimates in multicylinder or branched neuronal structures. J Neurophysiol 68:1401-20

Holstege, J. C. (1991) Ultrastructural evidence for GABAergic brain stem projections to spinal motoneurons in the rat. J Neurosci 11:159-67

Hongo, T., E. Jankowska, and A. Lundberg (1969) The rubrospinal tract. I. Effects on alpha-motoneurones innervating hindlimb muscles in cats. Exp Brain Res 7:344-64

Hori, Y., and K. Kanda (1996) Developmental alterations in NMDA receptor-mediated currents in neonatal rat spinal motoneurons. Neurosci Lett 205:99-102

Hounsgaard, J., H. Hultborn, B. Jespersen, and O. Kiehn (1984) Intrinsic membrane properties causing a bistable behavior of α-motoneurones. Exp Brain Res 55:391-394

Hounsgaard, J., H. Hultborn, B. Jespersen, and O. Kiehn (1988) Bistability of alpha-motoneurones in the decerebrate cat and in the acute spinal cat after intravenous 5-hydroxytryptophan. J Physiol (Lond) 405:345-67

Hounsgaard, J., and O. Kiehn (1985) Ca++ dependent bistability induced by serotonin in spinal motoneurons. Exp Brain Res 57:422-5

Hounsgaard, J., and O. Kiehn (1989) Serotonin-induced bistability of turtle motoneurones caused by a nifedipine-sensitive calcium plateau potential. J Physiol (Lond) 414:265-82

Hounsgaard, J., and O. Kiehn (1993) Calcium spikes and calcium plateaux evoked by differential polarization in dendrites of turtle motoneurones in vitro. J Physiol (Lond) 468:245-59

Hounsgaard, J., and I. Mintz (1988) Calcium conductance and firing properties of spinal motoneurones in the turtle. J Physiol (Lond) 398:591-603

Howe, J. R., and J. M. Ritchie (1992) Multiple kinetic components of sodium channel inactivation in rabbit Schwann cells. J Physiol (Lond) 455:529-566

Hsiao, C. F., C. A. DelNegro, P. R. Trueblood, and S. H. Chandler (1998) Ionic basis for serotonin-induced bistable membrane properties in guinea pig trigeminal motoneurons. J Neurophysiol 79:2847-2856

Hsiao, C. F., P. R. Trueblood, M. S. Levine, and S. H. Chandler (1997) Multiple effects of serotonin on membrane properties of trigeminal motoneurons in vitro. J Neurophysiol 77:2910-2924

Hultborn, H., R. Katz, and R. Mackel (1988) Distribution of recurrent inhibition within a motor nucleus. II. Amount of recurrent inhibition in motoneurons to fast and slow units. Acta Physiol Scand 134:363-374

Hultborn, H., and O. Kiehn (1992) Neuromodulation of vertebrate motor neuron membrane properties. Curr Opin Neurobiol 2:770-5

Hultborn, H., S. Lindstrom, and H. Wigstrom (1979) On the function of recurrent inhibition in the spinal cord. Exp Brain Res 37:399-403

Hunter, I., and M. J. Korenberg (1986) The identification of nonlinear biological systems: Wiener and Hammerstein cascade models. Biol Cybern 55:135-144

Iansek, R., and S. J. Redman (1973a) The amplitude, time course and charge of unitary excitatory post-synaptic potentials evoked in spinal motoneurone dendrites. J Physiol (Lond) 234:665-88

Iansek, R., and S. J. Redman (1973b) An analysis of the cable properties of spinal motoneurones using a brief intracellular current pulse. J Physiol (Lond) 234:613-36

Inoue, T., S. Itoh, M. Kobayashi, Y. Kang, R. Matsuo, S. Wakisaka, and T. Morimoto (1999) Serotonergic modulation of the hyperpolarizing spike afterpotential in rat jaw-closing motoneurons by PKA and PKC. J Neurophysiol 82:626-37

Jack, J. J., S. Miller, R. Porter, and S. J. Redman (1971) The time course of minimal excitatory post-synaptic potentials evoked in spinal motoneurones by group Ia afferent fibres. J Physiol (Lond) 215:353-380

Jack, J. J., and S. J. Redman (1971) An electrical description of the motoneurone, and its application to the analysis of synaptic potentials. J Physiol (Lond) 215:321-52

Jack, J. J., S. J. Redman, and K. Wong (1981) The components of synaptic potentials evoked in cat spinal motoneurones by impulses in single group Ia afferents. J Physiol (Lond) 321:65-96

Jack, J. J. B., D. Noble, and R. W. Tsien (1975) Electric Current Flow in Excitable Cells. Oxford: Clarendon Press, pp Pages

Jacobs, B. L., and C. A. Fornal (1993) 5-HT and motor control: a hypothesis. Trends Neurosci 6:346-352

Jankowska, E. (1992) Interneuonal relay in spinal pathways from proprioceptors. Prog Neurobiol 38:335-378

Jiang, Z. G., and N. J. Dun (1986) Presynaptic suppression of excitatory postsynaptic potentials in rat ventral horn neurons by muscarinic agonists. Brain Res 381:182-6

Johnston, D., J. C. Magee, C. M. Colbert, and B. R. Christie (1996) Active properties of neuronal dendrites. Annu Rev Neurosci 19:165-186

Jung, H.-Y., T. Mickus, and N. Spruston (1997) Prolonged sodium channel inactivation contributes to dendritic action potential attenuation in hippocampal pyramidal neurons. J Neurosci 17:6639-6646

Kalb, R. G., M. S. Lidow, M. J. Halsted, and S. Hockfield (1992) N-methyl-D-aspartate receptors are transiently expressed in the developing spinal cord ventral horn. Proc Natl Acad Sci USA 89:8502-6

Kanosue, K., M. Yoshida, K. Akazawa, and K. Fuji (1979) The Number of Active Motor Units and Their Firing Rates in Voluntary Contraction of Human Brachialis Muscle. Jap J Physiol 29:427-444

Katakura, N., and S. H. Chandler (1990) An iontophoretic analysis of the pharmacologic mechanisms responsible for trigeminal motoneuronal discharge during masticatory-like activity in the guinea pig. J Neurophysiol 63:356-69

Kawato, M. (1984) Cable properties of a neuron model with non-uniform membrane resistivity. J Theor Biol 111:149-69

Kellerth, J.-O., C.-H. Berthold, and S. Conradi (1979) Electron microscopic studies of serially sectioned cat spinal α-motoneurons: III. Motoneurons innervating fast-twitch (Type FR) units of the gastrocnemius muscle. J Comp Neurol 184:

Kellerth, J.-O., S. Conradi, and C.-H. Berthold (1983) Electron microscopic studies of serially sectioned cat spinal α-motoneurons: V. Motoneurons innervating fast-twitch (Type FF) units of the gastrocnemius muscle. J Comp Neurol 214:451-438

Kernell, D. (1965a) The adaptation and the relation between discharge frequency and current strength of cat lumbosacral motoneurones stimulated by long-lasting injected currents. Acta Physiol Scand 65:65-73

Kernell, D. (1965b) High frequency repetitive firing of cat lumbosacral motoneurones stimulated by long-lasting injected currents. Acta Physiol Scand 65:74-86

Kernell, D. (1965c) The limits of firing frequency in cat lumbosacral motoneurones possessing different time course of afterhyperpolarization. Acta Physiol Scand 65:87-100

Kernell, D. (1966) The repetitive discharge of motoneurones. In: R. Granit (eds.). Muscular afferents and Motor Control. Nobel Symp. I. Stockholm: Almqvist and Wiksell, pp 351-362

Kernell, D. (1968) The repetitive impulse discharge of a simple neurone model compared to that of spinal motoneurones. Brain Res 11:685-7

Kernell, D. (1970) Synaptic conductance changes and the repetitive impulse discharge of spinal motoneurones. Brain Res 15:291-294

Kernell, D. (1979) Rhythmic properties of motoneurones innervating muscle fibres of different speed in m. gastrocnemius medialis of the cat. Brain Res 160:159-62

Kernell, D. (1983) Functional properties of spinal motoneurons and gradation of muscle force. Adv Neurol 39:213-26

Kernell, D., O. Eerbeek, and B. A. Verhey (1983) Relation between isometric force and stimulus rate in cat's hindlimb motor units of different twitch contraction time. Exp Brain Res 50:220-7

Kernell, D., and A. W. Monster (1981) Threshold current for repetitive impulse firing in motoneurones innervating muscle fibres of different fatigue sensitivity in the cat. Brain Res 229:193-6

Kernell, D., and A. W. Monster (1982a) Motoneuron properties and motor fatigue. Exp Brain Res 46:197-204

Kernell, D., and A. W. Monster (1982b) Time course and properties of late adaptation in spinal motoneurones of the cat. Exp Brain Res 46:191-6

Kernell, D., and H. Sjoholm (1973) Repetitive impulse firing: comparisons between neurone models based on 'voltage clamp equations' and spinal motoneurones. Acta Physiol Scand 87:40-56

Kernell, D., and B. Zwaagstra (1981) Input conductance axonal conduction velocity and cell size among hindlimb motoneurones of the cat. Brain Res 204:311-26

Kernell, D., and B. Zwaagstra (1989a) Dendrites of cat's spinal motoneurones: relationship between stem diameter and predicted input conductance. J Physiol (Lond) 413:255-69

Kernell, D., and B. Zwaagstra (1989b) Size and remoteness: two relatively independent parameters of dendrites, as studied for spinal motoneurones of the cat. J Physiol (Lond) 413:233-54

Kiehn, O. (1991) Plateau potentials and active integration in the 'final common pathway' for motor behaviour. Trends Neurosci 14:68-73

Kiehn, O., and T. Eken (1997) Prolonged firing in motor units: evidence of plateau potentials in human motoneurons? J Neurophysiol 78:3061-8

Kiehn, O., and T. Eken (1998) Functional role of plateau potentials in vertebrate motor neurons. Cur Opin Neurobiol 8:746-752

Kiehn, O., J. Erdal, T. Eken, and T. Bruhn (1996) Selective depletion of spinal monoamines changes the rat soleus EMG from a tonic to a more phasic pattern. J Physiol (Lond). 492:173-184

Kiehn, O., and R. M. Harris-Warrick (1992) 5-HT modulation of hyperpolarization-activated inward current and calcium-dependent outward current in a crustacean motor neuron. J Neurophysiol 68:496-508

Kim, Y. I., and S. H. Chandler (1995) NMDA-induced burst discharge in guinea pig trigeminal motoneurons *in vitro*. J Neurophysiol 74:334-46

Kirkwood, P. A. (1979) On the use and interpretation of crosscorrelation measurements in the mammalian central nervous system. J Neurosci Methods 1:107-132

Kirkwood, P. A., and T. A. Sears (1978) The synaptic connexions to intercostal motoneurones as revealed by the average common excitation potential. J Physiol (Lond) 275:103-134

Knox, C. K. (1974) Cross-correlation functions for a neuronal model. Biophys J 14:567-582

Kobayashi, M., T. Inoue, R. Matsuo, Y. Masuda, O. Hidaka, Y. N. Kang, and T. Morimoto (1997) Role of calcium conductances on spike afterpotentials in rat trigeminal motoneurons. J Neurophysiol 77:3273-3283

Korogod, S. M., and I. B. Kulagina (1998) Geometry-induced features of current transfer in neuronal dendrites with tonically activated conductances. Biol Cybern 79:231-40

Krnjevi'c, K., and A. Lisiewicz (1972) Injections of calcium ions into spinal motoneurones. J Physiol (Lond) 225:363-390

Krnjevi'c, K., E. Puil, and R. Werman (1978) EGTA and motoneuronal afterpotentials. J Physiol (Lond) 275:199-223

Kuno, M., and J. T. Miyahara (1969) Non-linear summation of unit synaptic potentials in spinal motoneurons of the cat. J Physiol (Lond) 201:465-477

Lagerback, P. A., and B. Ulfhake (1987) Ultrastructural observations on beaded alpha-motoneuron dendrites. Acta Physiol Scand 129:61-6

Lape, R., and A. Nistri (1999) Voltage-activated K+ currents of hypoglossal motoneurons in a brain stem slice preparation from the neonatal rat. J Neurophysiol 81:140-8

Larkman, P. M., and J. S. Kelly (1992) Ionic mechanisms mediating 5-hydroxytryptamine- and noradrenaline-evoked depolarization of adult rat facial motoneurones. J Physiol (Lond) 456:473-90

Larkman, P. M., and J. S. Kelly (1998) Characterization of 5-HT-sensitive potassium conductances in neonatal rat facial motoneurones in vitro. J Physiol (Lond) 508:67-81

Larkum, M. E., M. G. Rioult, and H.-R. Luscher (1996) Propagation of action potentials in the dendrites of neurons from rat spinal cord slice cultures. J Neurophysiol 75:154-170

Lee, R. H., and C. J. Heckman (1996) Influence of voltage-sensitive dendritic conductances on bistable firing and effective synaptic current in cat spinal motoneurons in vivo. J Neurophysiol 76:2107-2110

Lee, R. H., and C. J. Heckman (1998a) Bistability in spinal motoneurons in vivo: Systematic variations in persistent inward currents. J Neurophysiol 80:583-593

Lee, R. H., and C. J. Heckman (1998b) Bistability in spinal motoneurons in vivo: Systematic variations in rhythmic firing patterns. J Neurophysiol 80:572-582

Lee, R. H., and C. J. Heckman (1999a) Enhancement of bistability in spinal motoneurons in vivo by the noradrenergic alpha1 agonist methoxamine. J Neurophysiol 81:2164-74

Lee, R. H., and C. J. Heckman (1999b) Paradoxical effect of QX-314 on persistent inward currents and bistable behavior in spinal motoneurons in vivo. J Neurophysiol 82:2518-27

Lee, R. H. and Heckman, C. J. (2000) Adjustable amplification of synaptic input in the dendrites of spinal motoneurons In vivo. J Neurosci 20: 6734-40

Lee, Y. W., and M. Schetzen (1965) Measurement of the Wiener kernels of a non-linear system by cross-correlation. International Journal of Control 2:237-254

Levine, E. S., W. J. Litto, and B. L. Jacobs (1990) Activity of cat locus coeruleus noradrenergic neurons during the defense reaction. Brain Res 531:189-195

Lindsay, A. D., and M.D. Binder (1991) Distribution of effective synaptic currents underlying recurrent inhibition in cat triceps surae motoneurons. J Neurophysiol 65:168-177

Lindsay, A. D., and J. L. Feldman (1993) Modulation of respiratory activity of neonatal rat phrenic motoneurones by serotonin. J Physiol (Lond) 461:213-33

Lipowsky, R., T. Gillessen, and C. Alzheimer (1996) Dendritic Na+ channels amplify EPSPs in hippocampal CA1 pyramical cells. J Neurophysiol 76:2181-2191

Lips, M. B., and B. U. Keller (1998) Endogenous calcium buffering in motoneurones of the nucleus hypoglossus from mouse. J Physiol (Lond) 511:105-17

Llinas, R. R. (1988) The intrinsic electrophysiological properties of mammalian neurons: Insights into central nervous system function. Science 242:1654-1664

London, M., C. Meunier, and I. Segev (1999) Signal transfer in passive dendrites with nonuniform membrane conductance. J Neurosci 19:8219-33

Luscher, H.-R., D. Thurbon, T. Hofstetter, and S. J. Redman (1997) Dendritic recording of action potentials and brief voltage transients in motoneurons of rat spinal cord slices. Soc Neurosci Abstr 23:1301

Lux, H. D., P. Schubert, and G. W. Kreutzberg (1970) Direct matching of morphological and electrophysiological data in cat spinal motoneurones. In: P. Anderson and J. K. S. Jansen (eds). Excitatory Synaptic Mechanisms. Oslo: Universitefsforlaget, pp 189-198

Madison, D. V., and R. A. Nicoll (1984) Control of the repetitive discharge of rat CA 1 pyramidal neurones in vitro. J Physiol (Lond) 354:319-31

Major, G., A. Larkman, P. Jonas, B. Sakmann, and J. J. B. Jack (1992) Detailed passive cable models of whole-cell recorded CA3 pyramidal neurons in rat hippocampal slices. J Neurosci 14:4613-4638

Maltenfort M.G., C.J. Heckman, and W.Z. Rymer (1998) Decorrelating actions of Renshaw interneurons on the firing of spinal motoneurons within a motor nucleus: a simulation study. J Neurophysiol 80:309-323

Marmarelis, P. Z., and V. Z. Marmarelis (1978) Analysis of Physiological Systems: The White Noise Approach. New York: Plenum, pp Pages

Mauritz, K. H., W. R. Schlue, D. W. Richter, and A. C. Nacimiento (1974) Membrane conductance course during spike intervals and repetitive firing in cat spinal motoneurons. Brain Res 76:223-233

Mayer, M. L., and G. L. Westbrook (1987) The physiology of excitatory amino acids in the vertebrate central nervous system. Progr Neurobiol 28:197-276

McBain, C. J., and M. L. Mayer (1994) N-methyl-D-aspartic acid receptor structure and function. Physiol Rev 74:723-60

McCobb, D. P., and K. G. Beam (1991) Action potential waveform voltage-clamp commands reveal striking differences in calcium entry via low and high voltage-activated calcium channels. Neuron 7:119-27

McLarnon, J. G. (1995) Potassium currents in motoneurones. Progr Neurobiol 47:513-531

Mel, B. W. (1994) Information processing in dendritic trees. Neural Computation 6:1031-1085

Midtgaard, J. (1994) Processing of Information from Different Sources–Spatial Synaptic Integration in the Dendrites of Vertebrate CNS Neurons. Trends in Neurosciences 17:166-173

Midtgaard, J. (1996) Active membrane properties and spatiotemporal synaptic integration in dendrites of vertebrate neurones. Acta Physiologica Scandinavica 157:395-401

Miller, J.F., K.D. Paul,, W.Z. Rymer and C.J. Heckman (1997) Intrathecal 2-amino-7-phophonohetanoic acid (AP-7) attenuates clasp knife reflex in decerebrate cat. Soc Neurosci Abstr 23: 1039

Monster, A. W., and H. Chan (1977) Isometric force production by motor units of extensor digitorum communis muscle in man. J Neurophysiol 40:1432-1443

Moore, J. A., and K. Appenteng (1991) The morphology and electrical geometry of rat jaw-elevator motoneurones. J Physiol (Lond) 440:325-43

Mosfeldt-Laursen, A., and J. C. Rekling (1989) Electrophysiological properties of hypoglossal motoneurons of guinea-pigs studied *in vitro*. Neuroscience 30:619-37

Munson, J. B. (1990) Synaptic inputs to type-identified motor units. In: M. D. Binder and L. M. Mendell (eds). The Segmental Motor System. New York: Oxford University Press, pp 291-307

Munson, J. B., J. W. Fleshman, and G. W. Sypert (1980) Properties of single-fiber spindle group II EPSPs in triceps surae motoneurons. J Neurophysiol 44:713-725

Munson, J. B., G. W. Sypert, J. E. Zengel, S. A. Lofton, and J. W. Fleshman (1982) Monosynaptic projections of individual spindle group II afferents to type-identified medial gastrocnemius motoneurons in the cat. J Neurophysiol 48:

Musick, J. R. (1999) Mechanisms of Spike-Frequency Adaptation in Hypoglossal Motoneurons. Ph.D. Dissertation, University of Washington

Neher, E. (1995) The use of fura-2 for estimating Ca buffers and Ca fluxes. Neuropharmacology 34:1423-1442

Nelson, P. G., and K. Frank (1964) La production du potentiel d'action etudee par la technique du voltage imposee sur le motoneurone du chat. Actual neurophysiol 5:15-35

Nelson, P. G., and H. D. Lux (1970) Some electrical measurements of motoneurone parameters. Biophys J 10:55-73

Nishimura, Y., P. C. Schwindt, and W. E. Crill (1989) Electrical properties of facial motoneurons in brainstem slices from guinea pig. Brain Res 502:127-42

Nistri, A., N. D. Fisher, and M. Gurnell (1990) Block by the neuropeptide TRH of an apparently novel K+ conductance of rat motoneurons. Neurosci Lett 120:25-30

O'Brien, J. A., J. S. Isaacson, and A. J. Berger (1997) NMDA and non-NMDA receptors are co-localized at excitatory synapses of rat hypoglossal motoneurons. Neurosci Lett 227:5-8

Oakley, J. C., P. C. Schwindt, and W. E. Crill (1999) Pruning the dendritic arbor of neocortical neurons with calcium plateaus: A gain control mechanism. Soc Neurosci Abstr 25:1741

Palecek, J., M. B. Lips, and B. U. Keller (1999) Calcium dynamics and buffering in motoneurones of the mouse spinal cord. J Physiol (Lond) 520:485-502

Palecek, J. I., G. Abdrachmanova, V. Vlachova, and L. Vyklick, Jr. (1999) Properties of NMDA receptors in rat spinal cord motoneurons. Eur J Neurosci 11:827-36

Parkis, M. A., D. A. Bayliss, and A. J. Berger (1995) Actions of norepinephrine on rat hypoglossal motoneurons. J Neurophysiol 74:1911-1919

Perkins, K. L., and R. K. S. Wong (1995) Intracellular QX-314 blocks the hyperpolarization-activated inward current Iq in hippocampal pyramidal cells. J Neurophysiol 73:911-915

Perrier, J. F., and J. Hounsgaard (1999) Ca(2+)-activated nonselective cationic current (I(CAN)) in turtle motoneurons. J Neurophysiol 82:730-5

Pierce, J. P., and L. M. Mendell (1993) Quantitative Ultrastructure of Ia Boutons in the Ventral Horn–Scaling and Positional Relationships. J Neurosci 13:4748-4763

Pinco, M., and L.-T. A. (1993) Synaptic excitation of alpha-motoneurons by dorsal root afferents in the neonatal rat spinal cord. J Neurophysiol 70:406-17

Pinter, M. J., R. L. Curtis, and M. J. Hosko (1983) Voltage threshold and excitability among variously sized cat hindlimb motoneurons. J Neurophysiol 50:644-57

Poliakov, A., R. K. Powers, and M. D. Binder (1997) Functional identification of the input-output transforms of motoneurones in the rat and cat. J Physiol (Lond) 504:401-424

Poliakov, A. V., R. K. Powers, A. Sawczuk, and M. D. Binder (1996) Effects of background noise on the response of rat and cat motoneurones to excitatory current transients. J Physiol (Lond) 495:143-157

Powers, R. K. (1993) A variable-threshold motoneuron model that incorporates time- and voltage-dependent potassium and calcium conductances. J Neurophysiol 70:246-62

Powers, R. K., and M. D. Binder (1985a) Determination of afferent fibers mediating oligosynaptic group I input to cat medial gastrocnemius motoneurons. J Neurophysiol 53:518-29

Powers, R. K., and M. D. Binder (1985b) Distribution of oligosynaptic group I input to the cat medial gastrocnemius motoneuron pool. J Neurophysiol 53:497-517

Powers, R. K., and M. D. Binder (1995) Effective synaptic current and motoneuron firing rate modulation. J Neurophysiol 74:793-801

Powers, R. K., and M. D. Binder (1999) Models of spike encoding and their use in the interpretation of motor unit recordings in man. In: Peripheral and Spinal Mechanism in the Neural Control of Movement. Progress in Brain Research Vol. 123. M.D. Binder (ed) Elevier: North-Holland, pp 83-98

Powers, R. K., and M. D. Binder (2000) Summation of effective synaptic currents and firing rate modulation in cat spinal motoneurons. J Neurophysiol 83:483-500

Powers, R. K., F. R. Robinson, M. A. Konodi, and M. D. Binder (1992) Effective synaptic current can be estimated from measurements of neuronal discharge. J Neurophysiol 68:964-8

Powers, R. K., F. R. Robinson, M. A. Konodi, and M. D. Binder (1993) Distribution of rubrospinal synaptic input to cat triceps surae motoneurons. J Neurophysiol 70:1460-1468

Powers, R. K., and W. Z. Rymer (1988) Effects of acute dorsal spinal hemisection on motoneuron discharge in the medial gastrocnemius of the decerebrate cat. J Neurophysiol 59:1540-1556

Powers, R. K., A. Sawczuk, J. R. Musick, and M. D. Binder (1999) Multiple mechanisms of spike-frequency adaptation in motoneurones. J Physiol (Paris) 93:101-114

Powers, R. K. D. B., and M. D. Binder (1996) Experimental evaluation of input-output models of motoneuron discharge. J Neurophysiol 75:367-379

Prather, J. F., Powers, R. K., and T. C. Cope (2001) Amplification and linear summation of synaptic effects on motoneuron firing rate. J Neurophysiol 85:43-53

Rajaofetra, N., J. L. Ridet, P. Poulat, M. Larlier, F. Sandillon, M. Geffard, and A. Privat (1992) Immunocytochemical mapping of noradrenergic projections to the rat spinal cord with an antiserum against noradrenaline. J Neurocytol 21:481-94

Rall, W. (1959) Branching dendritic trees and motoneuron membrane resistivity. Exp Neurol 1:491-527

Rall, W. (1964) Theoretical significance of dendritic trees for neuronal input-output relations. In: R. Reiss (eds). Neural Theory and Modelling. Stanford, CA: Stanford University Press, pp 73-97

Rall, W. (1967) Distinguishing theoretical synaptic potentials computed for different soma-dendritic distributions of synaptic input. J Neurophysiol 30:1138-68

Rall, W. (1977) Core conductor theory and cable properties of neurons. In: (eds.). Handbook of Physiology, The Nervous System, Cellular Biology of Neurons. Bethesda, MD: American Physiological Society, pp 39-97

Rall, W., R. E. Burke, W. R. Holmes, J. J. Jack, S. J. Redman, and I. Segev (1992) Matching dendritic neuron models to experimental data. Physiol Rev 86:

Rall, W., R. E. Burke, T. G. Smith, P. G. Nelson, and K. Frank (1967) Dendritic location of synapses and possible mechanisms for the monosynaptic EPSP in motoneurons. J Neurophysiol 30:1169-93

Rall, W., and J. Rinzel (1973) Branch input resistance and steady attenuation for input to one branch of a dendritic neuron model. Biophys J 13:648-87

Ramirez-Leon, V., and B. Ulfhake (1993) GABA-Like Immunoreactive Innervation and Dendro-Dendritic Contacts in the Ventrolateral Dendritic Bundle in the Cat S1 Spinal Cord Segment – An Electron Microscopic Study Exp Brain Res 97:1-12

Redman, S. (1976) A quantitative approach to the integrative function of dendrites. In: R. Porter (eds.). International Review of Physiology: Neurophysiology. Baltimore: University Park Press, pp 1-36

Redman, S., and B. Walmsley (1983a) Amplitude fluctuations in synaptic potentials evoked in cat spinal motoneurones at identified group Ia synapses. J Physiol (Lond) 343:135-145

Redman, S. J., and B. Walmsley (1983b) The time course of synaptic potentials evoked in cat spinal motoneurones at identified group Ia synapses. J Physiol (Lond) 343:117-33

Rekling, J. C. (1990) Excitatory effects of thyrotropin-releasing hormone in hypoglossal motoneurons. Brain Res 510:175-179

Rekling, J. C., and J. L. Feldman (1997) Calcium-dependent plateau potentials in rostral ambiguus neurons in the newborn mouse brain stem *in vitro*. J Neurophysiol 78:2483-2492

Reckling, J.C., G.D. Funk, D.A. Bayliss, X-W. Dong and J.L. Feldman (20000) Synaptic control of motoneuronal excitability. Physiological Reviews 80: 767-852.

Reyes, A. D., and E. E. Fetz (1993) Effects of Transient Depolarizing Potentials on the Firing Rate of Cat Neocortical Neurons. J Neurophysiol 69:1673-1683

Richter, D. W., W. R. Schlue, K. H. Mauritz, and A. C. Nacimiento (1974) Comparison of membrane properties of the cell body and the initial part of the axon of phasic motoneurones in the spinal cord of the cat. Exp Brain Res 21:193-206

Rinzel, J., and W. Rall (1974) Transient response in a dendritic neuron model for current injected at one branch. Biophys J 14:759-90

Rose, P. K. (1981) Distribution of dendrites from biventer cervicis and complexus motoneurons stained intracellularly with horseradish peroxidase in the adult cat. J Comp Neurol 197:395-409

Rose, P. K., and P. Brennan (1989) Somatic shunts in neck motoneurons of the cat. Soc Neurosci Abstr 15:922

Rose, P. K., and S. Cushing (1999) Non-linear summation of synaptic currents on spinal motoneurons: Lessons from simulations of the behavior of anatomically realistic models. In: M. D. Binder (eds). Peripheral and Spinal Mechanisms in the Neural Control of Movement. Amsterdam: Elsevier, pp 99-107

Rose, P. K., S. A. Keirstead, and S. J. Vanner (1985) A quantitative analysis of the geometry of cat motoneurons innervating neck and shoulder muscles. J Comp Neurol 239:89-107

Rose, P. K., and H. M. Neuber (1991) Morphology and frequency of axon terminals on the somata, proximal dendrites, and distal dendrites of dorsal neck motoneurons in the cat. J Comp Neurol 307:259-80

Rose, P. K., and S. J. Vanner (1988) Differences in somatic and dendritic specific membrane resistivity of spinal motoneurons: an electrophysiological study of neck and shoulder motoneurons in the cat. J Neurophysiol 60:149-66

Rossignol, S., C. Chau, E. Brustein, N. Giroux, L. Bouyer, H. Barbeau, and T. A. Reader (1998) Pharmacological activation and modulation of the central pattern generator for locomotion in the cat. Ann NY Acad Sci 860:346-59

Rudy, B. (1981) Inactivation in Myxicola giant axons responsible for slow and accumulative adaptation phenomena. J Physiol (Lond) 312:531-49

Rudy, B. (1988) Diversity and ubiquity of K channels. Neurosci 25:729-749

Safronov, B. V., and W. Vogel (1995) Single voltage-activated Na+ and K+ channels in the somata of rat motoneurones. J Physiol (Lond) 487:91-106

Safronov, B. V., and W. Vogel (1996) Properties and functions of Na^+-activated K^+ channels in the soma of rat motoneurones. J Physiol (Lond) 497:727-734

Sah, P. (1992) Role of calcium influx and buffering in the kinetics of Ca(2+)-activated K+ current in rat vagal motoneurons. J Neurophysiol 68:2237-47

Sah, P. (1996) Ca2+-activated K+ currents in neurones: Types, physiological roles and modulation. Trends in Neurosciences 19:150-154

Saha, S., K. Appenteng, and T. F. Batten (1991) Light and electron microscopical localisation of 5-HT-immunoreactive boutons in the rat trigeminal motor nucleus. Brain Res 559:145-8

Sakai, H. M. (1992) White-noise analysis in neurophysiology. Physiological Reviews 72:491-505

Sawczuk, A., R. K. Powers, and M. D. Binder (1995a) Intrinsic properties of motoneurons: Implications for muscle fatigue. In: S. Gandevia, R. M. Enoka, A. J. McComas, D. J. Stuart and C. K. Thomas (eds). Fatigue: Neural and Muscular Mechanisms. New York: Plenum Press, pp 123-134

Sawczuk, A., R. K. Powers, and M. D. Binder (1995b) Spike frequency adaptation studied in hypoglossal motoneurons of the rat. J Neurophysiol 73:1799-1810

Sawczuk, A., R. K. Powers, and M. D. Binder (1997) Contribution of outward currents to spike-frequency adaptation in hypoglossal motoneurons of the rat. J Neurophysiol 78:2246-2253

Schlue, W. R., D. W. Richter, K. H. Mauritz, and A. C. Nacimiento (1974) Mechanisms of accommodation to linearly rising currents in cat spinal motoneurones. J Neurophysiol 37:310-315

Schmidt, B. J., S. Hochman, and J. N. MacLean (1998) NMDA receptor-mediated oscillatory properties: potential role in rhythm generation in the mammalian spinal cord. Ann NY Acad Sci 860:189-202

Schwindt, P., and W. E. Crill (1977) A persistent negative resistance in cat lumbar motoneurons. Brain Res 120:173-8

Schwindt, P. C. (1973) Membrane-potential trajectories underlying motoneuron rhythmic firing at high rates. J Neurophysiol 36:434-9

Schwindt, P. C., and W. H. Calvin (1972) Membrane potential trajectories between spikes underlying motoneuron rhythmic firing. J Neurophysiol 35:311-325

Schwindt, P. C., and W. H. Calvin (1973a) Equivalence of synaptic and injected current in determining the membrane potential trajectory during motoneuron rhythmic firing. Brain Res 59:389-94

Schwindt, P. C., and W. H. Calvin (1973b) Nature of conductances underlying rhythmic firing in cat spinal motoneurons. J Neurophysiol 36:955-73

Schwindt, P. C., and W. E. Crill (1980a) Effects of barium on cat spinal motoneurons studied by voltage clamp. J Neurophysiol 44:827-46

Schwindt, P. C., and W. E. Crill (1980b) Properties of a persistent inward current in normal and TEA-injected motoneurons. J Neurophysiol 43:1700-24

Schwindt, P. C., and W. E. Crill (1980c) Role of a persistent inward current in motoneuron bursting during spinal seizures. J Neurophysiol 43:1296-1318

Schwindt, P. C., and W. E. Crill (1982) Factors influencing motoneuron rhythmic firing: results from a voltage-clamp study. J Neurophysiol 48:875-90

Schwindt, P. C., and W. E. Crill (1996) Equivalence of amplified current flowing from dendrite to soma measured by alteration of repetitive firing and by voltage clamp in layer 5 pyramidal neurons. J Neurophysiol 76:3731-3739

Schwindt, P. C., W. J. Spain, and W. E. Crill (1989) Long-lasting reduction of excitability by a sodium-dependent potassium current in cat neocortical neurons. J Neurophysiol 61:233-44

Schwindt, P. C., W. J. Spain, R. C. Foehring, M. C. Chubb, and W. E. Crill (1988) Slow conductances in neurons from cat sensorimotor cortex in vitro and their role in slow excitability changes. J Neurophysiol 59:450-67

Scroggs, R. S., S. M. Todorovic, E. G. Anderson, and A. P. Fox (1994) Variation in IH, IIR, and ILEAK between acutely isolated adult rat dorsal root ganglion neurons of different size. J Neurophysiol 71:271-279

Shapovalov, A. I. (1972) Extrapyramidal monosynaptic and disynaptic control of mammalian alpha-motoneurons. Brain Res 40:105-115

Shinoda, Y., T. Ohgaki, T. Futami, and Y. Sugiuchi (1988a) Structural basis for three-dimensional coding in the vestibulospinal reflex. Ann New York Acad Science 545:216-227

Shinoda, Y., T. Ohgaki, T. Futami, and Y. Sugiuchi (1988b) Vestibular projections to the spinal cord: the morphology of single vestibulospinal axons. Prog Brain Res 76:17-27

Skydsgaard, M., and J. Hounsgaard (1994) Spatial integration of local transmitter responses in motoneurones of the turtle spinal cord in vitro. J Physiol (Lond) 479:233-246

Skydsgaard, M., and J. Hounsgaard (1996) Multiple actions of iontophoretically applied serotonin on motorneurones in the turtle spinal cord in vitro. Acta Physiol Scand 158:301-310

Spencer, A. N., J. Przysiezniak, J. Acosta-Urquidi, and T. A. Basarsky (1989) Presynaptic spike broadening reduces junctional potential amplitude. Nature 340:636-8

Spielmann, J. M., Y. Laouris, M. A. Nordstrom, G. A. Robinson, R. M. Reinking, and D. G. Stuart (1993) Adaptation of cat motoneurons to sustained and intermittent extracellular activation. J Physiol (Lond) 464:

Spruston, N., and D. Johnston (1992) Perforated patch-clamp analysis of the passive membrane properties of three classes of hippocampal neurons. J Neurophysiol 67:508-529

Staley, K. J., T. S. Otis, and I. Mody (1992) Membrane properties of dentate gyrus granule cells: comparison of sharp microelectrode and whole-cell recordings. J Neurophysiol 67:1346-58

Stauffer, E. K., D. G. Watt, A. Taylor, R. M. Reinking, and D. G. Stuart (1976) Analysis of muscle receptor connections by spike-triggered averaging. 2. Spindle group II afferents. J Neurophysiol 39:1393-1402

Stefani, E., and A. B. Steinbach (1969) Resting potential and electrical properties of frog slow muscle fibers. Effect of different external solutions. J Physiol (Lond) 203:383-401

Stein, R. B., and R. Bertoldi (1981) The size principle: a synthesis of neurophysiological data.In: J.E. Desmedt (ed). Progress in Clinical Neurophysiology Basel: Karger, pp 85-96

Storm, J. F. (1990) Potassium currents in hippocampal pyramidal cells. Prog Brain Res 83:161-87

Streit, J., and H.-R. Luscher (1992) Miniature excitatory postsynaptic potentials in embryonic motoneurons grown in slice cultures of spinal cord, dorsal root ganglia and skeletal muscle. Exp Brain Res 89:453-458

Stuart, G., and N. Spruston (1998) Determinants of voltage attenuation in neocortical pyramidal neuron dendrites. J Neurosci 18:3501-3510

Svirskis, G., A. Gutman, and J. Hounsgaard (1997) Detection of a membrane shunt by DC field polarization during intracellular and whole cell recording. J Neurophysiol 77:579-586

Svirskis, G., and J. Hounsgaard (1997) Depolarization-induced facilitation of a plateau-generating current in ventral horn neurons in the turtle spinal cord. J Neurophysiol 78:1740-1742

Svirskis, G., and J. Hounsgaard (1998) Transmitter regulation of plateau properties in turtle motoneurons. J Neurophysiol 79:45-50

Takahashi, T. (1990) Membrane currents in visually identified motoneurones of neonatal rat spinal cord. J Physiol (Lond) 423:27-46

Talbot, M. J., and R. J. Sayer (1996) Intracellular QX-314 inhibits calcium currents in hippocampal CA1 pyramidal neurons. J Neurophysiol 76:2120-2124

Talley, E. M., N. N. Sadr, and D. A. Bayliss (1997) Postnatal development of serotonergic innervation, 5-HT1A receptor expression, and 5-HT responses in rat motoneurons. J Neurosci 17:4473-85

Tanji, J., and M. Kato (1973) Firing rate of individual motor units in voluntary contraction of abductor digiti minimi muscle in man. Exp Neurol 40:771-783

Thurbon, D., H. R. Luscher, T. Hofstetter, and S. J. Redman (1998) Passive electrical properties of ventral horn neurons in rat spinal cord slices. J Neurophysiol 79:2485-2502

Traub, R. D. (1977) Motoneurons of different geometry and the size principle. Biol Cybern 25:163-176

Traub, R. D., and R. Llinas (1977) The spatial distribution of ionic conductances in normal and axotomized motorneurons. Neuroscience 2:829-849

Trueblood, P. R., M. S. Levine, and S. H. Chandler (1996) Dual-component excitatory amino acid-mediated responses in trigeminal motoneurons and their modulation by serotonin *in vitro*. J Neurophysiol 76:2461-73

Turker, K. S., and R. K. Powers (1999) Effects of large excitatory and inhibitory inputs on motoneuron discharge rate and probability. J Neurophysiology 82:829-840

Turman, J. E., Jr., J. Ajdari, and S. H. Chandler (1999) NMDA receptor NR1 and NR2A/B subunit expression in trigeminal neurons during early postnatal development. J Comp Neurol 409:237-49

Ulfhake, B., and S. Cullheim (1988) Postnatal development of cat hind limb motoneurons. III: Changes in size of motoneurons supplying the triceps surae muscle. J Comp Neurol 278:103-20

Ulfhake, B., and J. O. Kellerth (1981) A quantitative light microscopic study of the dendrites of cat spinal alpha-motoneurons after intracellular staining with horse-radish peroxidase. J Comp Neurol 202:571-83

Ulfhake, B., and J. O. Kellerth (1984) Electrophysiological and morphological meas-urements in cat gastrocnemius and soleus alpha-motoneurones. Brain Res 307:167-79

Ulrich, D., R. Quadroni, and H. R. Luscher (1994) Electrotonic structure of motoneu-rons in spinal cord slice cultures: A comparison of compartmental and equiva-lent cylinder models. J Neurophysiol 72:861-871

Umemiya, M., and A. J. Berger (1994) Properties and function of low- and high-voltage-activated Ca2+ channels in hypoglossal motoneurons. J Neurosci 14:5652-5660

Urban, N. N., and G. Barrionuevo (1998) Active summation of excitatory postsynap-tic potentials in hippocampal CA3 pyramidal neurons. Proc Natl Acad Sci USA 95:11450-5

Vallbo, A. B. (1964) Accommodation related to the inactivation of the sodium per-meability in single myelinated nerve fibres from Xenopus laevis. Acta Physiol Scand 61:429-444

Viana, F., D. A. Bayliss, and A. J. Berger (1993a) Calcium conductances and their role in the firing behavior of neonatal rat hypoglossal motoneurons. J Neurophysiol 69:2137-49

Viana, F., D. A. Bayliss, and A. J. Berger (1993b) Multiple potassium conductances and their role in action potential repolarization and repetitive firing behavior of neonatal rat hypoglossal motoneurons. J Neurophysiol 69:2150-63

Wallen, P., J. T. Buchanan, S. Grillner, R. H. Hill, J. Christenson, and T. Hokfelt (1989) Effects of 5-hydroxytryptamine on the afterhyperpolarization, spike fre-quency regulation, and oscillatory membrane properties in lamprey spinal cord neurons. J Neurophysiol 61:759-68

Wallen, P., O. Ekeberg, A. Lansner, L. Brodin, H. Traven, and S. Grillner (1992) A computer-based model for realistic simulations of neural networks. II. The seg-mental network generating locomotor rhythmicity in the lamprey. J Neuro-physiol 68:1939-50

Wallen, P., and S. Grillner (1987) N-methyl-D-aspartate receptor-induced, inherent oscillatory activity in neurons active during fictive locomotion in the lamprey. J Neurosci 7:2745-55

Walton, K., and B. P. Fulton (1986) Ionic mechanisms underlying the firing proper-ties of rat neonatal motoneurons studied *in vitro*. Neuroscience 19:669-83

Wessel, R., W. B. Kristan, Jr., and D. Kleinfeld (1999) Supralinear summation of synaptic inputs by an invertebrate neuron: dendritic gain is mediated by an "inward rectifier" K(+) current. J Neurosci 19:5875-88

Westcott, S. L. (1993) Comparison of vestibulospinal synaptic input and Ia afferent synaptic input in cat triceps surae motoneurons. PhD, University of Washington.

Westcott, S. L., R. K. Powers, F. R. Robinson, and M. D. Binder (1995) Distribution of vestibulospinal input to cat triceps surae motoneurons. Exp Brain Res 107:1-8

White, S. R., and S. J. Fung (1989) Serotonin depolarizes cat spinal motoneurons in situ and decreases motoneuron afterhyperpolarizing potentials. Brain Res 502:205-13

White, S. R., S. J. Fung, D. A. Jackson, and K. M. Imel (1996) Serotonin, norepineph-rine and associated neuropeptides: effects on somatic motoneuron excitability. Progr Brain Res 107:183-199

Woodbury, J. W., and H. D. Patton (1952) Electrical activity of single spinal cord elements. Cold Spring Harbor Symp Quant Biol 17:185-188

Wu, S. Y., M. Y. Wang, and N. J. Dun (1991) Serotonin via presynaptic 5-HT1 receptors attenuates synaptic transmission to immature rat motoneurons *in vitro*. Brain Res 554:111-21

Zengel, J. E., S. A. Reid, G. W. Sypert, and J. B. Munson (1985) Membrane electrical properties and prediction of motor-unit type of medial gastrocnemius motoneurons in the cat. J Neurophysiol 53:1323-44

Zhang, L., and K. Krnjevi'c (1986) Effects of 4-aminopyridine on the action potential and the after-hyperpolarization of cat spinal motoneurons. Can J Physiol Pharmacol 64:1402-1406

Zhang, L., and K. Krnjevi'c (1988) Intracellular injection of Ca2+ chelator does not affect spike repolarization of cat spinal motoneurons. Brain Res 462:174-80

Zieglgansberger, W., and J. Champagnat (1979) Cat spinal motoneurones exhibit topographic sensitivity to glutamate and glycine. Brain Res 160:95-104

Ziskind-Conhaim, L. (1990) NMDA receptors mediate poly- and monosynaptic potentials in motoneurons of rat embryos. J Neurosci 10:125-35

Ziskind-Conhaim, L., B. S. Seebach, and B. X. Gao (1993) Changes in serotonin-induced potentials during spinal cord development. J Neurophysiol 69:1338-49

Zwaagstra, B., and D. Kernell (1980a) The duration of after-hyperpolarization in hindlimb alpha motoneurones of different sizes in the cat. Neurosci Lett 19:303-7

Zwaagstra, B., and D. Kernell (1980b) Sizes of soma and stem dendrites in intracellularly labelled alpha-motoneurones of the cat. Brain Res 204:295-309

Editor-in-charge: Professor L.M. Mendell